Microwave Ring Circuits and Related Structures

Microwave Ring Circuits and Related Structures

Second Edition

KAI CHANG
LUNG-HWA HSIEH

WILEY-INTERSCIENCE

A JOHN WILEY & SONS, INC., PUBLICATION

Published by John Wiley & Sons, Inc., Hoboken, New Jersey.
Published simultaneously in Canada.

Library of Congress Cataloging-in-Publication Data:

Chang, Kai, 1948–
Microwave ring circuits and related structures / Kai Chang, Lung-Hwa Hsieh.—2nd ed.
p. cm.—(Wiley series in microwave and optical engineering)
Includes bibliographical references and index.
ISBN 0-471-44474-X (cloth)
1. Microwave circuits. 2. Microwave antennas. I. Hsieh, Lung-Hwa. II. Title. III. Series.

TK7876.C439 2004
621.381′32—dc22

2003056885

10 9 8 7 6 5 4 3 2 1

Contents

Preface

For the past three decades, the ring resonator has been widely used in measurements, filters, oscillators, mixers, couplers, power dividers/combiners, antennas, frequency selective surfaces, and so forth. Recently, many new analyses, models, and applications of the ring resonators have been reported. To meet the needs for students and engineers, the first edition of the book has been updated by adding the latest material for ring circuits and applications. Also, all of the attractive features of the first edition have remained in the second edition. The objectives of the book are to introduce the analyses and models of the ring resonators and to apply them to the applications of filters, antennas, oscillators, couplers, and so on.

The revised book covers ring resonators built in various transmission lines such as microstrip, slotline, coplanar waveguide, and waveguide. Introduction on analysis, modeling, coupling methods, and perturbation methods is included. In the theory chapter, a new transmission-line analysis pointing out a literature error of the one-port ring circuit is added and can be used to analyze any shapes of the microstrip ring resonator. Moreover, using the same analyses, the ring resonator can be represented in terms of a lumped-element *G*, *L*, *C* circuit. After these theories and analyses, the updated applications of ring circuits in filters, couplers, antennas, oscillators, and tunable ring resonators are described. Especially, there is an abundance of new applications in bandpass and bandstop filters. These applications are supported by real circuit demonstrations. Extensive additions are given in the filter and coupler design and applications.

The book is based on the dissertations/theses and many papers published by graduate students: Lung-Hwa Hsieh, Tae-Yeoul Yun, Hooman Tehrani, Chien-Hsun Ho, T. Scott Martin, Ganesh K. Goplakrishnan, Julio A. Navarro, Richard E. Miller, James L. Klein, James M. Carroll, and Zhengping Ding. Dr. Cheng-Cheh Yu, Chun-Lei Wang, Lu Fan and F. Wang, Visiting Scholars or Research Associates of the Electromagnetics and Microwave Laboratory,

Texas A&M University, have also contributed to this research. The additional materials are mainly based on the recent publications by Lung-Hwa Hsieh, Tae-Yeooul Yun, Hooman Tehrani, and Cheng-Cheh Yu. The book will also included many recent publications by others. Finally, we would like to express thanks to our family for their encouragement and support.

KAI CHANG
LUNG-HWA HSIEH

College Station, Texas

CHAPTER ONE

Introduction

1.1 BACKGROUND AND APPLICATIONS

The microstrip ring resonator was first proposed by P. Troughton in 1969 for the measurements of the phase velocity and dispersive characteristics of a microstrip line. In the first 10 years most applications were concentrated on the measurements of characteristics of discontinuities of microstrip lines. Sophisticated field analyses were developed to give accurate modeling and prediction of a ring resonator. In the 1980s, applications using ring circuits as antennas, and frequency-selective surfaces emerged. Microwave circuits using rings for filters, oscillators, mixers, baluns, and couplers were also reported. Some unique properties and excellent performances have been demonstrated using ring circuits built in coplanar waveguides and slotlines. The integration with various solid-state devices was also realized to perform tuning, switching, amplification, oscillation, and optoelectronic functions.

The ring resonator is a simple circuit. The structure would only support waves that have an integral multiple of the guided wavelength equal to the mean circumference. The circuit is simple and easy to build. For such a simple circuit, however, many more complicated circuits can be created by cutting a slit, adding a notch, cascading two or more rings, implementing some solid-state devices, integrating with multiple input and output lines, and so on. These circuits give various applications. It is believed that the variations and applications of ring circuits have not yet been exhausted and many new circuits will certainly come out in the future.

Microwave Ring Circuits and Related Structures, Second Edition,
by Kai Chang and Lung-Hwa Hsieh
ISBN 0-471-44474-X

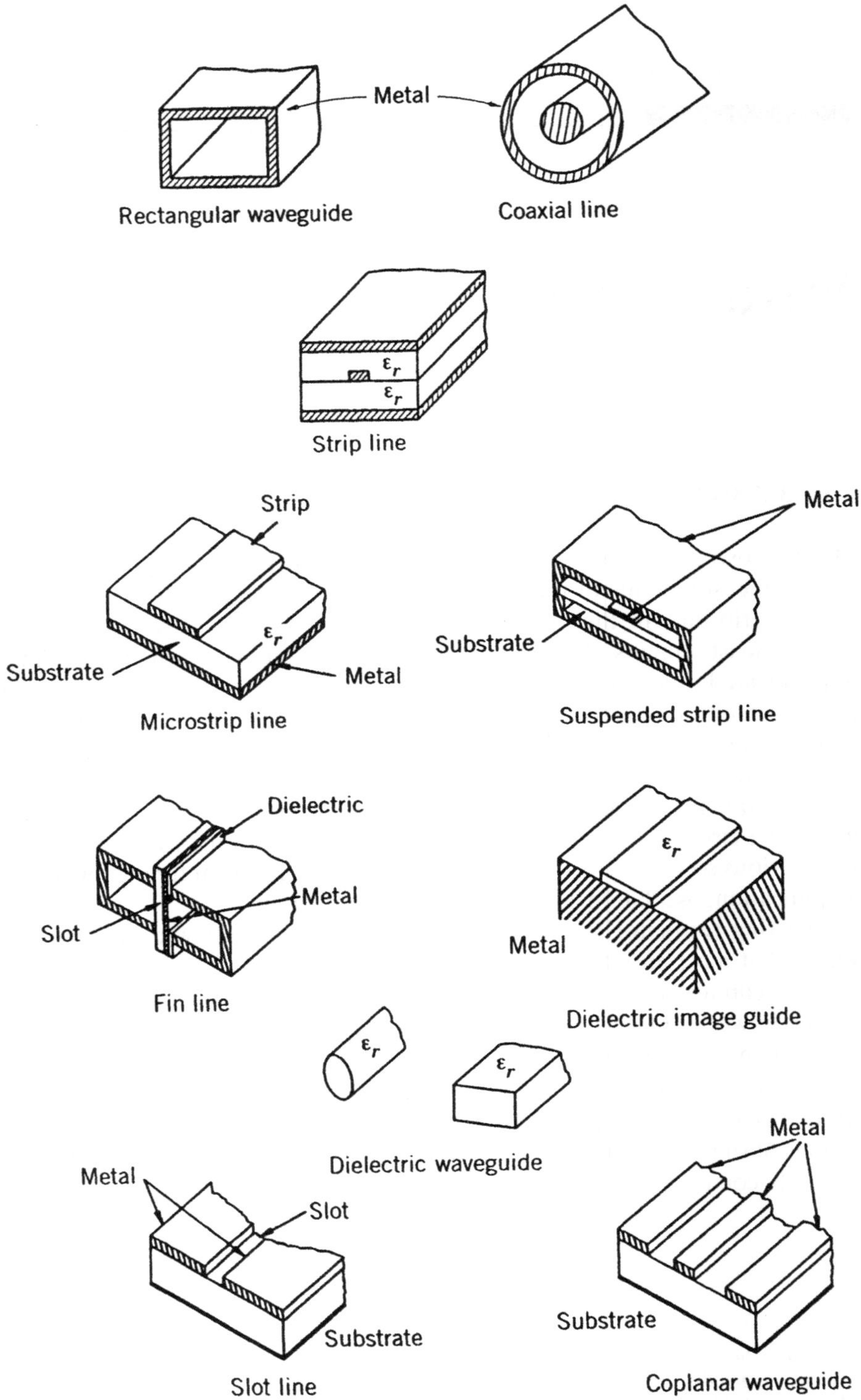

FIGURE 1.1 Various transmission lines and waveguides.

TABLE 1.1 Comparison of Guiding Media and Waveguides

Item	Useful Frequency (GHz)	Impedance Level (Ω)	Cross-Sectional Dimensions	Q Factor	Power Rating	Active Device Mounting	Potential for Low-Cost Production
Rectangular waveguide	<300	100–500	Moderate to large	High	High	Easy	Poor
Coaxial line	<50	10–100	Moderate	Moderate	Moderate	Fair	Poor
Strip line	<10	10–100	Moderate	Low	Low	Fair	Good
Microstrip line	≤100	10–100	Small	Low	Low	Easy	Good
Suspended strip line	≤150	20–150	Small	Moderate	Low	Easy	Fair
Fin line	≤150	20–400	Moderate	Moderate	Low	Easy	Fair
Slot line	≤60	60–200	Small	Low	Low	Fair	Good
Coplanar waveguide	≤60	40–150	Small	Low	Low	Fair	Good
Image guide	<300	30–30	Moderate	High	Low	Poor	Good
Dielectric guide	<300	20–50	Moderate	High	Low	Poor	Fair

1.2 TRANSMISSION LINES AND WAVEGUIDES

Many transmission lines and waveguides have been used for microwave and millimeter-wave frequencies. Figure 1.1 shows some of these lines and Table 1.1 summarizes their properties. Among them, the rectangular waveguide, coaxial line, and microstrip line are the most commonly used. Coaxial line has no cutoff frequency, can be made flexible, and can operate from dc to microwave or millimeter-wave frequencies. Rectangular waveguide has a cutoff frequency and low insertion loss, but it is bulky and requires precision machining. Microstrip line is the most commonly used in microwave integrated circuits (MIC) and monolithic microwave integrated circuits (MMIC). It has many advantages, which include low cost, small size, no critical machining, no cutoff frequency, ease of active device integration, use of pbotolithographic method for circuit production, good repeatability and reproducibility, and ease of mass production. In addition, coplanar waveguide and slotline can be the alternatives to microstrip line for some applications due to their uniplanar nature. In microstrip, the stripline and ground plane are located on opposite sides of the substrate. A hole is needed to be drilled for grounding or mounting solid-state devices in shunt. In the uniplanar circuits such as coplanar waveguide and slotline, the ground plane and circuit are located on the same side of the substrate, avoiding any circuit drilling or via holes.

Ring circuits can be built on all these transmission lines and waveguides. The selection of transmission lines and waveguides depends on applications and operating frequency ranges. Most ring circuits realized so far are in microstrip line, rectangular waveguide, coplanar waveguide, and stotline.

1.3 ORGANIZATION OF THE BOOK

This book is organized into 12 chapters. Chapters 2 and 3 give some general descriptions of a simple model, field analyses, a transmission-line model, modes, perturbation methods, and coupling methods of ring resonators. Chapters 4 and 5 discuss how electronically tunable and switchable ring resonators are made by incorporating varactor and PIN diodes into the ring circuits. Chapters 6, 7, 8, 9, and 10 present the applications of ring resonators to microwave measurements, filters, couplers, and magic-Ts. Chapter 11 gives a brief discussion of ring antennas, frequency selective surfaces, and active antennas. The last chapter (Chapter 12) summarizes applications for ring circuits in mixers, oscillators, optoelectronics, and metamaterials.

CHAPTER TWO

Analysis and Modeling of Ring Resonators

2.1 INTRODUCTION

This chapter gives a brief review of the methods used to analyze and model a ring resonator. The major goal of these analyses is to determine the resonant frequencies of various modes. Field analyses generally give accurate and rigorous results, but they are complicated and difficult to use. Circuit analyses are simple and can model the ring circuits with variations and discontinuities.

The field analysis "magnetic-wall model" for microstrip ring resonators was first introduced in 1971 by Wolff and Knoppik [1]. In 1976, Owens improved the magnetic-wall model [2]. A rigorous solution was presented by Pintzos and Pregla in 1978 based on the stationary principle [3]. Wu and Rosenbaum obtained the mode chart for the fields in the magnetic-wall model [4]. Sharma and Bhat [5] carried out a numerical solution using the spectral domain method. Wolff and Tripathi used perturbation analysis to design the open- and closed-ring microstrip resonators [6, 7]. So far, only the annular ring resonator has the field theory derivation for its frequency modes. For the square or meander ring resonators, it is difficult to use the magnetic-wall model to obtain the frequency modes of these ring resonators because of their complex boundary conditions. Also, the magnetic-wall model does not explain the dual-mode behavior very well, especially for ring resonators with complex boundary conditions.

The field analyses based on electromagnetic field theory are complicated and difficult to implement in a computer-aided-design (CAD) environment. Chang et al. [8] first proposed a straightforward but reasonably accurate transmission-

Microwave Ring Circuits and Related Structures, Second Edition,
by Kai Chang and Lung-Hwa Hsieh
ISBN 0-471-44474-X

line method that can include gap discontinuities and devices mounted along the ring. Gopalakrishnan and Chang [9] further improved the method with a distributed transmission-line method that included factors affecting resonances such as the microstrip dispersion, the curvature of the resonator, and various perturbations. The distributed transmission-line method can easily accommodate many solid-state devices, notches, gaps and various discontinuities along the circumference of the ring structure. Recently, Hsieh and Chang [10] used a simple transmission-line model unaffected by boundary conditions to calculate the frequency modes of ring resonators of any general shape such as annular, square, or meander. Moreover, it corrects an error in literature concerning the frequency modes of the one-port ring resonator [11]. Also, it can be used to describe the dual-mode behavior of the ring resonator that the magnetic-wall model cannot address well, especially for a ring resonator with complicated boundary conditions. In addition, they used the transmission-line model to extract the equivalent lumped element circuits for the closed- and open-loop ring resonators [12]. The unloaded Qs of the ring resonators can be calculated from the equivalent lumped elements G, L, and C. These simple expressions introduce an easy method for analyzing ring resonators in filters and provide, for the first time, a means of predicting their unloaded Q.

2.2 SIMPLE MODEL

The ring resonator is merely a transmission line formed in a closed loop. The basic circuit consists of the feed lines, coupling gaps, and the resonator. Figure 2.1 shows one possible circuit arrangement. Power is coupled into and out of the resonator through feed lines and coupling gaps. If the distance between the feed lines and the resonator is large, then the coupling gaps do not affect the resonant frequencies of the ring. This type of coupling is referred to in the literature as "loose coupling." Loose coupling is a manifestation of the

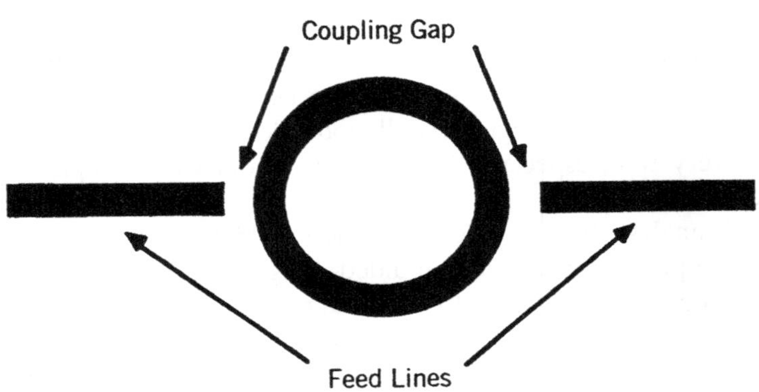

FIGURE 2.1 The microstrip ring resonator.

negligibly small capacitance of the coupling gap. If the feed lines are moved closer to the resonator, however, the coupling becomes tight and the gap capacitances become appreciable. This causes the resonant frequencies of the circuit to deviate from the intrinsic resonant frequencies of the ring. Hence, to accurately model the ring resonator, the capacitances of the coupling gaps should be considered. The effects of the coupling gaps are discussed in Chapter 3.

When the mean circumference of the ring resonator is equal to an integral multiple of a guided wavelength, resonance is established. This may be expressed as

$$2\pi r = n\lambda_g, \qquad \text{for } n = 1, 2, 3, \ldots \tag{2.1}$$

where r is the mean radius of the ring that equals the average of the outer and inner radii, λ_g is the guided wavelength, and n is the mode number. This relation is valid for the loose coupling case, as it does not take into account the coupling gap effects. From this equation, the resonant frequencies for different modes can be calculated since λ_g is frequency dependent. For the first mode, the maxima of field occur at the coupling gap locations, and nulls occur 90° from the coupling gap locations.

2.3 FIELD ANALYSES

Field analyses based on electromagnetic field theory have been reported in the literature [1–7]. This section briefly summarizes some of these methods described in [13].

2.3.1 Magnetic-Wall Model

One of the drawbacks of using the ring resonator is the effect of curvature. The effect of curvature cannot be explained by the straight-line approximation

$$2\pi r = n\lambda_g \tag{2.2}$$

To quantify the effects of curvature on the resonant frequency, Wolff and Knoppik [1] made some preliminary tests. They found that the influence of curvature becomes large if substrate materials with small relative permittivities and lines with small impedances are used. Under these conditions the widths of the lines become large and a mean radius is not well-defined. If small rings are used, then the effects become even more dramatic because of the increased curvature.

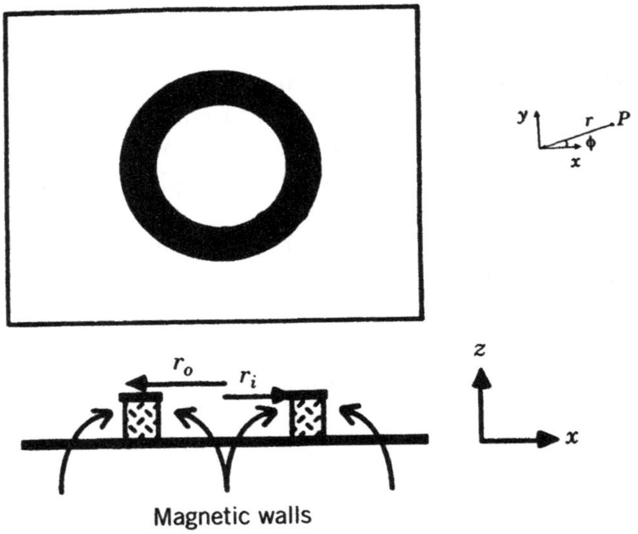

FIGURE 2.2 Magnetic-wall model of the ring resonator.

They concluded that a new theory that takes the curvature of the ring into account was needed. At the time there was no exact theory for the resonator for the dispersive effects on a microstrip line. They therefore assumed a magnetic-wall model for the resonator and used a frequency-dependent ε_{eff} to calculate the resonant frequencies.

The magnetic-wall model considered the ring as a cavity resonator with electric walls on the top and bottom and magnetic walls on the sides as shown in Figure 2.2. The electromagnetic fields are considered to be confined to the dielectric volume between the perfectly conducting ground plane and the ring conductor. It is assumed that there is no z-dependency ($\partial/\partial z = 0$) and that the fields are transverse magnetic (TM) to z direction. A solution of Maxwell's equations in cylindrical coordinates is

$$E_z = \{AJ_n(kr) + BN_n(kr)\}\cos(n\phi) \tag{2.3}$$

$$H_r = \frac{n}{j\omega\mu_0 r}\{AJ_n(kr) + BN_n(kr)\}\sin(n\phi) \tag{2.4}$$

$$H_\phi = \frac{k}{j\omega\mu_0}\{AJ'_n(kr) + BN'_n(kr)\}\cos(n\phi) \tag{2.5}$$

where A and B are constants, k is the wave number, ω is the angular frequency, J_n is a Bessel function of the first kind of order n, and N_n is a Bessel function of the second kind and order n. J'_n and N'_n are the derivatives of the Bessel functions with respect to the argument (kr).

The boundary conditions to be applied are

$$H_\phi = 0 \quad \text{at} \quad r = r_0$$

$$H_\phi = 0 \quad \text{at} \quad r = r_i$$

were r_0 and r_i are the outer and inner radii of the ring, respectively. Application of the boundary condition leads to the eigenvalue equation

$$J'_n(kr_0)N'_n(kr_i) - J'_n(kr_i)N'_n(kr_0) = 0 \tag{2.6}$$

where

$$k = \omega\sqrt{\varepsilon_0\varepsilon_r\mu_0} \tag{2.7}$$

Given r_0 and r_i, then Equation (2.6) can be solved for k. By using (2.7) the resonant frequency can be found.

The use of the magnetic-wall model eigenequation eliminates the error due to the mean radius approximation and includes the effect of curvature of the microstrip line. By using this analysis Wolff and Knoppik compared experimental and theoretical results in calculating the resonant frequency of the ring resonator. They achieved increased accuracy over Equation (2.2). Any errors that still remained were attributed to the fringing edge effects of the microstrip line.

2.3.2 Degenerate Modes of the Resonator

Using the magnetic-wall model it can be shown that the microstrip ring resonator actually supports two degenerate modes [14]. Degenerate modes in microwave cavity resonators are modes that coexist independently of each other. In mathematical terms this means that the modes are orthogonal to each other. One example of degeneracy is a circularly standing wave. This is the sum of two linearly polarized waves that are orthogonal and exist independently of each other.

Recall that the solution to the fields of the magnetic-wall model must satisfy the Maxwell's equations and boundary conditions. One proposed solution was given in Equations (2.3)–(2.5). The other set of solutions also satisfies the boundary conditions

$$E_z = \{AJ_n(kr) + BN_n(kr)\}\sin(n\phi) \tag{2.8}$$

$$H_r = -\frac{n}{j\omega\mu_0 r}\{AJ_n(kr) + BN_n(kr)\}\cos(n\phi) \tag{2.9}$$

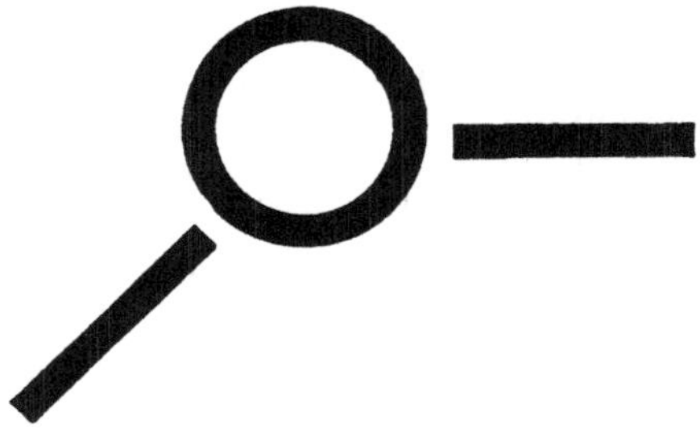

(*a*)

(*b*)

FIGURE 2.3 (*a*) Ring with asymmetrical feed lines, and (*b*) ring with a notch.

$$H_\phi = \frac{k}{j\omega\mu_0}\{AJ'_n(kr) + BN'_n(kr)\}\sin(n\phi) \tag{2.10}$$

The only difference between the field components of Equations (2.8)–(2.10) and (2.2)–(2.5) is that cosine as well as sine functions are solutions to the field dependence in the azimuthal direction, ϕ. Because sine and cosine functions are orthogonal functions, the solutions, (2.5) and (2.10), are also orthogonal. Both sets of solutions also have the same eigenvalue equation, (2.6). This means that two degenerate modes can exist at the resonance frequency. Because the modes are orthogonal, there is no coupling between them. The two modes can be interpreted as two waves, traveling clockwise and counterclockwise on the ring.

If circular symmetrical ring resonators are used with colinear feed lines, then only one of the modes will be excited. Wolff showed that if the coupling lines are arranged asymmetrically, as in Figure 2.3a, then both modes should be excited [14]. The slight splitting of the resonance frequency can be easily detected. Another way of exciting the two degenerate modes is to disturb the symmetry of the ring resonator. Wolff also demonstrated this by using a notch in the ring, as in Figure 2.3b [14].

Frequency splitting due to degenerate modes is undesirable in dispersion measurements. If both modes are excited due to an asymmetric circuit, the resonant frequency may be less distinct. To eliminate this source of error, care should be taken to ensure that the feed lines are perfectly colinear and the ring line width is constant.

2.3.3 Mode Chart for the Resonator

It has been established that the field components on the microstrip ring resonator are E_z, H_r, and H_ϕ. The resonant modes are a solution to the eigenequation

$$J_n'(kr_o)N_n'(kr_i) - J_n'(kr_i)N_n'(kr_o) = 0 \tag{2.11}$$

and may be denoted as TM_{nml}, where n is the azimuthal mode number, m is the root number for each n, and $l = 0$ because $\partial/\partial z = 0$. Close examination of Equation (2.11) reveals that for narrow microstrip widths, as r_l approaches r_o, the equation reduces to

$$[(kr_o)^2 - n^2]\{J_{n-1}(kr_o)N_n(kr_o) - N_{n-1}(kr_o)J_n(kr_o)\} = 0 \tag{2.12}$$

The second term of Equation (2.12) is nonzero, and therefore

$$(kr_o)^2 - n^2 = 0 \tag{2.13}$$

Substituting $k = 2\pi/\lambda_g$ and rearranging yields the well-known equation

$$n\lambda_g = 2\pi r_o$$

which gives the resonances of the TM_{n10} modes.

Wu and Rosenbaum presented a mode chart for the resonant frequencies of the various TM_{nm0} modes as a function of the ring line width [4]. They also pointed out that Equation (2.11) is the same equation that must be satisfied for the transverse electric (TE) modes in coaxial waveguides. The fields on the microstrip ring resonator are actually the duals of the TE modes in the coaxial waveguide.

From the mode chart of Wu and Rosenbaum, two important observations can be made [4]. As the normalized ring width, ring width/ring radius, (w/R) is increased, higher-order modes are excited. This occurs when the ring width reaches half the guided wavelength, and is similar to transverse resonance on a microstrip line. To avoid the excitation of higher-order modes, a design criteria of $w/R < 0.2$ should be observed. The other observation is the increase of dispersion on narrow rings. If rings for which $w/R < 0.2$ are used, then dispersion becomes important for the modes of $n > 4$. Wide rings do not suffer the effects of dispersion as much as narrow rings.

2.3.4 Improvement of the Magnetic-Wall Model

The magnetic-wall model is a nonrigorous but reasonable solution to the curvature problem in the microstrip ring resonator. The main criticism of the model is that it does not take into account the fringing fields of the microstrip

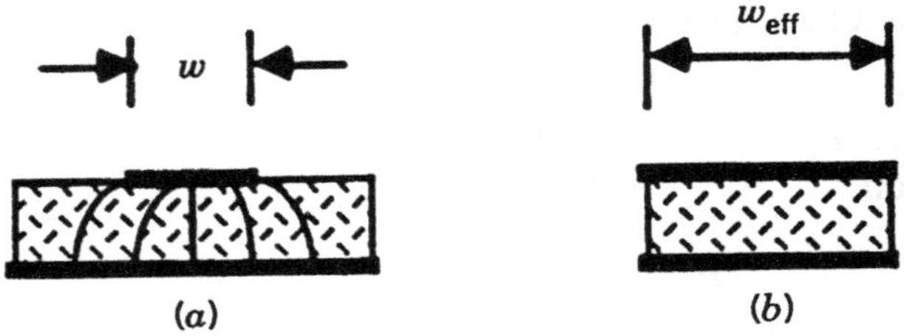

FIGURE 2.4 (*a*) Microstrip line and its electric fields, and (*b*) the planar waveguide model of a microstrip line.

line. In an attempt to take this into account, the substrate relative permittivity is made equal to the frequency-dependent effective relative permittivity, $\varepsilon_{eff}(f)$, while retaining the same line width, w. Owens argued that this increases the discrepancy between the quasi-static properties of the model and the microstrip ring that it represents [2]. He further argued that dispersion characteristics obtained in this way were still curvature-dependent. He proposed to correct this inconsistency by using the planar waveguide model for the microstrip line.

The planar waveguide model is similar to the magnetic-wall model of the ring resonator. In this model the width of the parallel conducting plates, $w_{eff}(f)$, is a function of frequency (see Fig. 2.4). The separation between the plates is equal to the distance between the microstrip line and its ground plane. Magnetic walls enclose the substrate with a permittivity of ε_{eff}. The following equations are used to calculate the effective line width:

$$w_{eff}(ff) = w + \frac{w_{eff}(0) - w}{1 + (f/f_p)^2} \tag{2.14}$$

where

$$w_{eff}(0) = \frac{h\eta_0}{Z_0\sqrt{\varepsilon_{eff}(0)}} \tag{2.15}$$

and

$$f_p = \frac{c}{w_{eff}(0)\sqrt{\varepsilon_{eff}(0)}} \tag{2.16}$$

where h is the substrate thickness, Z_0 is the characteristic impedance, η_0 is the free space impedance, and c is the speed of light in a vacuum [15, 16].

To apply the planar waveguide model to the ring resonator, the inner and outer radii of the ring, r_i and r_o, respectively, are compensated to give

$$R_o = \frac{1}{2}[(r_o + r_i) + w_{eff}(f)] \tag{2.17}$$

$$R_i = \frac{1}{2}[(r_o + r_i) - w_{\text{eff}}(f)] \tag{2.18}$$

where R_o and R_i are the radii using the new model. To find the resonant frequencies of the structure, solve for the eigenvalues of Equation (2.11).

Experimental results for this model compare quite accurately with known theoretical results. The results obtained for ε_{eff} were not curvature dependent as in the other models.

2.3.5 Simplified Eigenequation

The eigenequation for the magnetic-wall model can be solved numerically to determine the resonant frequency of a given circuit. The numerical solution is a tedious and time-consuming process that would make implementation into CAD inefficient. Therefore closed-form expressions for the technically interesting modes have been derived by Khilla [17]. The solution is as follows:

For the TM_{n10} modes

$$kR_e = (A1_n + A2_n)\frac{(\sin \pi X)^{B1_n}}{X^{B2_n}} \frac{(\cos \pi X/2)^{B3_n}}{(1-X)^{B4_n}} + A3_n(1-X) \tag{2.19}$$

For the TM_{010} mode and $0.5 < X \leq 1$

$$kR_e = 1.9159X^{0.0847}(1-\tan\alpha)^{0.3312} + (\tan\alpha)^{1.75} \tag{2.20}$$

where

$$X = \frac{0.5w_{\text{eff}}}{R_e}$$

$$\alpha = \frac{\pi(1-X)}{2}$$

$$R_e = \frac{(R_i + R_o)}{2}$$

and w_{eff}, R_o, and R_i are calculated from Equations (2.14), (2.17), and (2.18), respectively. The constants $A1_n$, $A2_n$, $A3_n$, $B1_n$, $B2_n$, $B3_n$, and $B4_n$ are given in Table 2.1. The accuracy is reported within ±0.4%.

TABLE 2.1 Constants for the Simplified Eigenequation

n	$A1_n$	$A2_n$	$A3_n$	$B1_n$	$B2_n$	$B3_n$	$B4_n$
1	0.9206	0.0493	0.0794	−0.4129	−1.0773	5.9931	4.5163
2	1.5271	1.42E − 4	0.4729	6.3852	5.6221	−1.9139	3.8091
3	2.1005	4.42E − 6	0.8995	10.6240	9.6195	−8.3029	1.8957

2.3.6 A Rigorous Solution

The magnetic-wall model is a nonrigorous method of analysis for the ring resonator. This method requires that either a frequency-dependent ε_{eff} or a frequency-dependent line width be used to describe the edge effects. This method adequately predicts the resonant frequency of not only the dominant modes but also higher-order modes; beyond this is may have limited applicability.

A rigorous solution based on the variational or stationary principle was developed by Pintzos and Pregla in 1978 [3]. A stationary expression was established for the resonant frequency of the dominant mode by means of the "reaction concept" of electromagnetic theory [18]. The reaction of a field $\mathbf{E}^a$, $\mathbf{H}^a$, on a source $\mathbf{J}^b$, $\mathbf{M}^b$ in a volume V is defined as

$$\langle a,b\rangle = \int_V (\mathbf{E}^a\cdot\mathbf{J}^b - \mathbf{H}^a\cdot\mathbf{M}^b)dV \tag{2.21}$$

In the case of a resonant structure, the self-reaction $\langle a, a\rangle$, the reaction of a field on its own source, is zero because the true field at resonance is source-free [19].

An approximate expression for the self-reaction can be derived using a trial field and source. By equating this to the correct reaction, a stationary formula for the resonant frequency can be obtained [19]. The only source is the trial current $\mathbf{J}_s$ on the microstrip line. The field associated with such a current can be considered a trial field as well. The self-reaction can now be defined as

$$\langle a,a\rangle = \int_V \mathbf{E}_{tr}\cdot\mathbf{J}_{tr}dV = 0 \tag{2.22}$$

Solving Equation (2.22) is the emphasis of the approach.

The fields existing in the structure can be expressed in terms of the vector potentials $\mathbf{A} = \mathbf{u}_z\Psi^E$ and $\mathbf{F} = \mathbf{u}_z\Psi^H$ by means of the following relations:

$$\mathbf{E} = -\nabla\times\mathbf{F} + \frac{1}{j\omega\varepsilon}\nabla\times\nabla\times\mathbf{A} \tag{2.23}$$

$$\mathbf{H} = -\nabla\times\mathbf{A} + \frac{1}{j\omega\mu}\nabla\times\nabla\times\mathbf{F} \tag{2.24}$$

The scalar potentials Ψ^E, Ψ^H satisfy the Helmholtz equation

$$\nabla\Psi^E + k_i^2\Psi^E = 0 \tag{2.25}$$

$$\nabla\Psi^H + k_i^2\Psi^H = 0 \tag{2.26}$$

and

$$k_i^2 = k_0^2\varepsilon_{ri}, \tag{2.27}$$

where $i = 1, 2$ and designates the subregions 1 (substrate) and 2 (air).

The solution of Equations (2.23) and (2.24) can be represented in the form of the Fourier–Bessel integrals for each region:

In the dielectric

$$\Psi^E = \sin(n\phi)\int_0^\infty A_n(k_\rho)\cosh(\gamma_1 z)k_\rho J_n(k_\rho \rho)dk_\rho \tag{2.28}$$

$$\Psi^H = \cos(n\phi)\int_0^\infty B_n(k_\rho)\sinh(\gamma_1 z)k_\rho J_n(k_\rho \rho)dk_\rho \tag{2.29}$$

In the air

$$\Psi^E = \sin(n\phi)\int_0^\infty C_n(k_\rho)e^{-\gamma_2(z-t)}k_\rho J_n(k_\rho \rho)dk_\rho \tag{2.30}$$

$$\Psi^E = \cos(n\phi)\int_0^\infty D_n(k_\rho)e^{-\gamma_2(z-t)}k_\rho J_n(k_\rho \rho)dk_\rho \tag{2.31}$$

where

$$r_i^2 + k_0^2\varepsilon_{ri} = k_\rho^2 \tag{2.32}$$

By applying the boundary conditions at the interface $z = t$, the coefficients A_n, B_n, C_n, and D_n can be determined. The continuity boundary conditions are as follows:

$$E_{\rho,1} = E_{\rho,2}$$

$$E_{\phi,1} = E_{\phi,2}$$

$$H_{\rho,1} - H_{\rho,2} = -I_\phi(\rho,\phi)$$

$$H_{\phi,1} - H_{\phi,2} = -I_\phi(\rho,\phi)$$

where $I_\rho(\rho, \phi)$ and $I_\phi(\rho, \phi)$ are the components of the sheet current density $\mathbf{J}_{tr}$ in the ρ and ϕ directions, respectively.

After the coefficients A_n, B_n, C_n, and D_n are expressed in terms of the trial current distribution on the surface, the expression for $\mathbf{E}_{tr}$ can be formed from Equation (2.23). Equation (2.22) can then be solved for the solution. Because the ρ component of the current is usually small when compared to the ϕ current component, it can be neglected. This results in

$$\langle a,a\rangle = \int_0^\infty E_{\phi,i}(\rho, z=t)I_\phi(\rho)\rho\, d\rho = 0 \tag{2.33}$$

for the stationary expression. This can be solved to determine the resonant frequency of the structure. Although many steps were omitted in the procedure explanation, the general idea of the method is presented.

Because this is a variational method, a crude approximation to the current distribution can be made. The trial fields due to this trial current distribution can be determined, as can the resonance. This method is a rigorous solution of the microstrip ring resonator, but in some ways is less desirable than the magnetic-wall model. The stationary formula is dependent on the trial current distribution. For the lower-older modes, the current density may be easily determine, but for the higher-order modes, the current density may be difficult to estimate. This would eliminate the analysis of higher-order modes. The stationary formula also requires quite a bit more computational effort, which may not be justified by the marginal increase in accuracy.

2.4 TRANSMISSION-LINE MODEL

It has been established that, although the ring has been studied extensively, there is a need for a new analysis technique. The magnetic-wall model is limited in that only the effects of varying the circuit parameters and dimensions can be studied. The rigorous solution using the stationary method is also limited due to its extensive computational time and difficulty in application. To extend the study of the microstrip ring resonator, the transmission-line analysis has been proposed [8, 13]. In the transmission-line approach, the resonator is represented by its equivalent circuit. Basic circuit analysis techniques can be used to determine the input impedance. From the input impedance the resonant frequency can be determined. This analysis technique allows various microwave circuits that use the ring resonator to be studied. The effect of the coupling gap on the resonant frequency can also be studied (see Chap. 3).

Application of the transmission-line method hinges on the ability to accurately model the ring resonator with an equivalent circuit. An equivalent circuit for the ring resonator is proposed [8, 13] in this section. The feed lines, coupling gap, and resonant structure are modeled and pieced together to form an overall equivalent circuit, and the equivalent circuit is verified with experimental results.

2.4.1 Coupling Gap Equivalent Circuit

The coupling gap is probably best modeled by an end-to-side gap. The end-to-side coupling is shown in Figure 2.5. This discontinuity is a difficult problem to solve because it cannot be reduced to a two-dimensional problem. The coupling gap of the resonator must thus be approximated by an end-to-end coupling gap. The end-to-end coupling gap is shown in Figure 2.6. The validity for this approximation has to be determined by experimental results.

The evaluation of the capacitance due to a microstrip gap has been treated by Farrar and Adams [20], Maeda [21], and Silvester and Benedek [22]. The capacitance associated with the discontinuities can be evaluated by finding the excess charge distribution near the discontinuity. The different methods used to find the charge distribution are the matrix inversion method [20], variational

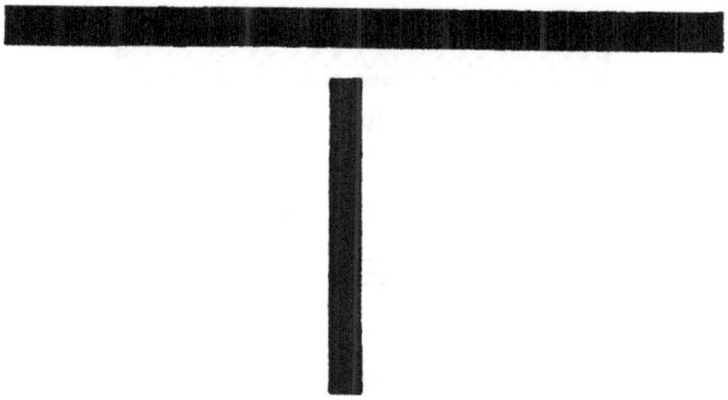

FIGURE 2.5 End-to-side coupling.

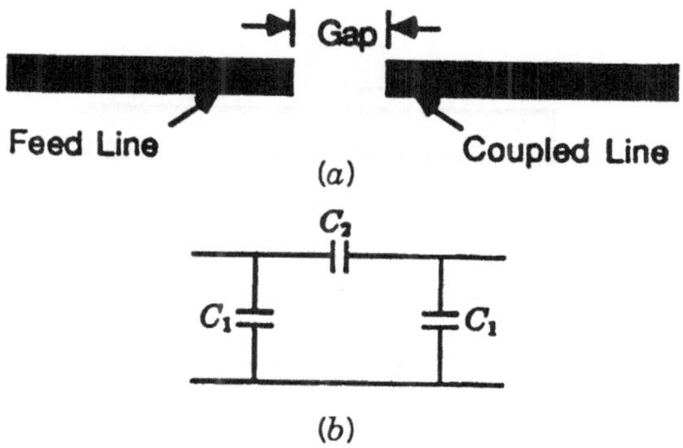

FIGURE 2.6 (*a*) End-to-end coupling, and (*b*) the equivalent circuit for the end-to-end coupling.

method [21], and use of line sources with charge reversal [22]. The matrix inversion and variational methods both involve the subtraction of two nearly equal numbers. Round-off error can become significant when two nearly equal large numbers are subtracted [23]. This subtraction could cause the matrix inversion and variational methods to suffer from computational errors. The charge-reversal method overcomes the round-off error difficulty and leads to increased accuracy. We now describe the method of charge reversal.

The proposed equivalent circuit for the microstrip gap is a symmetric two-port π-network shown in Figure 2.6. The capacitance C_2 is due to the charge buildup between the two microstrip lines. The capacitance C_1 is due to the fringing fields at the open circuits. There are two possible excitation conditions at the gap, even and odd. The symmetric excitation results in the capacitance C_{even}. The equivalent circuit for the symmetric excitation is shown in Figure 2.7. The antisymmetric excitation results in the capacitance C_{odd}. The equivalent circuit for the antisymmetric excitation is shown in Figure 2.8. The method of charge reversal is used to calculate C_{odd} and C_{even}. C_1 and C_2 can be computed from the following equation:

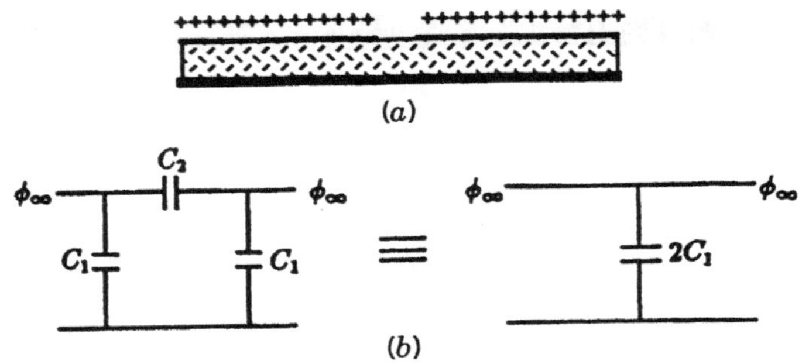

FIGURE 2.7 (*a*) Symmetric excitation of the coupling gap, and (*b*) the equivalent circuit.

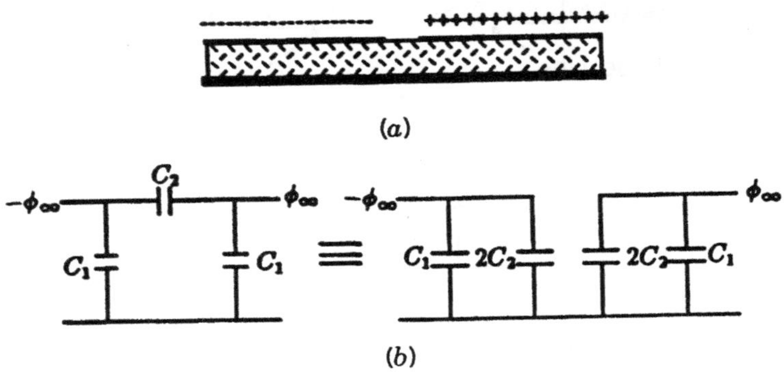

FIGURE 2.8 (*a*) Antisymmetric excitation of the coupling gap, and (*b*) the equivalent circuit.

$$C_1 = \frac{1}{2} C_{\text{even}} \tag{2.34}$$

$$C_2 = \frac{1}{2}\left(C_{\text{odd}} - \frac{1}{2} C_{\text{even}}\right) \tag{2.35}$$

The problem remains to obtain C_{even} and C_{odd}. If we let $\phi_\infty(P)$ be the potential due to an infinitely extending microstrip line with a corresponding charge-density distribution $\sigma_\infty(P')$, then

$$\phi_\infty(P) = \int \sigma_\infty(P') G_\infty(P; P') dP' \tag{2.36}$$

where $G_\infty(P; P')$ is the Green's function for the infinite microstrip. Now if we let $\phi_\xi(P)$ be the potential associated with a charge distribution $\sigma_\infty(P')$ for $z \geq \xi$ and $-\sigma_\infty(P')$ for $z < \xi$, then

$$\phi_\xi(P) = \int \sigma_\infty(P')G_\xi(P;P')dP' \tag{2.37}$$

where $G_\xi(P; P')$ is the Green's function for the charge distribution with polarity reversal at $z = \xi$.

Using Equations (2.36) and (2.37), three cases of line charges can be formed: infinite extending line charge, charge reversal at $s/2$, and charge reversal at $-s/2$. The infinite extending line charge is represented by Equation (2.36) and shown in Figure 2.9b. According to Equation (2.37), line charges with charge reversals at $s/2$ and $-s/2$ are governed by

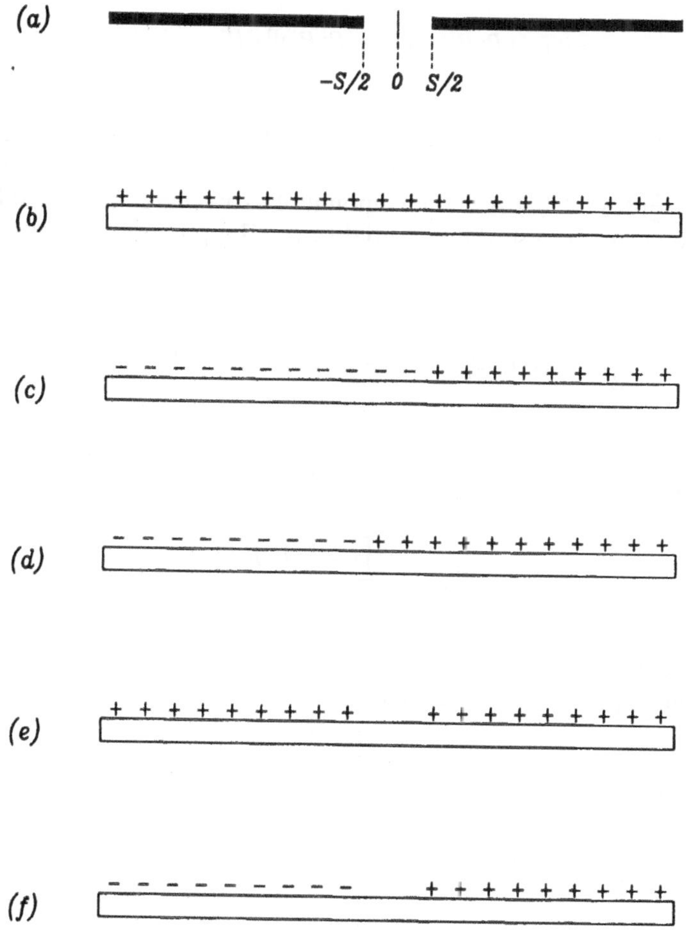

FIGURE 2.9 Formulation of the microstrip gap in terms of line charges. (*a*) Microstrip with a gap. (*b*) G_∞; infinitely extending line charge. (*c*) $G_{s/2}$; charge reversal at $s/2$. (*d*) $G_{-s/2}$; charge reversal at $-s/2$. (*e*) $G_{even} = G_\infty + 1/2(G_{s/2} - G_{-s/2})$. (*f*) $G_{odd} = 1/2(G_{s/2} + G_{-s/2})$

$$\frac{1}{2}\phi_{s/2}(P)=\frac{1}{2}\int\sigma_{\infty}(P')G_{s/2}(P;P')dP' \tag{2.38}$$

$$\frac{1}{2}\phi_{-s/2}(P)=\frac{1}{2}\int\sigma_{\infty}(P')G_{-s/2}(P;P')dP' \tag{2.39}$$

and shown in Figures 2.9c and 2.9d.

If we superposition these lines by adding Equations (2.37) and (2.38) and subtracting Equation (2.39), the result is

$$\phi_{\infty}(P)+\frac{1}{2}\{\phi_{s/2}(P)-\phi_{-s/2}(P)\}=\int\sigma_{\infty}(P')\{G_{\infty}(P;P') \\ +\frac{1}{2}[G_{s/2}(P;P')-G_{-s/2}(P;P')]\}dP' \tag{2.40}$$

Equation (2.40) represent the charge distribution of the symmetric excitation represented by Figure 2.9e. Note that on the strips the potential is not ϕ_{∞} but rather $\phi_{\infty}+\frac{1}{2}[\phi_{s/2}-\phi_{-s/2}]$. A certain amount of extra charge ϕ_e^{even} must be added to the two strips to raise the potential to ϕ_{∞}. The potential corresponding to the extra charge is

$$\frac{1}{2}\{\phi_{s/2}(P)-\phi_{-s/2}(P)\}=\int\sigma_e^{\text{even}}(P')G^{\text{even}}(P;P')dP' \tag{2.41}$$

Noting that the excess charge $\sigma_e^{\text{even}}(P')$ is responsible for the discontinuity capacitance C_{even} and solving Equation (2.41) for $\sigma_e^{\text{even}}(P')$ results in

$$C_{\text{even}}=\frac{2\int\sigma_e^{\text{even}}(P')dP'}{\phi_{\infty}} \tag{2.42}$$

To evaluate C_{odd}, we use a similar process. Subtracting Equations (2.38) and (2.39) from Equation (2.37) results in

$$\phi_{\infty}(P)-\frac{1}{2}\{\phi_{s/2}(P)+\phi_{-s/2}(P)\} \\ =\int\sigma_{\infty}(P')\{G_{\infty}(P;P')-\frac{1}{2}[G_{s/2}(P;P')+G_{-s/2}(P;P')]\}dP' \tag{2.43}$$

which represents the charge distribution of the asymmetrical excitation shown in Figure 2.9f. A certain amount of charge is needed to raise the potential to ϕ_{∞} for $z>s/2$ and lower the potential to $-\phi_{\infty}$ for $z<-s/2$. The extra charge is σ_e^{odd} and $-\sigma_e^{\text{odd}}$. The corresponding integral equation is

$$\phi_\infty - \frac{1}{2}\{\phi_{s/2}(P) + \phi_{-s/2}(P)\} = \int \sigma_e^{\text{odd}}(P')G^{\text{odd}}(P;P')dP' \tag{2.44}$$

and C_{odd} is evaluated from

$$C_{\text{odd}} = \frac{2\int \sigma_e^{\text{odd}}(P')dP'}{\phi_\infty} \tag{2.45}$$

Using the concepts outlined earlier, Silvester and Benedek calculated the capacitance for a gap in a microstrip line [22]. The Green's functions for the microstrip line are obtained by considering the multiple images of a line charge when placed parallel to a dielectric slab [24]. Equations (2.42) and (2.43) permit the solutions for excess charge density and excess capacitance directly. Subtraction of two nearly equal large quantities is avoided.

Numerical results for C_{odd} and C_{even} are avilable in the form of graphs the have been plotted for some discrete values of parameters [22]. The coupling capacitance C_2 decreases with an increase in gap spacing, and for infinite spacing, C_2 should approach zero. The shunt capacitance C_1 should equal the end capacitance of an open-ended line for an infinite spacing. Difficulty arises when capacitance values are needed for parameters that have not been graphed. The number of available graphs is limited, and interpolation methods between these discrete values are not given. To solve this problem, Garg and Bahl [25] have taken the numerical results of Silvester and Benedek [22] and obtained closed-form expressions for C_{odd} and C_{even}. The closed-form expressions were obtained by using polynomial approximations of the available numerical results. The numerical results for C_1 and C_2 are as follows [25, 26]:

$$C_1 = \frac{1}{2}C_{\text{even}}$$
$$C_2 = \frac{1}{2}\left(C_{\text{odd}} - \frac{1}{2}C_{\text{even}}\right) \tag{2.46}$$

where

$$\frac{C_{\text{odd}}}{w} = \left(\frac{s}{w}\right)^{m_o} e^{k_o} \quad [pF/m] \tag{2.47}$$

$$\frac{C_{\text{even}}}{w} = 12\left(\frac{s}{w}\right)^{m_e} e^{k_e} \quad [pF/m] \tag{2.48}$$

applicable for $\varepsilon_r = 9.6$ and $0.5 \le w/h \le 2.0$, where

$$\left.\begin{aligned} m_o &= (w/h)(0.619\log(w/h) - 0.3853) \\ k_o &= 4.26 - 1.453\log(w/h) \end{aligned}\right\} \quad 0.1 \le s/w \le 1.0 \tag{2.49}$$

$$\left.\begin{aligned} m_e &= 0.8675 \\ k_e &= 2.043(w/h)^{0.12} \end{aligned}\right\} \quad 0.1 \le s/w \le 0.3 \tag{2.50}$$

$$\left.\begin{aligned} m_e &= \left(1.565/(w/h)^{0.16}\right) - 1 \\ k_e &= 1.97 - (0.03/(w/h)) \end{aligned}\right\} \quad 0.3 \le s/w \le 1.0 \tag{2.51}$$

Note that there is an error in the calculation of C_{even} from the equation given in [25, 26]. The correct expression is shown here in Equation (2.48). The values of C_{even} and C_{odd} for other values of ε_r in the range $2.5 \le \varepsilon_r \le 15$ can be calculated by using the following scaling factors:

$$C_{\text{even}}(\varepsilon_r) = C_{\text{even}}(9.6)\left(\frac{\varepsilon_r}{9.6}\right)^{0.9} \tag{2.52}$$

$$C_{\text{odd}}(\varepsilon_r) = C_{\text{odd}}(9.6)\left(\frac{\varepsilon_r}{9.6}\right)^{0.8} \tag{2.53}$$

In the expressions for C_{odd} and C_{even}, w is the strip width, h is the substrate height, and s is the gap width. The expressions for the capacitances are quoted to an accuracy of 7% for the mentioned ranges. An example of the capacitance values that can be expected is shown in Figure 2.10.

2.4.2 Transmission-Line Equivalent Circuit

The ring resonator can be modeled by its transmission-line equivalent circuit. In filter analysis it is a common practice to employ a lumped-parameter-equivalent, two-port network for a particular length of transmission line. It is assumed that the length and impedance of the line represented is known. The general T-network is chosen for the analysis and shown in Figure 2.11. The lumped parameters, Z_a and Z_b, are expressed as follows:

$$Z_a = Z_0 \tanh\frac{\gamma l}{2} \tag{2.54}$$

$$Z_b = \frac{Z_0}{\sinh \gamma l} \tag{2.55}$$

where γ is the propagation constant, l is the length of line represented, and Z_0 is the characteristic impedance of the line.

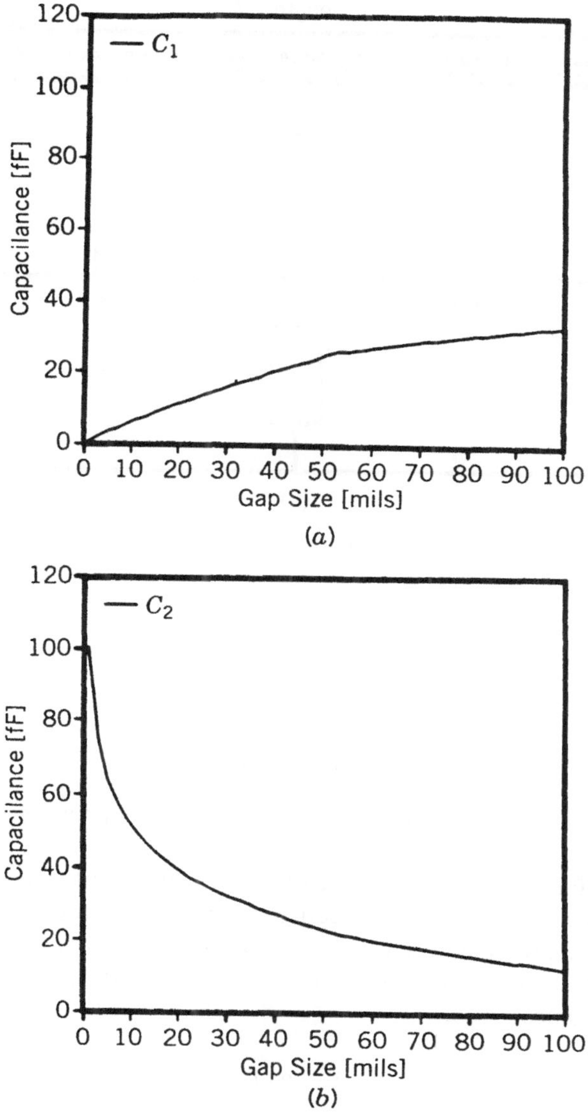

FIGURE 2.10 Coupling capacitance (*a*) C_1 and (*b*) C_2 for w = 2.3495 mm, h = 0.762 mm, and $\varepsilon_r = 2.2$.

A transmission line can be characterized by four quantities: a resistance R along the line, an inductance L along the line, a conductance G shunting the line, and a capacitance C shunting the line. From these primary constants the propagation of the wave along a line can be characterized by the complex propagation constant γ as

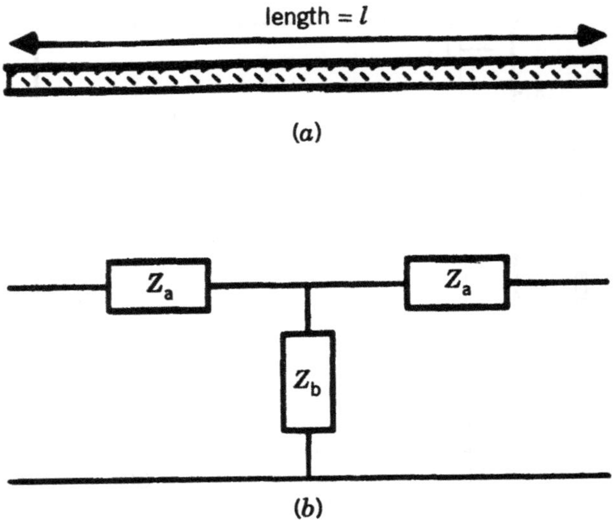

FIGURE 2.11 (*a*) Transmission line of length *l* and (*b*) the T-network equivalent.

$$\gamma = \sqrt{(R + j\omega L)(G + j\omega C)} \tag{2.56}$$

or

$$\gamma = \alpha + j\beta \tag{2.57}$$

where α = the attenuation constant and β = the phase constant (wave-number).

In most RF transmission lines the effects due to L and C tend to dominate, because of the relatively high inductive reactance and capacitive susceptibility. These lines are generally referred to as "loss-free" lines. If loss-free lines are assumed, then R and G in Equation (2.56) become negligible, and the equation becomes

$$\gamma \approx j\omega\sqrt{LC} \tag{2.58}$$

or

$$\gamma \approx j\beta \tag{2.59}$$

Substituting for γ in Equations (2.54) and (2.55) yields the T-network parameters for loss-free lines:

$$Z_a = jZ_0 \tan\frac{\beta l}{2} \tag{2.60}$$

$$Z_b = -jZ_0 \csc \beta l \tag{2.61}$$

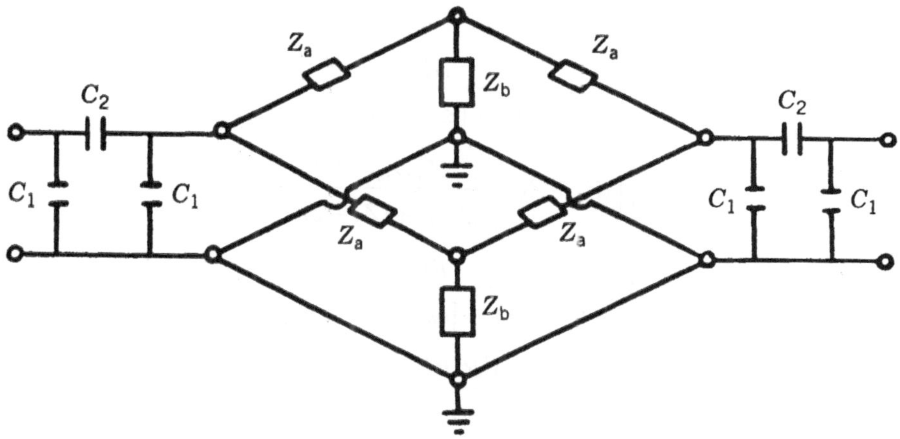

FIGURE 2.12 Equivalent circuit for the ring resonator [8]. (Permission from IEEE.)

Equations (2.60) and (2.61) are used for equivalent-circuit analysis.

2.4.3 Ring Equivalent Circuit and Input Impedance

The coupling gap and transmission line of the ring resonator have been modeled by their lumped-parameter equivalent circuit. The total equivalent circuit can now be pieced together to form a two-port network like that shown in Figure 2.12. The circuit can be reduced to a one-port circuit by terminating one of the two ports with an arbitrary impedance. The terminating impedance should correspond to the impedance of the feed lines. The feed lines will normally have an impedance equal to the impedance of the test equipment that they connect to. The standard for microwave measurements is 50 Ω.

Because of the symmetry of the circuit, the input impedance can be found by simplifying parallel and series combinations. The input impedance is expressed as [8, 13]:

$$R_{in} = \frac{C(C_1+C_2)[(C_1+C_2)-\omega D(C_1^2+2C_1C_2)]}{[(C_1+C_2)-\omega D(C_1^2+2C_1C_2)]^2+[\omega C(C_1^2+2C_1C_2)]^2} + \frac{[D(C_1+C_2)-\omega^{-1}][\omega C(C_1^2+2C_1C_2)]}{[(C_1+C_2)-\omega D(C_1^2+2C_1C_2)]^2+[\omega C(C_1^2+2C_1C_2)]^2} \tag{2.62}$$

$$X_{in} = \frac{[D(C_1+C_2)-\omega^{-1}][(C_1+C_2)-\omega D(C_1^2+2C_1C_2)]}{[(C_1+C_2)-\omega D(C_1^2+2C_1C_2)]^2+[\omega C(C_1^2+2C_1C_2)]^2} + \frac{[D(C_1+C_2)-\omega C^2(C_1+C_2)(C_1^2+2C_1C_2)]}{[(C_1+C_2)-\omega D(C_1^2+2C_1C_2)]^2+[\omega C(C_1^2+2C_1C_2)]^2} \tag{2.63}$$

where

$$C = \frac{AZ_b^2}{(2A)^2 + (Z_a - 2B - Z_b)^2}$$

$$D = \frac{1}{2}\left[(Z_a - Z_b) - \frac{Z_b^2(Z_a - 2B - Z_b)}{(2A)^2 + (Z_a - 2B - Z_b)^2}\right]$$

$$A = \frac{RC_2^2}{(C_1 + C_2)^2 + [\omega R(C_1^2 + 2C_1C_2)]^2}$$

$$B = \frac{(C_1 + C_2) + \omega^2 R^2 (C_1^2 + 2C_1C_2)(C_1 + C_2)}{\omega(C_1 + C_2)^2 + \omega[\omega R(C_1^2 + 2C_1C_2)]^2}$$

where R is the terminated load. The input impedance is

$$Z_{\text{in}} = R_{\text{in}} + jX_{\text{in}} \tag{2.64}$$

The equivalent circuit of the ring can also be modeled using commercially available software such as *Touchstone* or *Supercompact*. The resonant frequency for the circuit is defined as the frequency that makes the impedance seen by the source purely resistive. In other words, the circuit resonates when $X_{\text{in}} = 0$.

Using Equations (2.62) and (2.63), the impedance can be plotted as a function of frequency. The normalized imaginary (X_{in}/Z_0) and real impedance (R_{in}/Z_0) are shown for an arbitrary circuit in Figure 2.13. As can be seen from the imaginary impedance (see Fig. 2.13b), there are two resonance points ($X_{\text{in}} = 0$), f_s and f_p. The resonance f_s is a series resonance. At the frequency f_s, the imaginary impedance is equal to zero, and the real impedance has a normalized value of 1 (see Fig. 2.13a). The resonance f_p is a parallel resonance point. In the imaginary impedance, f_p is an asymptote that is approached from positive infinity and negative infinity. The resonance f_p is a parallel resonance point. At f_p the real impedance has a maximum value. The circuit Q of the ring resonator can be shown to be directly related to the size of the coupling gap. As the size of the gap is increased, the series and parallel resonance points become closer together and the Q is increased. The difference between f_s and f_p as a function of gap size is shown in Figure 2.14. We will see later in the experimental results that as the coupling gap is increased, the circuit Q is increased. The impedance function of a microstrip ring is similar to the piezoelectric quartz crystal [27]. The crystal also has parallel and series resonance points. The high Q in the crystal is a result of the low impedance at f_s followed by the high impedance at f_p. The same is true for the ring resonator. The close resonance points result in a steeper attenuation slope before and after the resonant frequency than with conventional resonator filters.

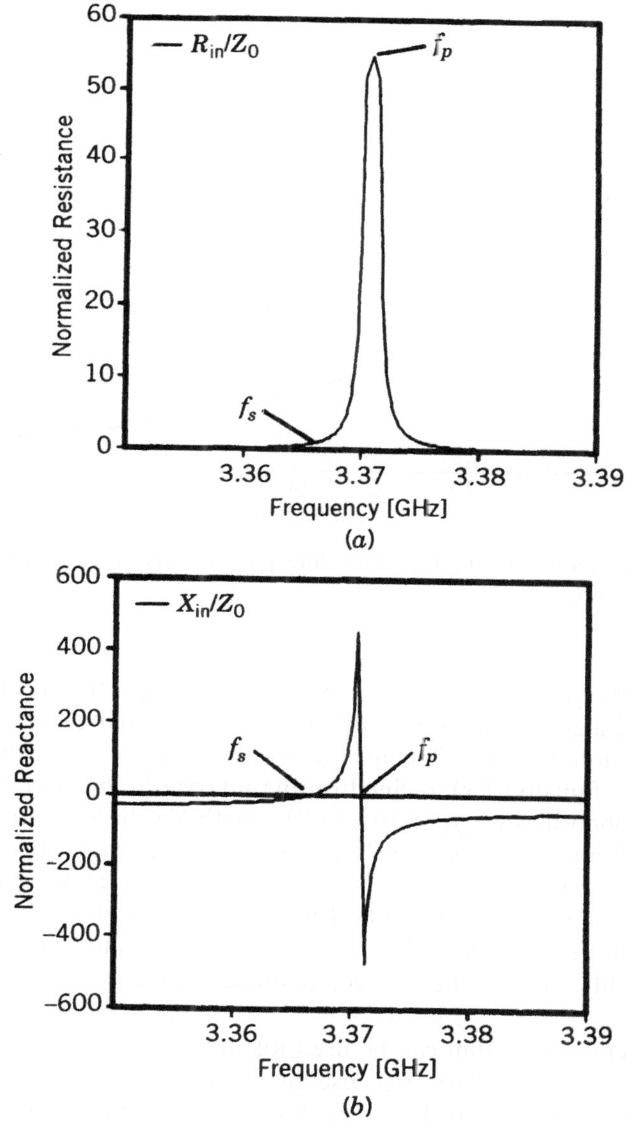

FIGURE 2.13 Normalized input (*a*) resistance and (*b*) reactance for a ring with $\varepsilon_r = 2.2$, $h = 0.762$ mm, $w = 2.34954$ mm, gap = 0.520 mm, and $r = 10.2959$ mm.

2.4.4 Frequency Solution

The solution of Equation (2.63) for the resonance condition $X_{in} = 0$ is merely a root-finding problem [8, 13]. There are several methods available to solve this problem, each of which has its advantages and disadvantages. The bisection method was chosen for the analysis. Other methods may offer greater

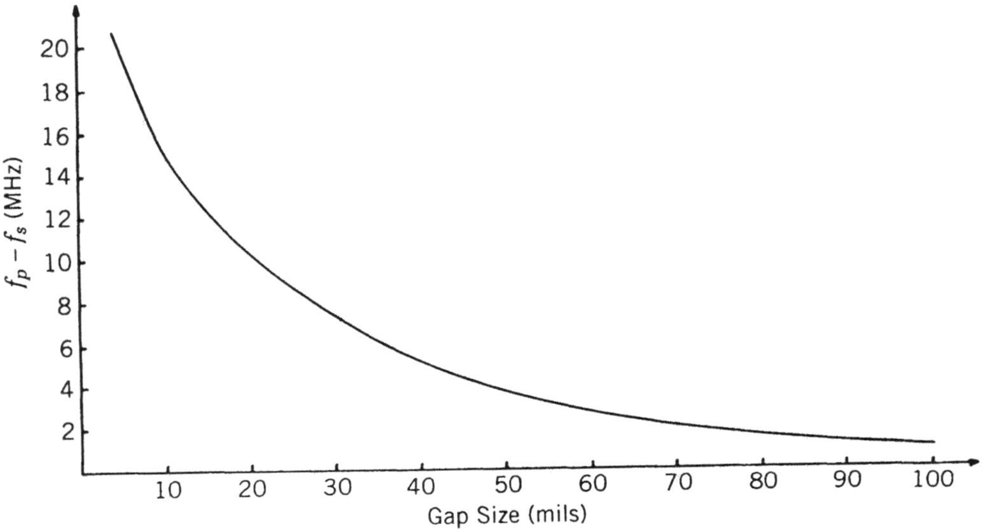

FIGURE 2.14 Difference of the series and parallel resonance frequencies for an increasing gap size.

rates of convergence, but they cannot converge unless the function is well-behaved and a good approximation is used for the initial guess.

The bisection method will converge for all continuous functions. Suppose a continuous function $f(x)$, defined on the interval $[a, b]$, is given, with $f(a)$ and $f(b)$ of opposite sign ($f(a)f(b) < 0$). The method calls for the interval $[a, b]$ to be halved into two subintervals, $[a, p]$ and $[p, b]$, where $p = \frac{1}{2}(a+b)$. The function is evaluated at point p and each subinterval is again checked for opposite signs ($f(a)f(p) < 0$ or $f(p)f(b) < 0$). The interval that contains opposite signs is again halved. This procedure is repeated until the interval being checked is smaller than a given tolerance or the solution is determined exactly.

The bisection algorithm can be used for the solution of the resonant frequency from Equation (2.63). Because this equation has two solutions that are close together, special care has to be taken so that only the desired root is obtained. It would be inconsistent to allow the algorithm to solve for f_s one time and f_p another time. To avoid this inconsistency the root can be found by a moving interval that always approaches from the same side. The interval $[a, b]$ is made smaller than the difference $f_p - f_s$. To find the series resonance the initial guess f_0 is made smaller than f_s. The interval to be checked, $[a, b]$, is then started at f_0($a = f_0$ and $b = f_0 + (b - a)$). If no solution is found in that interval ($f(a)f(b) > 0$), it is moved such that $a = f_0 + (b - a)$ and $b = f_0 + 2(b - a)$. The interval is gradually moved until the solution lies within it. When the solution is known to lie within the interval, the bisection algorithm is used to determine the solution.

TABLE 2.2 Data for the Circuits Used to Verify the Circuit Model

Circuit	Substrate	Relative Permittivity	Height (mm)	Width (mm)	Mean Radius (mm)	Gap (mm)
1	6010	10.5	0.635	0.602	6.984	0.077
2	6010	10.5	0.254	0.279	2.451	0.066
3	5880	2.2	0.254	0.838	4.900	0.069

TABLE 2.3 Resonant Frequencies for the Circuits of Table 2.2

Circuit	Resonant Frequency (GHz)	
	$n = 1$	$n = 2$
1	2.56	5.00
2	7.19	14.28
3	7.10	14.13

2.4.5 Model Verification

The transmission-line approach allows analysis of the ring resonators loaded with discontinuities or solid-state devices. This analysis will not be valid if the circuit model does not accurately represent the ring. This proposed equivalent circuit should be verified by experimental results [13]. The most obvious assumption is the use of the end-to-end model for the coupling gap.

Rings were designed on RT/Duroid 6010 and 5880. The data for the circuits are in Table 2.2.

Experimental data on the resonant frequencies was measured for the first two resonant modes ($n = 1$ and $n = 2$). These experimental resonant frequencies are recorded in Table 2.3. The experimental data is then compared with the theoretical resonant frequencies obtained using the transmission-line method (upper half of Table 2.4), and the magnetic-wall model [2] (lower half). It can be seen that the transmission-line method accurately predicts the resonant frequency to within 1%. This is comparable to the results obtained from the magnetic-wall model calculations in [2].

2.4.6 Frequency Modes for Ring Resonators [10]

Unlike the conventional magnetic-wall model, a simple transmission-line model unaffected by boundary conditions is used to calculate the frequency modes of ring resonators of any general shape such as annular, square, or meander. Figure 2.15 shows the configurations of the one-port square and annular ring resonators. For a ring of any general shape, the total length l may be divided into l_1 and l_2 sections. In the case of the square ring, each section

TABLE 2.4 A Comparison of Table 2.3 and the Theoretical Results from (upper) the Transmission Line Method and (lower) the Magnetic-Wall Model

	Frequency Error (%)	
Circuit	$n = 1$	$n = 2$
1	0.79	0.60
2	0.28	0.49
3	0.56	0.28

	Frequency Error (%)	
Circuit	$n = 1$	$n = 2$
1	0.78	0.89
2	0.07	0.37
2	0.63	0.38

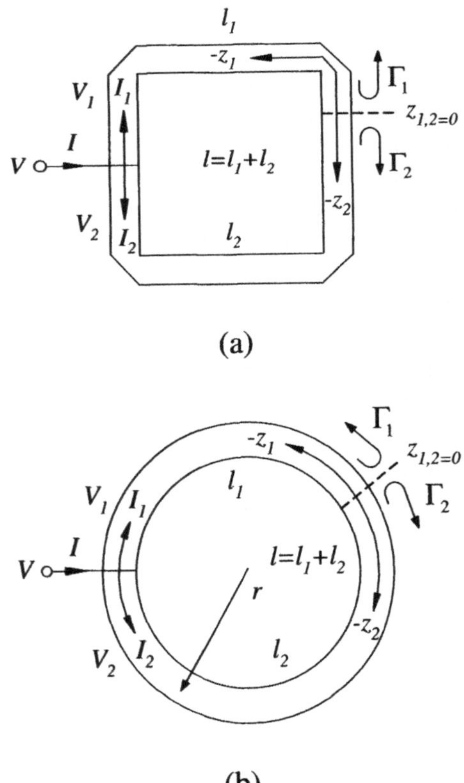

FIGURE 2.15 The configurations of one-port (a) square and (b) annular ring resonators [10].

is considered to be a transmission line. z_1 and z_2 are the coordinates corresponding to sections l_1 and l_2, respectively. The ring is fed by the source voltage V at somewhere with $z_{1,2} < 0$. The positions of the zero point of $z_{1,2}$ and the voltage V are arbitrarily chosen on the ring.

For a lossless transmission line, the voltages and currents for the two sections are given as follows:

$$V_{1,2}(z_{1,2}) = V_o^+\left(e^{-j\beta z_{1,2}} + \Gamma_{1,2}(0)e^{j\beta z_{1,2}}\right) \tag{2.65a}$$

$$I_{1,2}(z_{1,2}) = \frac{V_o^+}{Z_o}\left(e^{-j\beta z_{1,2}} - \Gamma_{1,2}(0)e^{j\beta z_{1,2}}\right) \tag{2.65b}$$

where $V_o^+ e^{-j\beta z_{1,2}}$ is the incident wave propagating in the $+z_{1,2}$ direction, $V_o^+\Gamma_{1,2}(0)e^{j\beta z_{1,2}}$ is the reflected wave propagating in the $-z_{1,2}$ direction, $\Gamma_{1,2}(0)$ is the reflection coefficient at $z_{1,2} = 0$, and Z_0 is the characteristic impedance of the ring.

When a resonance occurs, standing waves set up on the ring. The shortest length of the ring resonator that supports these standing waves can be obtained from the positions of the maximum values of these standing waves. These positions can be calculated from the derivatives of the voltages and currents in Equation (2.65). The derivatives of the voltages are

$$\frac{\partial V_{1,2}(z_{1,2})}{\partial z_{1,2}} = -j\beta V_o^+\left(e^{-j\beta z_{1,2}} - \Gamma_{1,2}(0)e^{j\beta z_{1,2}}\right) \tag{2.66}$$

Letting $\left.\frac{\partial V_{1,2}(z_{1,2})}{\partial z_{1,2}}\right|_{z_{1,2}=0} = 0$, the reflection coefficients can be found as

$$\Gamma_{1,2}(0) = 1 \tag{2.67}$$

Substituting $\Gamma_{1,2}(0) = 1$ into Equation (2.65), the voltages and currents can be obtained as

$$V_{1,2}(z_{1,2}) = 2V_o^+ \cos(\beta z_{1,2}) \tag{2.68a}$$

$$I_{1,2}(z_{1,2}) = -\frac{j2V_o^+}{Z_o}\sin(\beta z_{1,2}) \tag{2.68b}$$

Based on Equation (2.68), the absolute values of voltage and current standing waves on each section l_1 and l_2 are shown in Figure 2.16.

Inspecting Figure 2.16, the standing waves repeat for multiples of $\lambda_g/2$ on the each section of the ring. Thus, to support standing waves, the shortest length of each section on the ring has to be $\lambda_g/2$. which can be treated as the fundamental mode of the ring. For higher order modes,

$$l_{1,2} = n\frac{\lambda_g}{2} \qquad \text{for } n = 1, 2, 3, \ldots \tag{2.69}$$

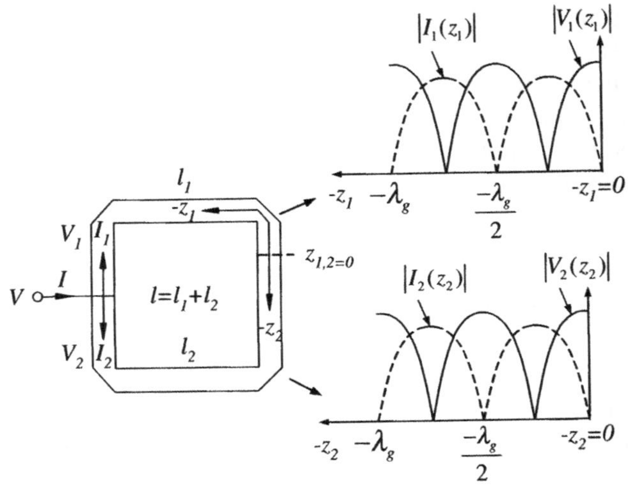

FIGURE 2.16 Standing waves on each section of the square ring resonator [10].

where n is the mode number. Therefore, the total length of the square ring resonator is

$$l = l_1 + l_2 = nl_g \tag{2.70}$$

or in terms of the annular ring resonator with a mean radius r as shown in Figure 2.15b,

$$l = n\lambda_g = 2\pi r \tag{2.71}$$

Equation (2.70) shows a general expression for frequency modes and may be applied to any configuration of microstrip ring resonators, including those shown in [28, 29].

2.4.7 An Error in Literature for One-Port Ring Circuit

In [11], one- and two-port ring resonators show different frequency modes. For a one-port ring resonator, as shown in Figure 2.17a, the frequency modes are given as

$$2\pi r = n\frac{\lambda_g}{2} \qquad n = 1, 2, 3, \ldots \tag{2.72a}$$

$$f_o = \frac{nc}{4\pi r\sqrt{\varepsilon_{eff}}} \tag{2.72b}$$

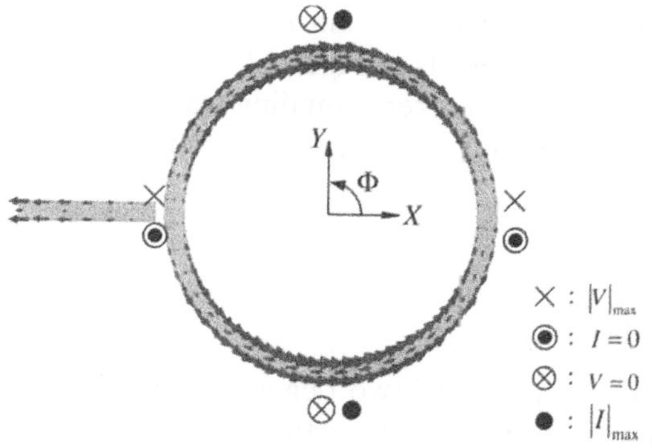

(a)

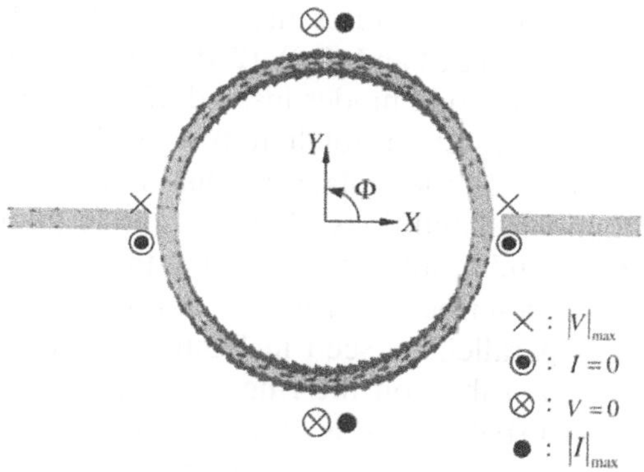

(b)

FIGURE 2.17 Simulated electrical current standing waves for (a) one- and (b) two-port ring resonators at $n = 1$ mode [10].

where f_o is the resonant frequencies. For the two-port ring resonator, as shown in Figure 2.17b, the frequency modes are

$$2\pi r = n\lambda_g \qquad n = 1, 2, 3, \ldots \tag{2.73a}$$

$$f_o = \frac{nc}{2\pi r\sqrt{\varepsilon_{eff}}} \tag{2.73b}$$

However, in Section 2.4.6, the one-port ring resonator has the same frequency modes given in Equation (2.71) as those of the two-port ring resonator given in Equation (2.73a). The results can also be investigated by EM simulation performed by the IE3D electromagnetic simulator based on the method of moment [30]. The ring resonators in Figure 2.17 are designed at fundamental mode at 2 GHz with dielectric constant $\varepsilon_r = 10.2$ and thickness $h = 50$ mil. As seen from the simulation results in Figure 2.17, both exhibit the same electrical current flows, which are current standing waves. Therefore, both one- and two-port ring resonators have the same frequency modes as given in Equations (2.71) or (2.73a).

2.4.8 Dual Mode

The dual mode is composed of two degenerate modes or splitting resonant frequencies that may be excited by perturbing stubs, notches, or asymmetrical feed lines. The dual mode follows from the solution of Maxwell's equations for the magnetic-wall model of the ring resonator in Equations (2.3)–(2.5) and (2.8)–(2.10). However, the ring resonator with a perturbing stub or notch at $\Phi = 45°$, 135°, 225°, or 315° generates the dual mode only for odd modes. Inspecting Equations (2.3)–(2.5) and (2.8)–(2.10), they cannot explain why the dual mode only happens for odd modes instead of even modes when the ring resonator has a perturbing stub or notch at $\Phi = 45°$, 135°, 225°, or 315°. Also, the magnetic-wall model cannot explain the dual mode of the ring resonator with complicate boundary conditions. This dual-mode phenomenon may be explained more simply and more generally using the transmission-line model of Section 2.4.6, which describes the ring resonator as two identical $\lambda_g/2$ resonators connected in parallel. As seen in Figure 2.17, two identical current standing waves are established on the ring resonator in parallel. If the ring does not have any perturbation and is excited by symmetrical feed lines, two identical resonators are excited and produce the same frequency response, which overlap each other. However, if one of the $\lambda_g/2$ resonators is perturbed out of balance with the other, two different frequency modes are excited and couple to each other. To investigate the dual-mode behavior, a perturbed square ring resonator is simulated in Figure 2.18. The perturbed square ring designed at fundamental mode of 2 GHz is fabricated on a RT/Duroid 6010.2 $\varepsilon_r = 10.2$ substrate with a thickness $h = 25$ mil.

Figure 2.18 shows the simulated electric currents on the square ring resonator with a perturbing stub at $\Phi = 45°$ for the $n = 1$ and the $n = 2$ modes. For the $n = 1$ mode, one of $\lambda_g/2$ resonators is perturbed so that the two $\lambda_g/2$ resonators do not balance each other. Thus, two splitting different resonant frequencies are generated. Figures 2.18a and 2.18b show the simulated electrical currents for the splitting resonant frequencies. Figure 2.19 illustrates the measured S21 confirming the splitting frequencies for the $n = 1$ mode around 2 GHz. Furthermore, for the $n = 2$ mode, Figure 2.18c shows the perturbing stub located at the position of zero voltage, which is a short circuit. Therefore, the perturbed stub does not disturb the resonator and both $\lambda_g/2$ resonators balance each other without frequency splitting. Measured results in Figure 2.19 has confirmed that the resonant frequency at the $n = 2$ mode of 4 GHz is not affected by the perturbation.

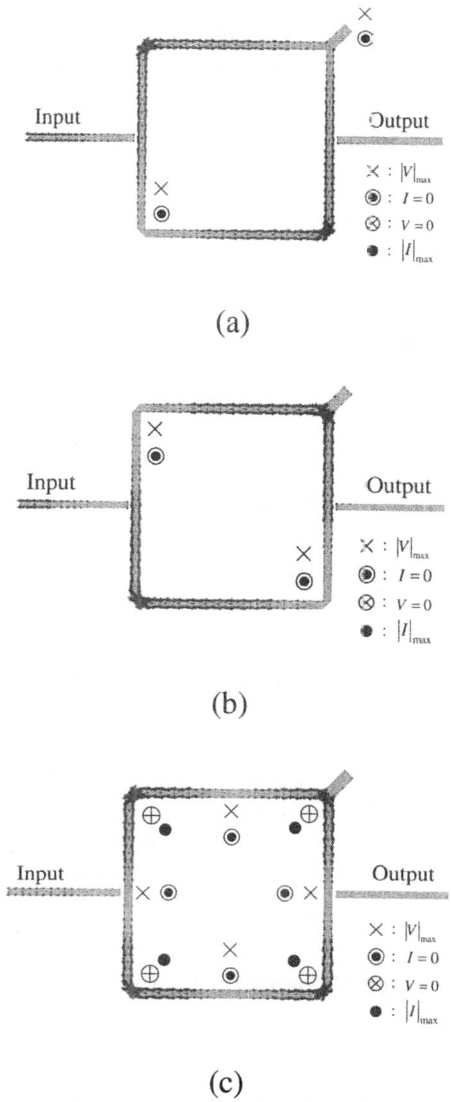

FIGURE 2.18 The simulated electrical currents of the square ring resonator with a perturbed stub at $\Phi = 45°$ for (a) the low splitting resonant frequency of $n = 1$ mode and (b) high splitting resonant frequency of mode $n = 1$, and (c) mode $n = 2$ [10].

2.5 RING EQUIVALENT CIRCUIT IN TERMS OF *G*, *L*, *C*

The basic operation of the ring resonator based on the magnetic-wall model was originally introduced by Wolff and Knoppik [1]. In addition, a simple mode chart of the ring was developed to describe the relation between the physical ring radius and resonant mode and frequency [4]. Although the mode

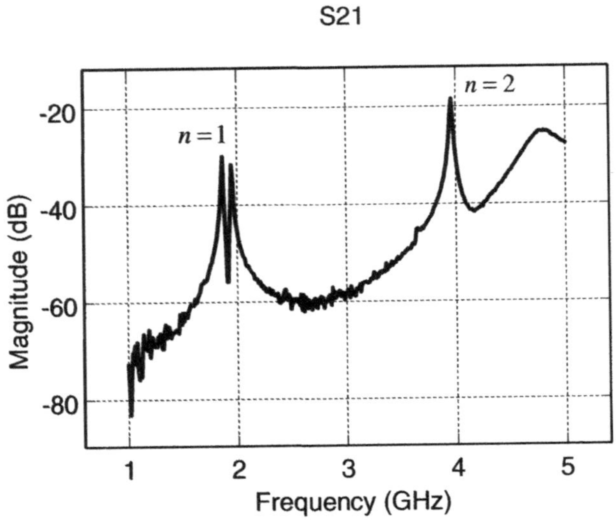

FIGURE 2.19 The measured results for modes $n = 1$ and 2 of the square ring resonator with a perturbed stub at $\Phi = 45°$ [10].

chart of the magnetic-wall model has been studied extensively, it provides only a limited description of the effects of the circuit parameters and dimensions. A further study on a ring resonator using the transmission-line model was introduced in Section 2.4. The transmission-line model used a T-network in terms of equivalent impedances to analyze a ring circuit. Although this model could predict the behavior of a ring resonator well, it could not provide a straightforward circuit view, such as equivalent lumped elements G, L, and C for the ring circuit.

2.5.1 Equivalent Lumped Elements for Closed- and Open-Loop Microstrip Ring Resonators [12]

As seen in Figure 2.20, the two-port network with an open circuit at port 2 ($i_2 = 0$) models a one-port network to find the equivalent input impedance through *ABCD* matrix and Y parameters operations [31].

The ring resonator is divided by input and output ports on arbitrary positions of the ring with two sections l_1 and l_2 to form a parallel circuit. For this parallel circuit, the overall Y parameters are given by

$$\begin{bmatrix} Y_{11} & Y_{12} \\ Y_{12} & Y_{22} \end{bmatrix} = \begin{bmatrix} Y_o[\coth(\gamma l_1) + \coth(\gamma l_2)] & -Y_o[\operatorname{csch}(\gamma l_1) + \operatorname{csch}(\gamma l_2)] \\ -Y_o[\operatorname{csch}(\gamma l_1) + \operatorname{csch}(\gamma l_2)] & Y_o[\coth(\gamma l_1) + \coth(\gamma l_2)] \end{bmatrix} \quad (2.74)$$

By setting i_2 to zero, the input impedance Z_{ic} of the ring can be found as follows:

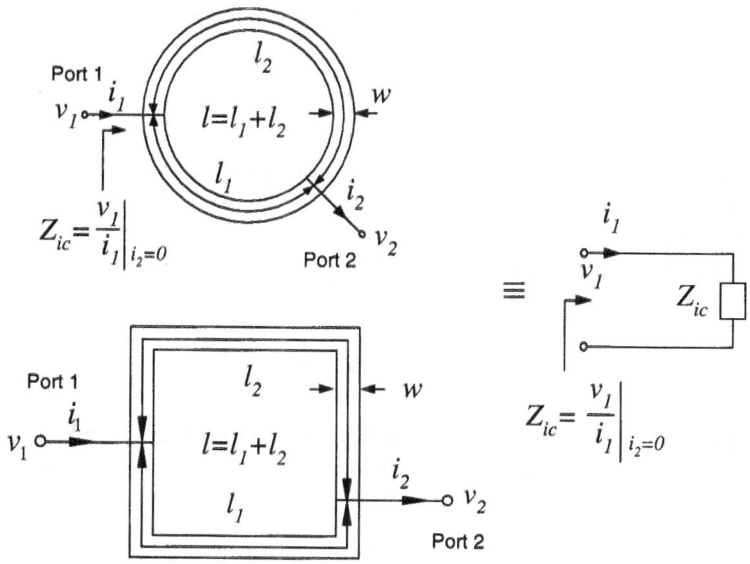

FIGURE 2.20 The input impedance of two-port network of the closed-loop ring resonator [12]. (Permission from IEEE.)

$$Z_{ic}=\frac{Y_{22}}{Y_{11}Y_{22}-Y_{12}Y_{21}}=\frac{Z_o}{2}\frac{1+j\tanh(\alpha l_g)\tan(\beta l_g)}{\tanh(\alpha l_g)+j\tan(\beta l_g)} \tag{2.75}$$

where $l_g = l/2 = \lambda_g/2$. Using some assumptions and derivations for αl_g and βl_g [12], the input impedance Z_{ic} can be approximated as

$$Z_{ic}\cong\frac{Z_a/2\alpha l_g}{1+j\pi\Delta\omega/\alpha l_g\omega_o} \tag{2.76}$$

For a general parallel G, L, C circuit, the input impedance is [32]

$$Z_i=\frac{1}{G+2j\Delta\omega C} \tag{2.77}$$

Comparing Equation (2.76) with Equation (2.77), the conductance of the equivalent circuit of the ring is

$$G_c=2\alpha l_g/Z_o=\alpha\lambda_g/Z_o \tag{2.78a}$$

and the capacitance of the equivalent circuit of the ring is

$$G_c=\pi/Z_o\omega_o \tag{2.78b}$$

The inductance of the equivalent circuit of the ring can be derived from $\omega_0 = 1/\sqrt{L_c C_c}$ and is given by

$$L_c = 1/\omega_o^2 C_c \tag{2.78c}$$

where G_c, L_c, and C_c stand for the equivalent conductance, inductance, and capacitance of the closed-loop ring resonator. Figure 2.21 shows the equivalent lumped element circuit of the ring in terms G_c, L_c, and C_c. Moreover, the unloaded Q of the ring resonator can be found from Equation (2.78) and the unloaded Q is

$$Q_{uc} = \omega_o C_c / G_c = \pi / \alpha\lambda_g \tag{2.79}$$

Figure 2.22a shows the configuration of open-circuited $\lambda_g/2$ microstrip ring resonators with annular and U shapes. As seen in Figure 2.22a, l_3 is the phys-

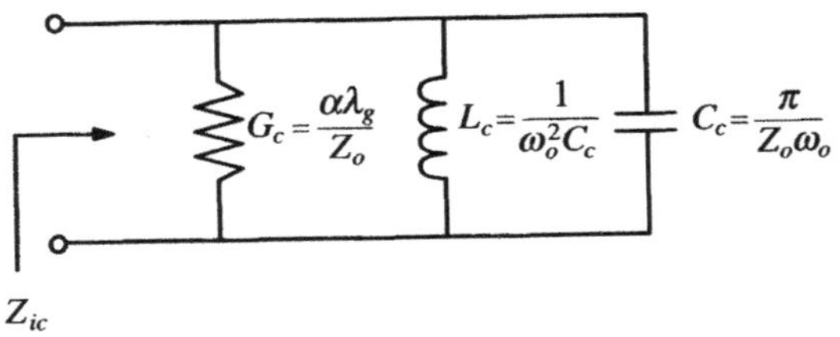

FIGURE 2.21 Equivalent elements G_c, L_c, and C_o of the closed-loop ring [12]. (Permission from IEEE.)

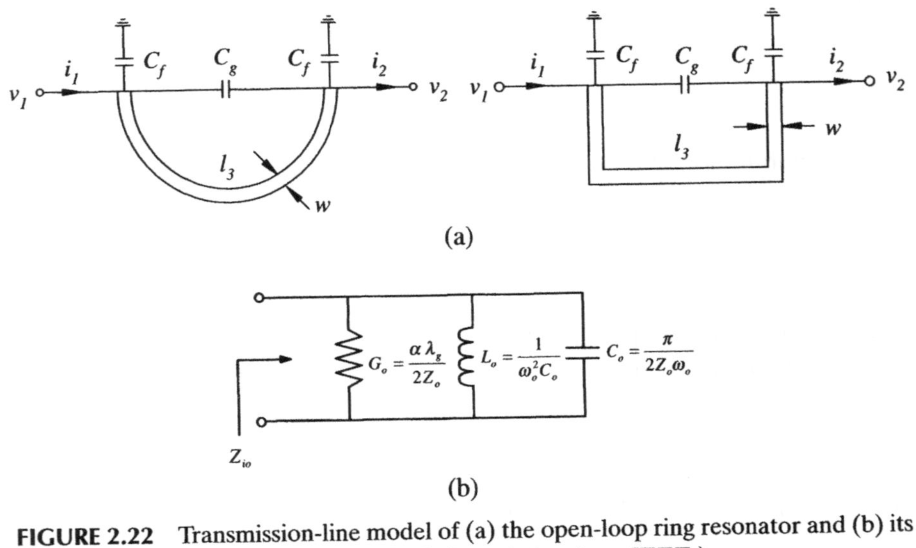

FIGURE 2.22 Transmission-line model of (a) the open-loop ring resonator and (b) its equivalent elements G_o, L_o, and C_o [12]. (Permission from IEEE.)

ical length of the ring, C_g is the gap capacitance, and C_f is the fringe capacitance caused by fringe field at the both ends of the ring. The fringe capacitance can be replaced by an equivalent length Δl [33]. Considering the open-end effect, the equivalent length of the ring is $l_3 + 2\Delta l = \lambda_g/2 = l_g$ for the fundamental mode.

Through the same derivations in Section 2.5.1, the input impedance Z_{io} of the open-loop ring can be approximated as

$$Z_{io} = \frac{Z_o/\alpha l_g}{1 + j\pi\Delta\omega/\alpha l_g \omega_g} \tag{2.80}$$

Comparing Equation (2.80) with Equation (2.77), the conductance, capacitance, and inductance of the equivalent circuit of the ring are

$$G_o = \alpha\lambda_g/2Z_o, \quad C_o = \pi/2Z_o\omega, \quad and \quad L_o = 1/\omega_o^2 C_o \tag{2.81}$$

The equivalent circuit in terms of G_o, L_o, and C_o is shown in Figure 2.22b. Moreover, the unloaded Q of the ring is given by

$$Q_{uo} = \omega_o C_o/G_o = \pi/\alpha\lambda_g \tag{2.82}$$

Inspecting the equivalent conductances, capacitances, and inductances of the closed- and open-loop ring resonators in Equations (2.78) and (2.81), the relations of the equivalent lumped elements *G, L, C* between these two rings can be found as follows:

$$G_c = 2G_o \qquad \text{for the same attenuation constant} \tag{2.83a}$$

$$C_c = 2C_o \quad and \quad L_c = L_c/2 \tag{2.83b}$$

In addition, observing the Equations (2.79) and (2.82), the unloaded Q of the closed- and open-loop ring resonators are equal, namely

$$Q_{uc} = Q_{uo} \qquad \text{for the same attenuation constant} \tag{2.84}$$

Equations (2.83a) and (2.84) sustain for the same losses condition of the closed- and the open-loop ring resonator. In practice, the total losses for the closed- and the open-loop ring resonator are not the same. In addition to the dielectric and conductor losses, the open-loop ring resonator has a radiation loss caused by the open ends [34]. Thus, total losses of the open-loop ring are larger than that of the closed-loop ring. Under this condition, Equations (2.83a) and (2.84) should be rewritten as follows:

$$Q_{uc} > Q_{uo} \quad \text{and} \quad G_c < 2G_o \tag{2.85}$$

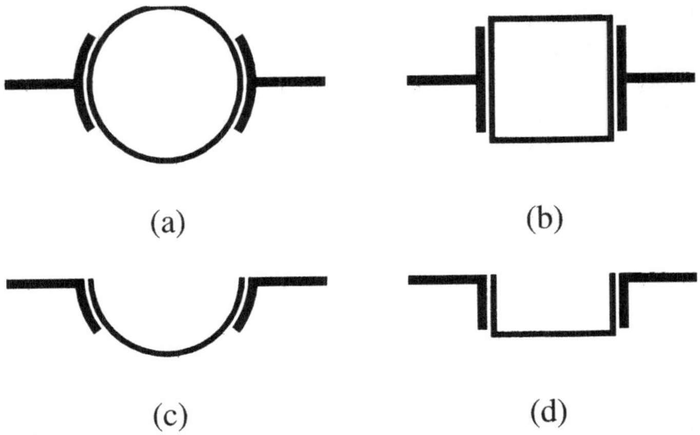

FIGURE 2.23 Layouts of the (a) annular, (b) square, (c) open-loop with the curvature effect, and (d) U-shaped open-loop ring resonators [12]. (Permission from IEEE.)

2.5.2 Calculated and Experimental Results

To verify the calculations of the unloaded Q and G, L, C of the closed- and open-loop ring resonators [12], four configurations of the closed- and open-loop ring resonators as shown in Figure 2.23 were designed at the fundamental mode of 2 GHz. The ring resonators were fabricated for two different dielectric constants: RT/Duriod 5870 with $\varepsilon_r = 2.33$, $h = 10$ mil, and $t = 0.7$ mil and RT/Duriod 6010.2 with $\varepsilon_r = 10.2$, $h = 10$ mil, and $t = 0.7$ mil, where ε_r is the relative dielectric constant, h is the substrate thickness, t is the foil thickness, and Δ is the surface roughness.

As seen in Tables 2.5 through 2.8, the measured unloaded Qs and equivalent lumped elements of the closed- and open-loop rings show good agreement with each other. The largest difference between the measured and calculated unloaded Q shown in Table 2.7 for the closed-loop square ring resonator is 5.7%.

2.6 DISTRIBUTED TRANSMISSION-LINE MODEL

The transmission-line model described in Section 2.4 is straightforward and provides reasonably accurate results for simple circuits at low frequencies. The method lends itself to CAD implementation, and circuits loaded with solid-state devices and discontinuities along the rings can be analyzed. However, the model is not accurate because the effects of the dispersive nature of the microstrip line and curvature of the ring resonator are neglected. A more accurate distributed transmission-line model has been proposed to overcome these problems [9, 35]. The model includes the losses and can deal with multiple devices, discontinuities, and feeds located at any place along the ring. This section summarizes this method based on [35].

TABLE 2.5 Unloaded Qs for the Parameters: $\varepsilon_r = 2.33$, $h = 10$ mil, $t = 0.7$ mil, $w = 0.567$ mm for a 60-ohms Line, $\Delta = 1.397\ \mu$m, and $\lambda_g = 108.398$ mm

Resonators	Annular Ring	Open-loop Ring with the Curvature Effect	Square Ring	U-Shape Open-loop Ring
Designed Resonant Frequency (GHz)	2	2	2	2
Measured Resonant Frequency (GHz)	1.963	1.964	1.977	1.983
Measured Insertion Loss L_{meas} (dB)	32.66	31.33	32.3	33.12
Measured 3-dB Bandwidth $BW_{3dB,means}$(MHz)	19	19.5	19	19.5
Measured Loaded Q	103.32	100.72	104.05	101.69
Measured Unloaded Q	105.78	103.53	106.64	103.98
Calculated Unloaded Q	103.35	102.41	103.35	102.41

TABLE 2.6 Equivalent Elements for the Parameters: $\varepsilon_r = 2.33$, $h = 10$ mil, $t = 0.7$ mil, $w = 0.567$ mm for a 60-ohms Line, $\Delta = 1.397\ \mu$m, and $\lambda_g = 108.398$ mm

Resonators	Annular Ring	Open-loop Ring with the Curvature Effect	Square Ring	U-Shape Open-loop Ring
Calculated α (dB/mm)	2.45×10^{-3}	2.43×10^{-3}	2.45×10^{-3}	2.43×10^{-3}
Calculated Conductance G (mS)	0.508	0.256	0.508	0.256
Calculated Capacitor C (pF)	4.17	2.08	4.17	2.08
Calculated Inductor L (nH)	1.52	3.04	1.52	3.04
Measured α_{meas} (dB/mm)	2.38×10^{-3}	2.43×10^{-3}	2.36×10^{-3}	2.42×10^{-3}
Measured Conductance G (mS)	0.495	0.253	0.49	0.252
Measured Capacitor C (pF)	4.25	2.12	4.22	2.1
Measured Inductor L (nH)	1.55	3.1	1.54	3.07

2.6.1 Microstrip Dispersion

When a radio frequency (RF) wave propagates down a microstrip line, both longitudinal and transverse currents are excited. These currents cause the normally independent longitudinal section electric (LSE) and longitudinal section magnetic (LSM) modes to couple, thereby producing a hybrid mode

TABLE 2.7 Unloaded *Q*s for the Parameters: $\varepsilon_r = 10.2$, $h = 10$ mil, $t = 0.7$ mil, $w = 0.589$ mm for a 30-ohms Line, $\Delta = 1.397\ \mu m$, and $\lambda_g = 55.295$ mm

Resonators	Annular Ring	Open-loop Ring with the Curvature Effect	Square Ring	U-Shape Open-loop Ring
Designed Resonant Frequency (GHz)	2	2	2	2
Measured Resonant Frequency (GHz)	1.974	1.968	2.03	2.03
Measured Insertion Loss L_{meas} (dB)	35.83	35.48	35.48	33.4
Measured 3-dB Bandwidth $BW_{3dB,mens}$ (MHz)	20.5	21	20.5	21
Measured Loaded Q	96.29	95.36	97.71	95.38
Measured Unloaded Q	97.87	96.99	99.38	97.46
Calculated Unloaded Q	93.65	93.21	93.65	93.21

TABLE 2.8 Equivalent Elements for the Parameters: $\varepsilon_r = 10.2$, $h = 10$ mil, $t = 0.7$ mil, $w = 0.589$ mm for a 30-ohms Line, $\Delta = 1.397\ \mu m$, and $\lambda_g = 55.295$ mm

Resonators	Annular Ring	Open-loop Ring with the Curvature Effect	Square Ring	U-Shape Open-loop Ring
Calculated α (dB/mm)	5.29×10^{-3}	5.27×10^{-3}	5.29×10^{-3}	5.27×10^{-3}
Calculated Conductance G (mS)	1.12	0.56	1.12	0.56
Calculated Capacitor C (pF)	8.33	4.17	8.33	4.17
Calculated Inductor L (nH)	0.76	1.52	0.76	1.52
Measured α_{meas} (dB/mm)	5.04×10^{-3}	5.09×10^{-3}	4.97×10^{-3}	5.06×10^{-3}
Measured Conductance G (mS)	1.06	0.54	1.05	0.54
Measured Capacitor C (pF)	8.44	4.23	8.21	4.11
Measured Inductor L (nH)	0.77	1.54	0.75	1.5

configuration [36]. The coupling increases with frequency, owing to the betten confinement of the fields to the dielectric at higher frequencies. This can be mathematically represented by introducing a frequency-dependent expression for the effective dielectric constant ($\varepsilon_{eff}(f)$). A nonlinear relation between the wave number and frequency is thus introduced, causing different frequencies to propagate at different velocities. This phenomenon is termed *microstrip dispersion.*

Kirschning and Jansen [37] proposed an accurate closed-form empirical relation for $\varepsilon_{\text{eff}}(f)$ that can readily be implemented into any CAD program. This is given by

$$\varepsilon_{eff}(f) = \varepsilon_r - \frac{\varepsilon_r - \varepsilon_e}{1 + P(f)} \tag{2.86}$$

where

$$P(f) = P_1 P_2 [(0.1844 + P_3 P_4) 10 fh]^{1.5763} \tag{2.87}$$

with

$$P_1 = 0.27488 + \left[0.6315 + \frac{0.525}{(1 + 0.157 fh)^{20}}\right]\frac{w}{h} - 0.065683 e^{-8.7513 w/h} \tag{2.88}$$

$$P_2 = 0.33622[1 - e^{-0.03442 \varepsilon_r}] \tag{2.89}$$

$$P_3 = 0.0363 e^{-4.6 w/h}\left[1 - e^{(-fh/3.87)^{4.97}}\right] \tag{2.90}$$

$$P_4 = 1 + 2.751\left[1 - e^{(-\varepsilon_r/15.916)^8}\right] \tag{2.91}$$

where f is the frequency in GHz; w and h are the microstrip width and height in cm, respectively; ε_r is the relative dielectric constant of the substrate; and ε_e is the static value of the effective dielectric constant, which is dependent on the geometry of the microstrip. In the limit $f \to 0$, $\varepsilon_{\text{eff}}(f) \to \varepsilon_e$. Here ε_e is given by

$$\varepsilon_e = \frac{\varepsilon_r + 1}{2} + \frac{\varepsilon_r - 1}{2} F\left(\frac{w}{h}\right) - \frac{\varepsilon_r - 1}{4.6} \frac{t/h}{\sqrt{w/h}} \tag{2.92}$$

where

$$F\left(\frac{w}{h}\right) = \begin{cases} \left(1 + \frac{12h}{w}\right)^{-0.5} + 0.04\left(1 - \frac{w}{h}\right)^2 & \text{if } \frac{w}{h} \le 1 \\ \left(1 + \frac{12h}{w}\right)^{-0.5} & \text{if } \frac{w}{h} \ge 1 \end{cases} \tag{2.93}$$

In the preceding equation, t denotes the thickness of the metal that constitutes the microstrip line. The accuracy of Equation (2.86) is better than 0.6% in the range $0.1 \le w/h \le 100$, and $1 \le \varepsilon_r \le 20$, and is valid up to about 60 GHz. This equation spans a fairly wide variety of frequencies and dielectric substrates, hence $\varepsilon_{\text{eff}}(f)$ can be evaluated very accurately.

2.6.2 Effect of Curvature

A curved microstrip line can be modeled as a cascade of sections of microstrip lines with chamfered bends. Illustrated in Figure 2.24a is a typical bend in a microstrip line for an arbitrary bend angle θ; also shown in the same figure are the reference planes that define the edges of the bend. The equivalent circuit of the bend, in the region restricted to the confines of the reference planes, is shown in Figure 2.24b. For optimum chamfer, the ratio of the width of the chamfered region b to the width of the microstrip line w is approximately 0.5 [38]. In the equivalent-circuit representation of the bend, the inductance L and capacitance C represent the inductance associated with the discontinuity and the capacitance to ground, respectively. Kirschning et al. [39] derived an empirical closed-form expression to represent the equivalent circuit of the bend. For optimum chamfer, the capacitance C (pf) and inductance L (nH) are given by

$$C = 0.001h\left(\frac{180-\theta}{90}\right)\left[(3.393\varepsilon_r + 0.62)\left(\frac{w}{h}\right)^2 + (7.6\varepsilon_r + 3.8)\left(\frac{w}{h}\right)\right] \quad (2.94)$$

$$L = 0.44h\left(\frac{180-\theta}{90}\right)\left[1 - 1.062e^{-0.177(w/h)^{0.947}}\right] \quad (2.95)$$

where h and ε_r are the thickness in mm and the dielectric constant, respectively, of the substrate; and θ is the angle of the chamfer in degrees, and in the limit $\theta \to 180$, $C, L \to 0$. This reduces to the straight-line case; hence there are no discontinuities. These equations are in general valid for w/h and ε_r in the ranges $0.2 \le w/h \le 6$, and $2 \le \varepsilon_r \le 13$. When $0.2 \le w/h \le 1$, the accuracy of the

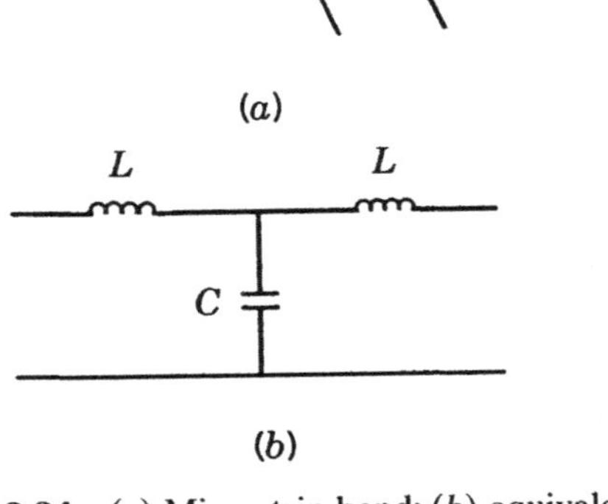

FIGURE 2.24 (*a*) Microstrip bend; (*b*) equivalent circuit.

model is within 0.3%. Since ring resonators are usually built on substrates with dielectric constants greater than 2, and since $w/h \leq 1$ for standard 50-Ω lines on high dielectric-constant substrates, this model can be applied to accurately model the curvature of conventional microstrip ring resonators. A more detailed account of the application of this model to the microstrip ring resonator is presented in the following.

2.6.3 Distributed-Circuit Model

The distributed ring circuit model is described in [9, 35]. The basic microstrip ring resonator is illustrated in Figure 2.25. Power is coupled into and out of the resonator via two feed lines located at diametrically opposite points. If the distance between the feed lines and the resonator is large, then the coupling gap does not affect the resonant frequencies of the ring. The resonator in this case is said to be "loosely coupled." Loose coupling is a manifestation of the negligibly small capacitance of the coupling gap. If the feed lines are moved closer to the resonator, however, the coupling becomes tight and the gap capacitance becomes appreciable. This causes the circuit's resonant frequency to deviate from the intrinsic resonant frequency of the ring. Hence, to accurately model the ring resonator, the capacitance of the coupling gap should be considered in conjunction with microstrip dispersion and curvature.

When the mean circumference of the ring resonator is equal to an integral multiple of a guided wavelength, resonance is established. This may be expressed as

$$2\pi r = n\lambda_g \qquad \text{for } n = 1, 2, \ldots \tag{2.96}$$

where r is the mean radius of the ring (i.e., $r = (r_i + r_o)/2$); λ_g is the guided wavelength; and n is the mode number. This relation is valid for the loose coupling case, as it does not take into consideration the effect of the coupling gap.

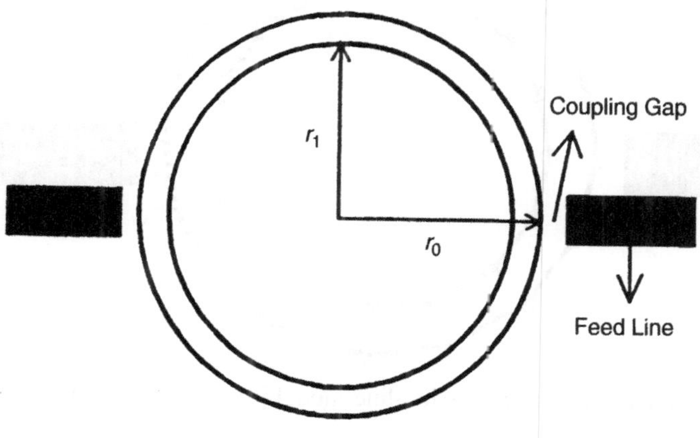

FIGURE 2.25 Layout of the microstrip ring resonator.

In order to apply the distributed transmission-line model, the mean radius of the ring resonator must be known. This may be estimated from Equation (2.96) as follows: For a given frequency and a dielectric material of known thickness, the dimensions of a 50-Ω line are estimated from a commercially available program called *Linecalc* [40]; the effective dielectric constant and guided wavelength are estimated from Equations (2.86) through (2.93). The value of the guided wavelength thus determined is substituted into Equation (2.96) to evaluate the mean radius of the ring. Although the resonant frequencies of an ideal ring resonator are independent of the characteristic impedance of the line that forms the closed loop, it is conventional to use lines whose characteristic impedance corresponds to 50 Ω.

The approach underlying the distributed transmission-line model is that the ring is analyzed as a polygon of n sides. This is illustrated in Figure 2.26 wherein the ring resonator is represented by a 16-sided polygon. In actuality, however, a 36-sided polygon was used, and it was found that any further increase in the number of sides did not improve the accuracy of the model. The sides of the polygon and the feed lines were modeled as sections of lossy microstrip transmission lines; the length of each side of the polygon was fixed to be on thirty-sixth of the ring's mean circumference. The discontinuities at the vertices of the polygon were modeled as optimally chamfered microstrip bends; for a 36-sided polygon, the bend angle θ is 170°. The gap between the feed lines and the resonator is modeled in accordance with Hammerstad's model [41] for the microstrip gap. Although this gap model is valid only for symmetric gaps, it was successfully applied to the asymmetric gaps between the feed lines and the resonator. In actuality, the small curvature of the ring over the region corresponding to the width of the feed lines makes the gaps appear symmetric. Further, when the ring is symmetrically excited, the maximum field points in

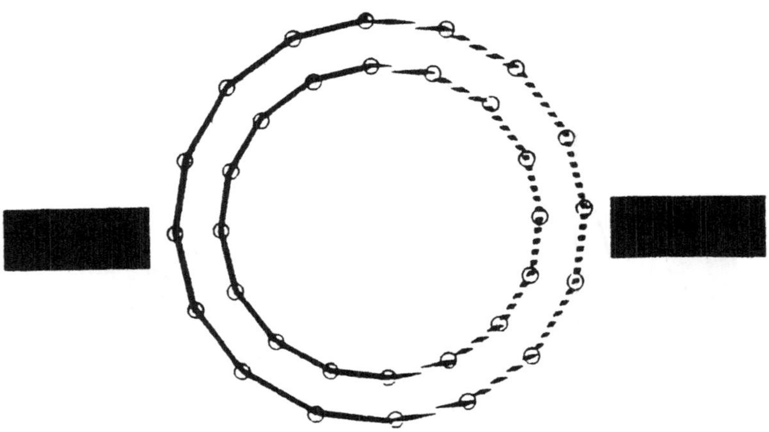

FIGURE 2.26 Distributed transmission-line model [9]. (Permission from *Electronics Letters.*)

both the feed lines and the resonator are collinear. This bears resemblance to a microstrip gap, and hence application of the symmetric-gap model is further justified.

Having represented the ring resonator as a cascade of sections of transmission lines with discontinuities, each section is modeled by its equivalent *ABCD*-matrix. The *ABCD*-matrix for a transmission line (TRL) of length l is

$$\begin{bmatrix} A & B \\ C & D \end{bmatrix}_{\text{TRL}} = \begin{bmatrix} \cosh \gamma l & jZ_0 \sinh \gamma l \\ \dfrac{\sinh \gamma l}{Z_0} & \cosh \gamma l \end{bmatrix} \tag{2.97}$$

where the propagation constant γ is given by

$$\gamma = \alpha + j\beta \tag{2.98}$$

where α and β are the attenuation and phase constants, respectively.

The attenuation constant of a microstrip line is given as follows [42]:

$$\alpha = \alpha_d + \alpha_c \tag{2.99a}$$

where $\alpha_d = 27.3 \dfrac{\varepsilon_r}{\varepsilon_r - 1} \dfrac{\varepsilon_{reff}(f) - 1}{\sqrt{\varepsilon_{reff}(f)}} \dfrac{\tan\delta}{\lambda_0}$ and α_c are dielectric and conductor attenuation constants, and $\varepsilon_{reff}(f)$ is the relative dielectric including the effects of dispersion [43]. The conductor attenuation constant α_c can be approximately expressed as [42]

$$w/h \le 1/2\pi$$

$$\alpha_c = \frac{8.68 R_{s1} M}{2\pi Z_o h}\left(1 + \frac{h}{w_{eff}} + \frac{hU}{\pi w_{eff}}\right) \quad dB/\text{unit length} \tag{2.99b}$$

$$1/2\pi \le w/h \le 2$$

$$\alpha_c = \frac{8.68 R_{s1} M}{2\pi Z_o h}\left(1 + \frac{h}{w_{eff}} + \frac{hV}{\pi w_{eff}}\right) \quad dB/\text{unit length} \tag{2.99c}$$

$$w/h \ge 2$$

$$\alpha_c = \frac{8.68 R_{s1}}{2\pi Z_o h}\left[\frac{w_{eff}}{h} + \frac{w_{eff}/\pi h}{\dfrac{w_{eff}}{2h} + 0.94}\right]\left(1 + \frac{h}{w_{eff}} + \frac{hV}{\pi w_{eff}}\right)$$
$$\times \left\{\frac{w_{eff}}{h} + \frac{2}{\pi}\ln\left[2\pi e\left(\frac{w_{eff}}{2h} + 0.94\right)\right]\right\}^{-2} \quad dB/\text{unit length} \tag{2.99d}$$

$$\text{with } M = 1 - \left(\frac{w_{eff}}{4h}\right)^2, \quad U = \ln\frac{4\pi w}{t} + \frac{t}{w}, \quad V = \ln\frac{2h}{t} + \frac{t}{w},$$

$$R_{s1} = R_s\left\{1 + \frac{2}{\pi}\tan^{-1}\left[1.4\left(\frac{\Delta}{\delta_s}\right)^2\right]\right\} \quad \text{and} \quad R_s = \sqrt{\frac{\pi f \mu_0}{\sigma}} \text{ [44]}$$

where R_{s1} is the surface-roughness resistance of the conductor, R_s is the surface resistance of the conductor, Δ is the surface roughness, $\delta_s = 1/R_s\sigma$ is the skin depth, σ is the conductivity of the microstrip line, f is the frequency, t is the microstrip thickness, w is the width of the microstrip line, and w_{eff} is the effective width of the microstrip line.

The phase constant β is defined as

$$\beta = \frac{2\pi}{\lambda_g} \tag{2.100}$$

where λ_g is the guided wavelength of the line. For discontinuities such as gaps, notches, or solid-state devices, the capacitances and inductances involved are modeled, respectively, as an admittance Y or an impedance Z. The $ABCD$-matrices for an admittance (Y) and impedance (Z) are

$$\begin{bmatrix} A & B \\ C & D \end{bmatrix}_Y = \begin{bmatrix} 1 & 0 \\ Y & 1 \end{bmatrix} \tag{2.101}$$

and

$$\begin{bmatrix} A & B \\ C & D \end{bmatrix}_Z = \begin{bmatrix} 1 & Z \\ 0 & 1 \end{bmatrix} \tag{2.102}$$

The flowchart for the modeling procedure is shown in Figure 2.27. Computation of the $ABCD$-matrix for the 36-sided polygon is done in two steps. First, the $ABCD$-matrix for each half of the polygon is computed by cascading the $ABCD$-matrices corresponding to the transmission lines and bends. This is transformed into a Y-matrix according to the following transformation:

$$\begin{bmatrix} Y_{11} & Y_{12} \\ Y_{21} & Y_{22} \end{bmatrix} = \frac{1}{Z_0}\begin{bmatrix} \dfrac{D}{B} & \dfrac{BC - AD}{B} \\ \dfrac{-1}{B} & \dfrac{A}{B} \end{bmatrix} \tag{2.103}$$

The Y-matrices for each half-section are then added, and the resultant matrix is transformed into an $ABCD$-matrix in accordance with the following transformation:

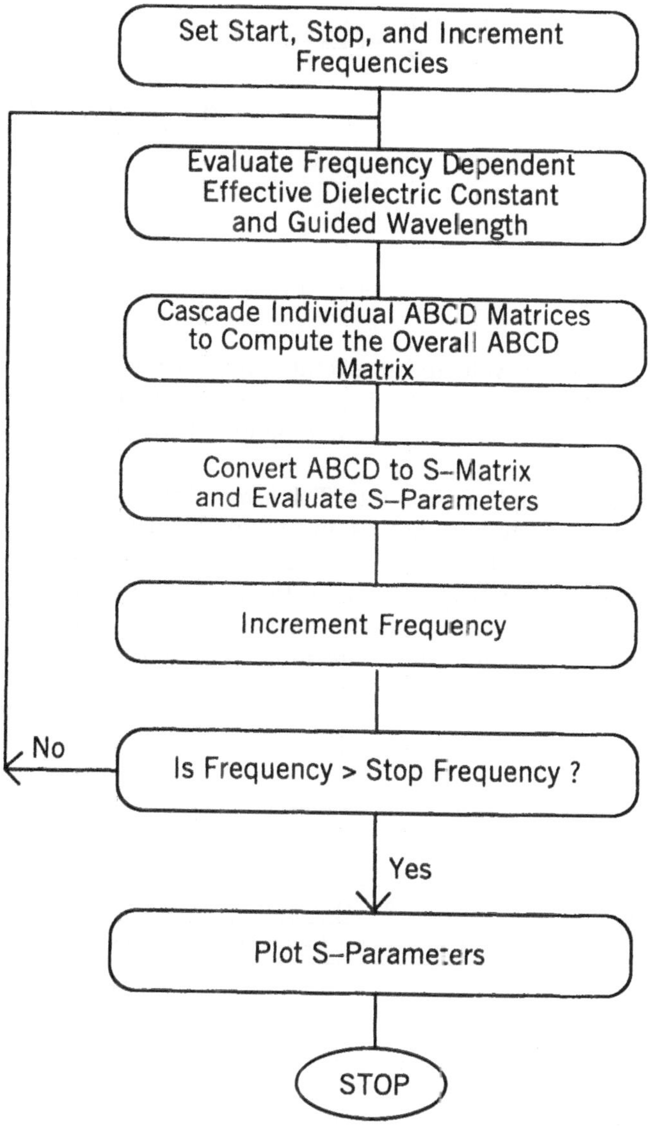

FIGURE 2.27 Flowchart for estimation of *S*-parameters.

$$\begin{bmatrix} A & B \\ C & D \end{bmatrix} = \frac{1}{Y_{21}} \begin{bmatrix} -Y_{22} & -1 \\ (Y_{12}Y_{21} - Y_{11}Y_{22}) & -Y_{11} \end{bmatrix} \tag{2.104}$$

The overall *ABCD*-matrix of the circuit is then computed by cascading the equivalent *ABCD*-matrices for the feed lines, gap, and the resonator; 50 Ω terminations are assumed at the input and output. The *ABCD*-matrix thus obtained is converted to an *S*-matrix based on the following transformation:

$$\begin{bmatrix} S_{11} & S_{12} \\ S_{21} & S_{22} \end{bmatrix} = \frac{1}{\Delta_s} \begin{bmatrix} A + \dfrac{B}{Z_0} - CZ_0 - D & 2(AD - BC) \\ 2 & -A + \dfrac{B}{Z_0} - CZ_0 + D \end{bmatrix} \tag{2.105}$$

where

$$\Delta_s = A + \frac{B}{Z_0} + CZ_0 + D \tag{2.106}$$

The S-parameters are evaluated at several frequency points over a swept range of frequencies. Resonant frequencies are then determined from plots of $|S_{21}|$ versus f, and resonance is said to occur at points where the insertion loss ($|S_{21}|$) is minimum.

The model was implemented into *Touchstone* [45] and was applied to the ring resonators reported by Pintzos and Pregla [3]. The results obtained are shown in Figure 2.28 for the first, third, and fifth modes. The agreement between theory and the experiments of Pintzos and Pregla is good even for rings whose mean radii are small; the smaller the mean radius, the wider the ring, and hence curvature effects are more pronounced. Thus, by virtue of

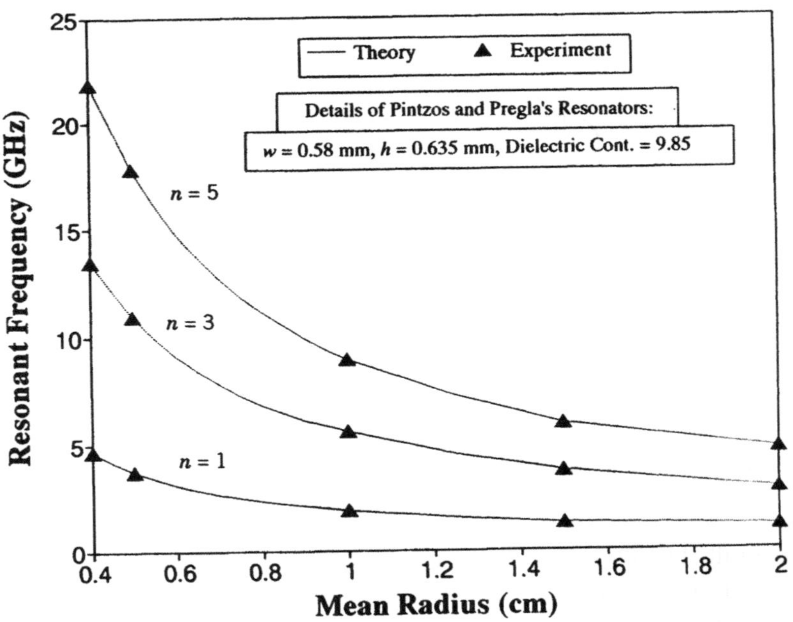

FIGURE 2.28 Mean radius of ring vs. resonant frequency.

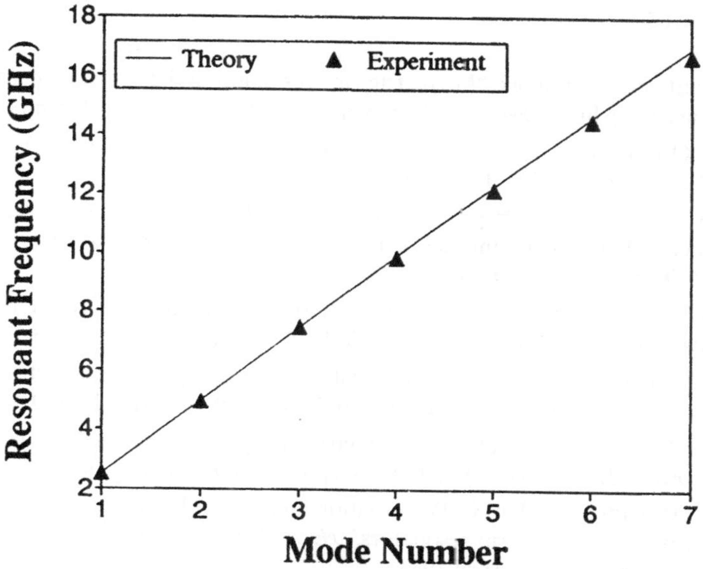

FIGURE 2.29 Mode number vs. resonant frequency.

being valid for resonators of small mean radii, the accuracy of the model is borne out even further.

Although the validity of the model was established by comparison with the results of Pintzos and Pregla [3], the size of the coupling gap was 4 mm, and hence the coupling was loose. In a circuit environment, however, it is desirable to have tight coupling to obtain a lower insertion loss. In this case, the coupling gap affects the intrinsic resonances of the ring, and hence circuits have to be simulated and tested in a tightly coupled environment, to validate the accuracy of the model. Toward this, ring resonators were fabricated on RT Duroid 6010 ($\varepsilon_r = 10.5$) substrates as per the following dimensions:

Substrate thickness = 0.635 mm
Line width = 0.573 mm
Coupling gap = 0.25 mm
Mean radius of the resonator = 7.213 mm

These ring circuits were tested using an HP 8510B automatic network analyzer. The measured resonant frequencies were compared with the distributed transmission-line model in Figure 2.29. As can be seen, the results compared quite well. It should be mentioned that the discrepancy between theory and experiment is of the order of 1% for modes greater than 5; this is attributed to the error margins associated with the discontinuity and dispersion models.

REFERENCES

[1] I. Wolff and N. Knoppik, "Microstrip ring resonator and dispersion measurements on microstrip lines," *Electron. Lett.*, Vol. 7, No. 26, pp. 779–781, December 30, 1971.

[2] R. P. Owens, "Curvature effect in microstrip ring resonators," *Electron. Lett.*, Vol. 12, No. 14, pp. 356–357, July 8, 1976.

[3] S. G. Pintzos and R. Pregla, "A simple method for computing the resonant frequencies of microstrip ring resonators," *IEEE Trans. Microwave Theory Tech.*, Vol. MTT-26, pp. 809–813, October 1978.

[4] Y. S. Wu and F. J. Rosenbaum, "Mode chart for microstrip ring resonators," *IEEE Trans. Microwave Theory Tech.*, Vol. MTT-21, pp. 487–489, July 1973.

[5] A. K. Sharma and B. Bhat, "Spectral domain analysis of microstrip ring resonators", *Arch. Elek. Übertragung*, Vol. 33, pp. 130–132, March 1979.

[6] I. Wolff and V. K. Tripathi, "The microstrip open-ring resonator," *IEEE Trans. Microwave Theory Tech.*, Vol. MTT-32, pp. 102–106, January 1984.

[7] V. K. Tripathi and I. Wolff, "Perturbation analysis and design equations for open- and close-ring microstrip resonators," *IEEE Trans. Microwave Theory Tech.*, Vol. MTT-32, pp. 405–409, April 1984.

[8] K. Chang, T. S. Martin, F. Wang, and J. L. Klein, "On the study of microstrip ring and varactor-tuned ring circuits," *IEEE Trans. Microwave Theory Tech.*, Vol. MTT-35, pp. 1288–1295, December 1987.

[9] G. K. Gopalakrishnan and K. Chang, "Bandpass characteristics of split-modes in asymmetric ring resonators", *Electron. Lett.*, Vol. 26, No. 12, pp. 774–775, June 7, 1990.

[10] L.-H. Hsieh and K. Chang, "Simple analysis of the frequency modes for microstrip ring resonators of any general shape and the correction of an error in literature," *Microwave Opt. Technol. Lett*, Vol. 38, pp. 209–213, August 2003.

[11] T. C. Edwards, "Microstrip measurements," *MTT-S Int. Microwave Symp. Dig.*, pp. 338–341, 1982.

[12] L.-H. Hsieh and K. Chang, "Equivalent lumped elements G, L, C, and unloaded Q's of closed- and open-loop ring resonators," *IEEE Trans. Microwave Theory Tech.*, Vol. 50, pp. 453–460, February 2002.

[13] T. S. Martin, "A Study of the Microstrip Ring Resonator and Its Applications", M.S. thesis, Texas A&M University, College Station, December 1987.

[14] I. Wolff, "Microwave bandpass filter using degenerate mode of a microstrip ring resonator," *Electron. Lett.*, Vol. 8, No. 12, pp. 302–303, June 15, 1972.

[15] G. Kompa and R. Mehran, "Planar-waveguide model for calculating microstrip components," *Electron. Lett.*, Vol. 11, No. 19, pp. 459–460, September 18, 1975.

[16] R. P. Owens, "Predicted frequency dependence of microstrip characteristic impedance using the planar-waveguide model." *Electron. Lett.*, Vol. 12, No. 11, pp. 269–270, May 27, 1976.

[17] A. M. Khilla, "Ring and disk resonator CAD model," *Microwave J.*, Vol. 27, pp. 91–105, November 1984.

[18] V. H. Rumsey, "The reaction concept in electromagnetic theory," *Phys. Rev.*, Vol. 94, pp. 1483–1491, June 1954.

[19] R. F. Harrington, *Time-Harmonic Electromagnetic Fields*, McGraw-Hill, New York, 1961.

[20] A. Farrar and A. T. Adams, "Matrix methods for microstrip three-dimensional problems," *IEEE Trans. Microwave Theory Tech.*, Vol. MTT-20, pp. 497–504, August 1972.

[21] M. Maeda, "An analysis of gap in microstrip transmission lines," *IEEE Trans. Microwave Theory Tech.*, Vol. MTT-20, pp. 390–396, June 1972.

[22] P. Silvester and P. Benedek, "Equivalent capacitance for microstrip gaps and steps," *IEEE Trans. Microwave Theory Tech.*, Vol. MTT-20, pp. 729–733, November 1972.

[23] A. Ralston and P. Rabinowitz, *A First Course in Numerical Analysis*, McGraw-Hill, New York, 1965.

[24] P. Silvester and P. Benedek, "Equivalent capacitance of microstrip open circuits," *IEEE Trans. Microwave Theory Tech.*, Vol. MTT-20, pp. 511–516, August 1972.

[25] R. Garg and I. J. Bahl, "Microstrip discontinuities," *Int. J. Electron.* Vol. 45, pp. 81–87, July 1978.

[26] K. C. Gupta, R. Garg, and I. J. Bahl, *Microstrip Lines and Slotlines*, Artech House, Dedham, Mass., 1979.

[27] I. M. Gottlieb, *Basic Oscillators*, Rider, New York, 1963.

[28] Z. M. Hejazi, P. S. Excell, and Z. Jiang, "Compact dual-mode filters for HTS satellite communication systems," *IEEE Microwave Guided Wave Lett.*, Vol. 8, pp. 275–277, August 1998.

[29] J. S. Hong and M. J. Lancaster, "Microstrip bandpass filter using degenerate modes of a novel meander loop resonator," *IEEE Microwave Guided Wave Lett.*, Vol. 5, pp. 371–372, November 1995.

[30] IE3D Version 9.16, Zeland Software Inc., Fremont, CA, July 2002.

[31] C.-C. Yu and K. Chang, "Transmission-line analysis of a capacitively coupled microstrip-ring resonator," *IEEE Trans. Microwave Theory Tech.*, Vol. 45, pp. 2018–2024, November 1997.

[32] D. M. Pozar, *Microwave Engineering*, 2nd ed. New York: Wiley, 1998, pp. 303–306.

[33] R. Garg and I. J. Bahl, "Microstrip discontinuities," *Int. J. Electron.*, Vol. 45, pp. 81–87, July 1978.

[34] E. Belohoubek and E. Denlinger, "Loss considerations for microstrip resonators," *IEEE Trans. Microwave Theory Tech.*, Vol. 23, pp. 522–526, June 1975.

[35] G. K. Gopalakrishnan, "Microwave and optoelectronic performance of hybrid and monolithic microstrip ring resonator circuits," Ph.D. dissertation, Texas A&M University, College Station, May 1991.

[36] P. Daly, "Hybrid-mode analysis of microstrip by finite-element methods," *IEEE Trans. Microwave Theory and Tech.*, Vol. MTT-19, pp. 19–25, January 1971.

[37] M. Kirschning and R. H. Jansen, "Accurate model for effective dielectric constant of microstrip and validity up to millimeter-wave frequencies," *Electron. Lett.*, Vol. 18, pp. 272–273, March 1982.

[38] T. C. Edwards, *Foundations for Microstrip Circuit Design*, Wiley, Chichester, England, 1981; 2d ed., 1992.

[39] M. Kirschning, R. H. Jansen, and M. H. L. Koster, "Measurement and computer-aided modeling of microstrip discontinuities by an improved resonator model," in 1983 *IEEE MTT-S International Microwave Symposium Digest*, pp. 495–497, June 1983.

[40] LINECALC, EEsof Inc., Westlake Village, Calif.

[41] E. Hammerstad, "Computer-aided design of microstrip couplers with accurate discontinuity models," in 1981 *IEEE MTT-S International Microwave Symposium Digest*, pp. 54–56, June 1981.

[42] R. A. Pucel, D. J. Masse', and C. P. Hartwig, "Losses in microstrip," *IEEE Trans. Microwave Theory Tech.*, Vol. 16, pp. 342–350, June 1968.

[43] M. Kobayashi, "A dispersion formula satisfying recent requirements in microstrip CAD," *IEEE Trans. Microwave Theory Tech.*, Vol. 36, pp. 1246–1250, August 1988.

[44] E. Hammerstad and O. Jensen, "Accurate models for microstrip computer-aided design," *IEEE MTT-S Int. Microwave Symp. Dig.*, pp. 407–409, 1980.

[45] TOUCHSTONE, EEsof Inc., Westlake Village, Calif.

CHAPTER THREE

Modes, Perturbations, and Coupling Methods of Ring Resonators

3.1 INTRODUCTION

According to the simple model and field analysis in Chapter 2, many modes can be supported by the ring resonators. All these modes satisfy the boundary conditions and can be excited if desired. The excitation of these modes depends on the perturbation and coupling methods. This chapter discusses the various mode phenomena, excitation techniques, perturbations, and coupling methods based on references 1 and 2.

The mode phenomenon of the annular ring element is caused by different types of excitation and perturbation. The resonant modes of the coupled annular ring circuit are divided into three groups according to the different types of excitation and perturbation: (1) regular mode, (2) forced mode (or excited mode), and (3) split mode. The operating principle and design rule of each mode are discussed in the following sections. The discussion is concentrated on microstrip rings. However, the theory applies to waveguide ring cavities and uniplanar rings.

3.2 REGULAR RESONANT MODES

A *regular mode* is obtained by applying symmetric input and output feedlines on the annular ring element [1, 2]. The resonant wavelengths of the regular mode are determined by Equation (2.1) and repeated here:

Microwave Ring Circuits and Related Structures, Second Edition,
by Kai Chang and Lung-Hwa Hsieh
ISBN 0-471-44474-X

$$2\pi r = n\lambda_g \tag{3.1}$$

where r is the mean radius of the annular ring element; λ_g is the guided wavelength; and n is the mode number. Some modifications of the curvature effect for this equation may result in a more accurate prediction of the resonant frequency [3].

The ring akin to any microwave resonator has both resonant and antiresonant frequencies. Basically, the ring comprises two half-wavelength linear resonators connected in parallel. Since the parallel connection alleviates problems related to radiation from open ends, the ring has a higher Q-factor compared to the linear resonator. Resonance occurs when standing waves are set up in the ring; this happens when its circumference is an integral multiple of the guided wavelength. To understand the basic phenomena underlying the operation of the ring, it is imperative to first understand its field configuration for the different modes. The absolute values of maximum field points for the first four modes are shown in Figure 3.1; the field is minimum midway between these points. In the absence of slits or other discontinuities, a maximum field point occurs where the feed line excites the resonator. This point is independent of the azimuthal position of the feed line that extracts microwave

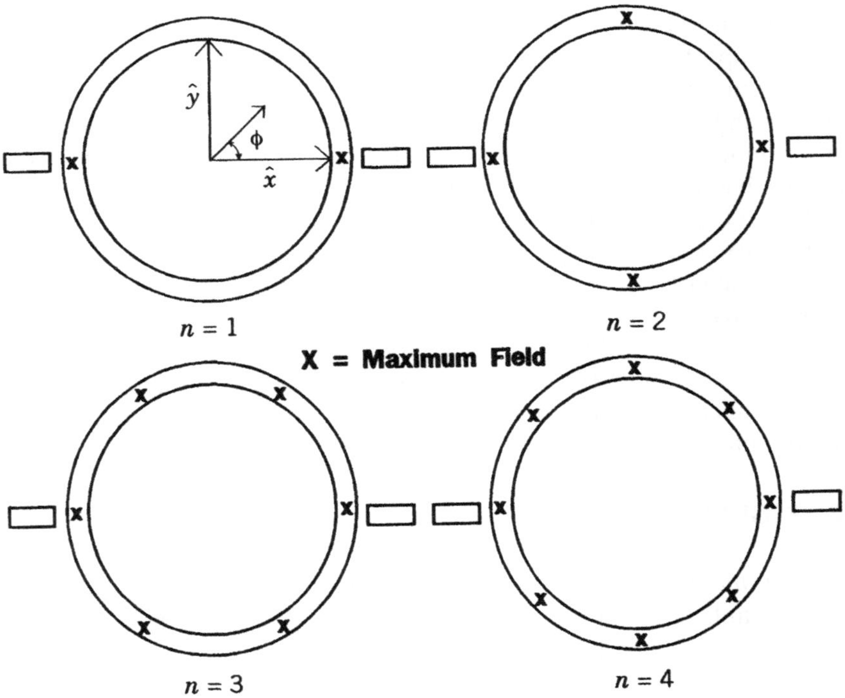

FIGURE 3.1 Maximum field points for the first four modes.

power. This is important from the standpoint of mode suppression. For example, it can be seen from Figure 3.1 that when the azimuthal angle $\phi = 90°$, there is a field minimum for the first and third modes. These modes and other higher-order odd modes can be suppressed if the feed line that extracts power is located at $\phi = 90°$. In the presence of discontinuities such as slits, the fields in the resonator readjust themselves, so as to first satisfy the boundary conditions caused by the slits in the resonator. In other words, if there are slits, then the maximum field point is not necessarily collinear with the feed line that excites the resonator.

The S_{21} characteristics of the first seven modes of the ring resonator whose dimensions were specified in Section 2.6 are shown in Figure 3.2. A Smith chart is shown in Figure 3.3. The frequencies corresponding to markers 1 through 5 in Figures 3.2 and 3.3 are the first five resonant frequencies of the ring. The loops in the Smith chart indicate resonance. They are associated with the fact that at resonance, the reactance X goes from either being inductive to capacitive or vice versa (i.e., the phase of the signal goes through zero). The fact that the loops are skewed from the $X = 0$ line in the chart is attributed to the finite reactances associated with the connectors and the feed lines. By using

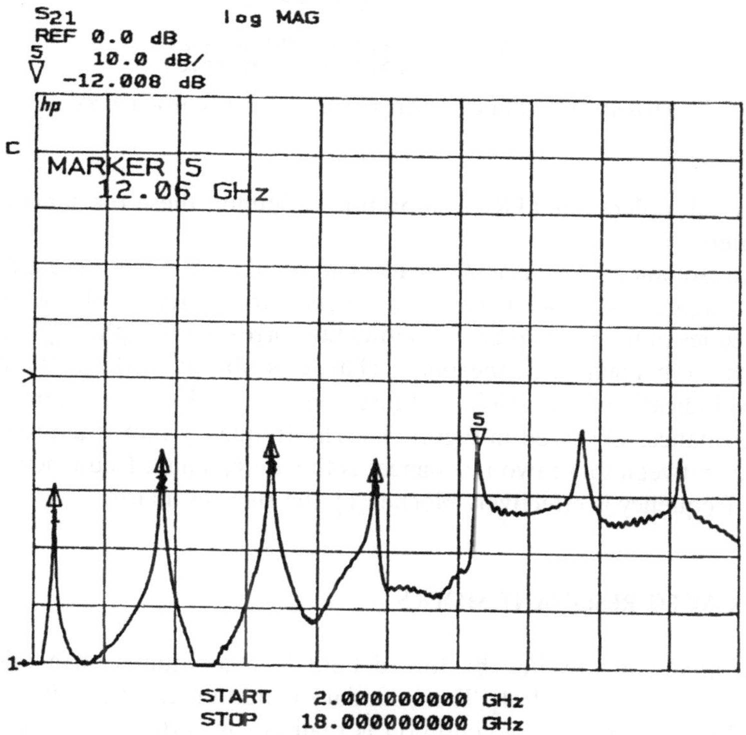

FIGURE 3.2 $|S_{21}|$ vs. frequency for the first seven resonances.

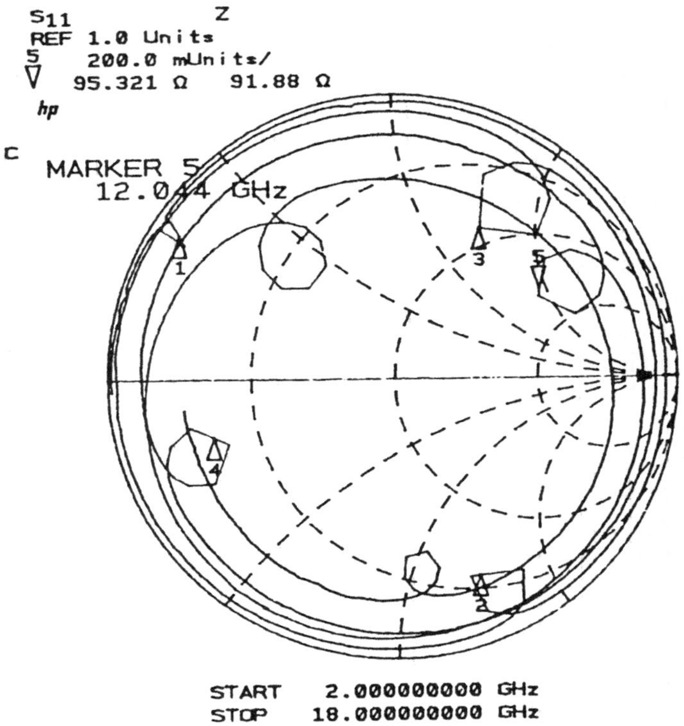

FIGURE 3.3 Smith's chart for the first seven resonances.

the through-reflect-line (TRL) calibration technique, these reactances can be calibrated.

Antiresonance in the ring resonator can be illustrated with the aid of two figures. Figure 3.4 shows a plot of $|S_{21}|$ versus frequency for the second and third modes of the ring; Figure 3.5 shows the corresponding phase plot. As can be seen from Figure 3.5, the phase change is abrupt at three frequencies, thereby indicating the possibility of three resonances. However, there are only two resonances observed in Figure 3.4. The frequency that is approximately midway between these two resonances is the antiresonant frequency; passage of this frequency is effectively blocked by the ring resonator.

3.3 FORCED RESONANT MODES

Forced modes are excited by forced boundary conditions on a microstrip annular ring element [1, 4]. The boundary condition can be either open or short. The open boundary condition is realized by cutting open slits on the annular ring element [5]. The shorted boundary condition show in Figure 3.6 is obtained by inserting a thin conductor sheet inside the substrate.

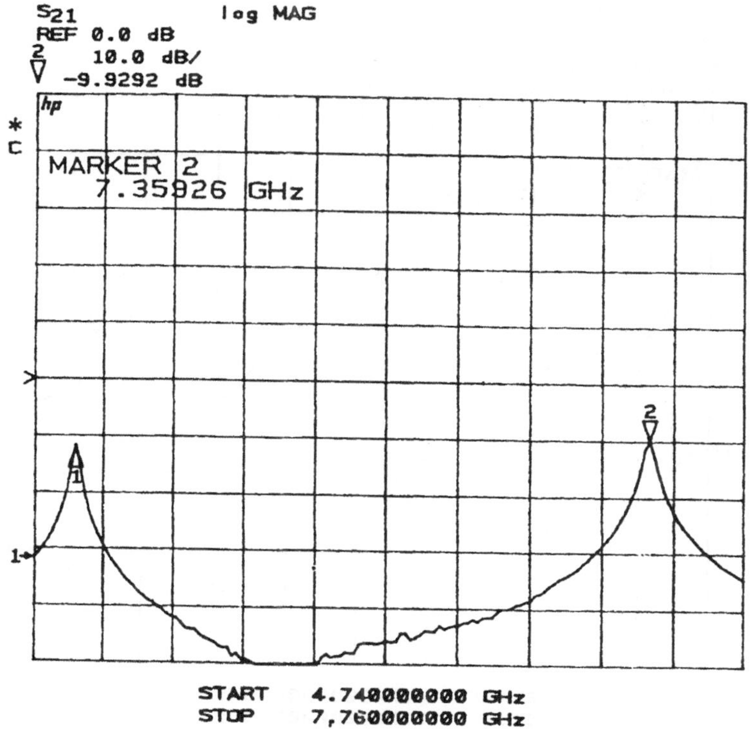

FIGURE 3.4 $|S_{21}|$ vs. frequency for the second and third resonances.

The short plane in Figure 3.6 is located at the annular angle of 90°. This boundary condition forces minima of the electric field to occur on both sides of the short plane. The standing-wave patterns for the first four resonant modes are illustrated in Figure 3.7. As shown in Figure 3.7, the standing-wave patterns with the even mode numbers result in minimum magnitudes at the input and output feed lines. This means that no energy is transferred between the input port and the output port. Therefore the resonant modes with even mode numbers cannot exist in this shorted *forced mode*. The theoretical and experimental results illustrated in Figure 3.8 and 3.9 agree with the standing-wave pattern analysis [4]. The theoretical analysis was based on the *distributed transmission-line model* [6]. Figure 3.8 shows that the even modes are nonexistent. The results agree with the prediction of standing-wave pattern analysis. The test circuit was built on a RT/Duroid 6010.5 substrate with the following dimensions:

Substrate thickness = 0.635 mm

Line width = 0.6 mm

Coupling gap = 0.1 mm

Ring radius = 6 mm

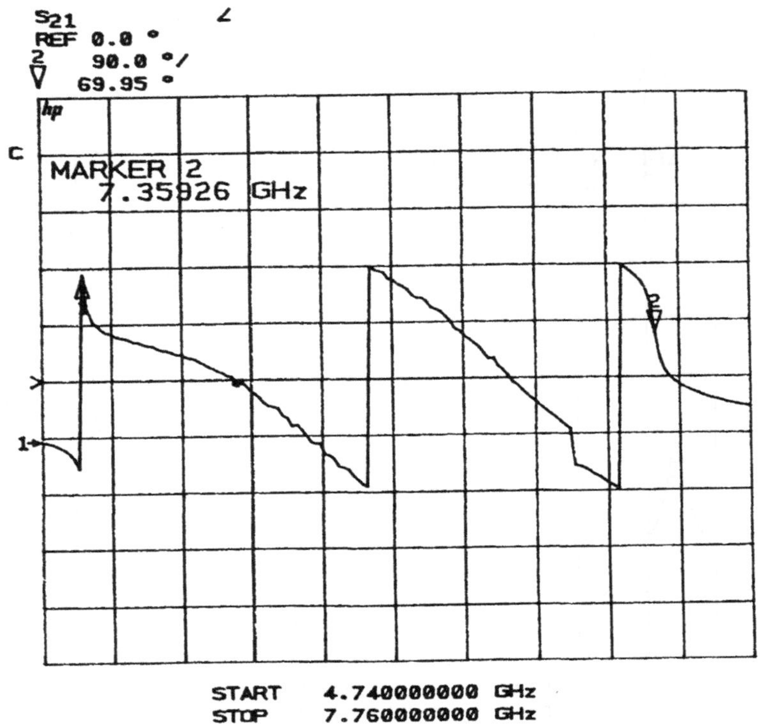

FIGURE 3.5 Phase of S_{21} vs. frequency for the second and third resonances.

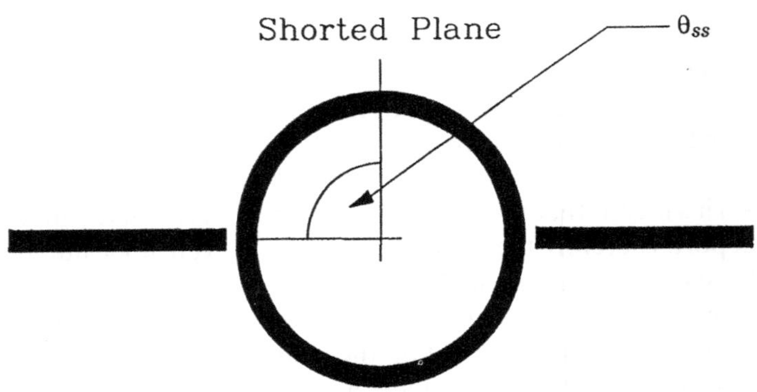

FIGURE 3.6 Coupled annular circuit with short plane at $\theta_{ss} = 90°$.

According to the preceding analysis, a general design rule for the single shorted boundary condition is summarized in the following:

Given an annular angle $\phi = \theta_{ss}$ of the shorted conductor sheet, the resonant modes that have integer mode number $n = m \cdot 90°/|\theta_{ss}|$, for $-90° \leq \theta_{ss} \leq 90°$, or $n = m \cdot 90°/|\theta_{ss} - 180°|$, for $90° \leq \theta_{ss} \leq 270°$, where $m = 2, 4, 6$, and so on, will not exist in

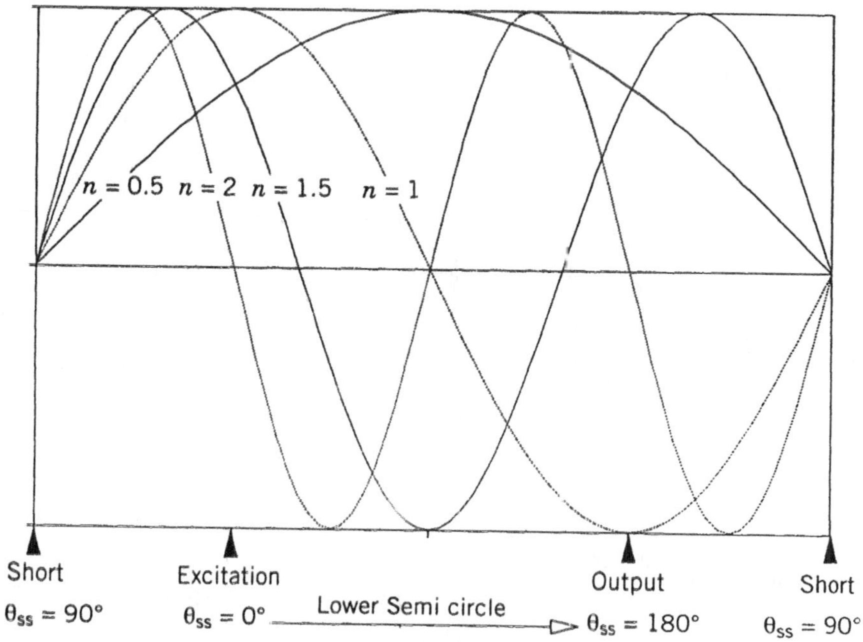

FIGURE 3.7 Standing wave patterns of the shorted forced mode.

the shorted forced mode. On the other hand, the half-wavelength resonant modes with mode number $v = m/2$, where $m = 1, 3, 5$, and so on, will be excited due to the shorted boundary condition. If the shorted conductor sheet is at 0° or 180° of the annular angle, then there is no energy transferred between the input port and the output port.

A similar design rule for the single open slit, as mentioned before, can also be summarized in the following:

Given an annular angle $\phi = \theta_{os}$ of the open slit, the resonant modes that have integer mode number $n = m \cdot 90°/|\theta_{oy}|$, for $-90° \leq \theta_{os} \leq 90°$, or $n = m \cdot 90°/|\theta_{os} - 180°|$, for $90° \leq \theta_{os} \leq 270°$, where $m = 1, 3, 5$, and soon, will not exist in the opened forced mode. On the other hand, the half-wavelength resonant modes with mode number $v = m/2$, where $m = 1, 3, 5$, and so on, will be excited due to the open boundary condition. If the open slit is at 0° or 180° of the annular angle, then the resonant modes are regular modes.

3.4 SPLIT RESONANT MODES

The *split resonant mode* was first reported by Wolff [7]. He used asymmetric feedlines or notch perturbation to obtain the *split resonant modes*. Besides these methods, two other new techniques can also be used to generate the *split*

(a)

(b)

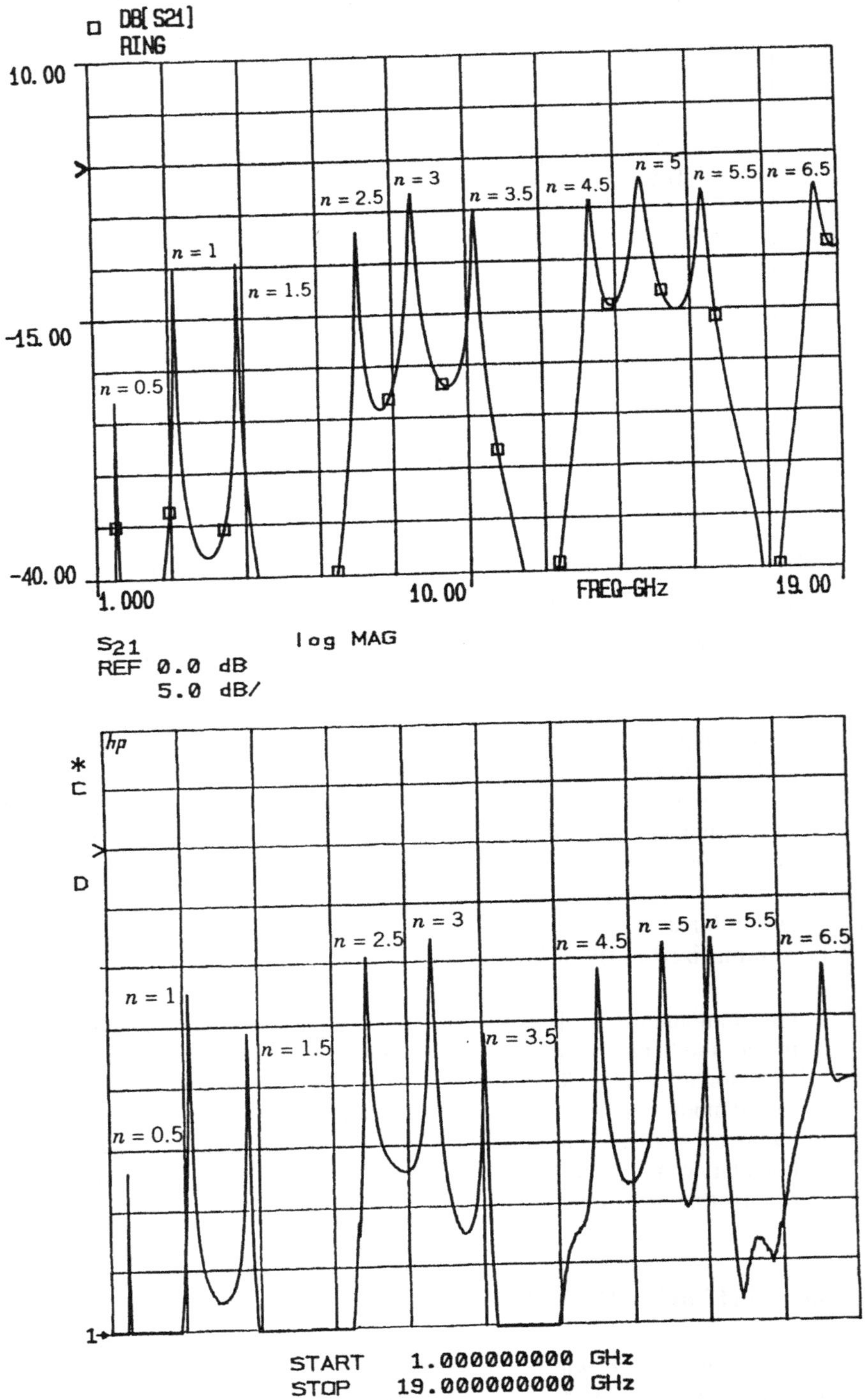

FIGURE 3.8 $|S_{21}|$ vs. frequency for the shorted forced mode: (a) theoretical result; (b) experimental result.

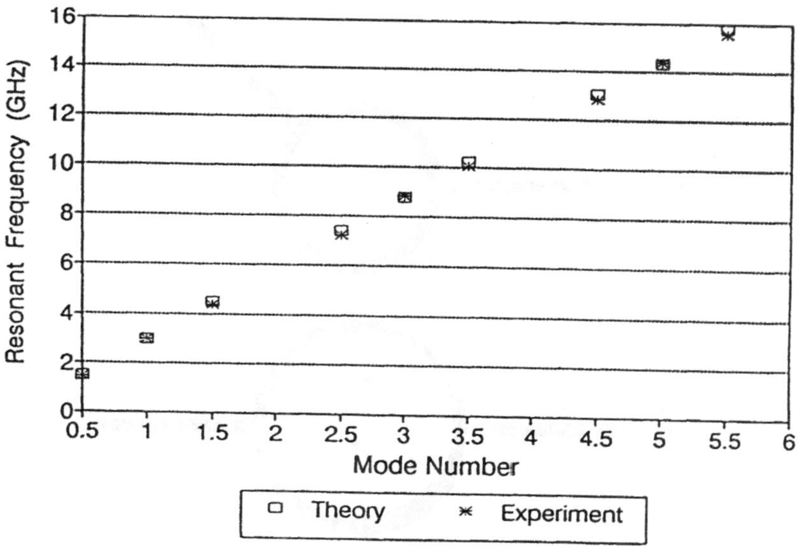

FIGURE 3.9 Resonant frequency vs. mode number for the shorted forced mode.

resonant mode. According to the different types of perturbation, the *split mode* can be classified into the following four types: (1) *coupled split mode* [7], (2) *local resonant split mode*, (3) *notch perturbation split mode* [7], and (4) *patch perturbation split mode*. Figure 3.10*a–d* illustrate the basic circuit structures for these four types of *split resonant modes* [4]. The following sections discuss the operating principle and design rule for each type of *split resonant modes* [1, 4].

3.4.1 Coupled Split Modes

The *coupled split mode*, as shown in Figure 3.10*a*, is generated by asymmetric feedlines [7]. The annular angle θ between the asymmetric feed lines determines the splitting frequency of the split mode [8]. The power transmission can be calculated as [8]:

$$|S_{21}|^2 = 4\cos^2\theta \frac{g_{11}^2 + b_{11}^2}{\left[(1+g_{11})^2 - b_{11}^2 - \cos^2\theta\right]^2 + 4b_{11}^2(1+g_{11})^2} \tag{3.2}$$

where g_{11} and b_{11} denote the normalized input conductance and susceptance, respectively, of a one-port annular ring resonator of the same size. The power transmission versus frequency is illustrated in Figure 3.11 [8]. As shown in Figure 3.11, the double-tuned characteristics are always found, except when $\theta = \pi$ or $\theta = \pi/2$ [8].

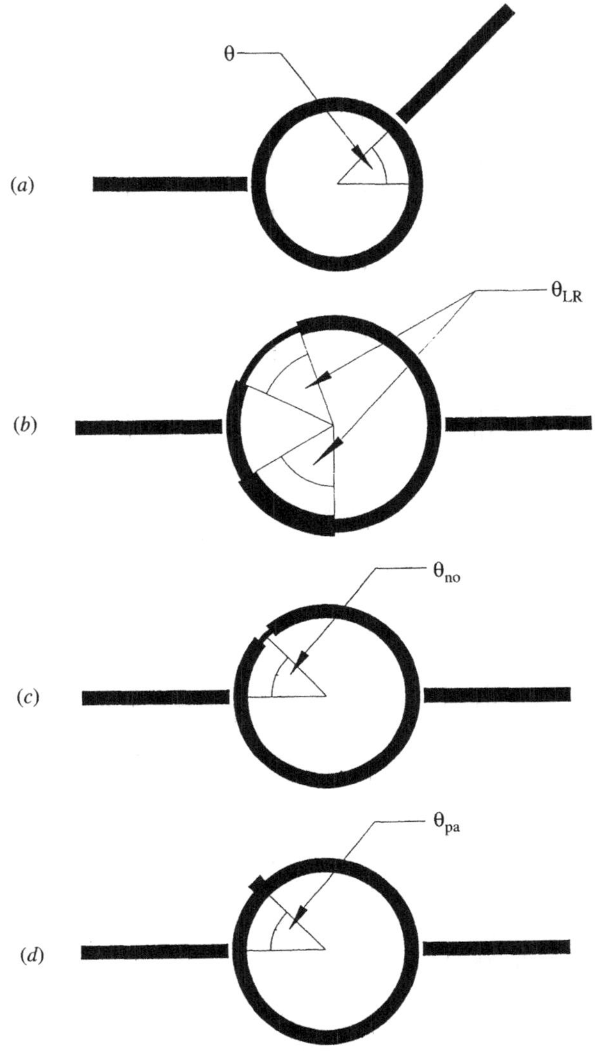

FIGURE 3.10 Four types of split modes: (*a*) coupled split mode; (*b*) local resonant split mode; (*c*) notch perturbation split mode; (*d*) patch perturbation split mode.

3.4.2 Local Resonant Split Modes

The *local resonant split mode*, as shown in Figure 3.10*b*, is excited by changing the impedance of one annular sector on the annular ring element. The high- or low-impedance sector will build up a local resonant boundary condition to store or split the energy of the different resonant modes. Figure 3.12 illustrates a coupled annular ring element with a 45° high-impedance *local resonant sector* (LRS). According to the standing-wave pattern analysis, only the

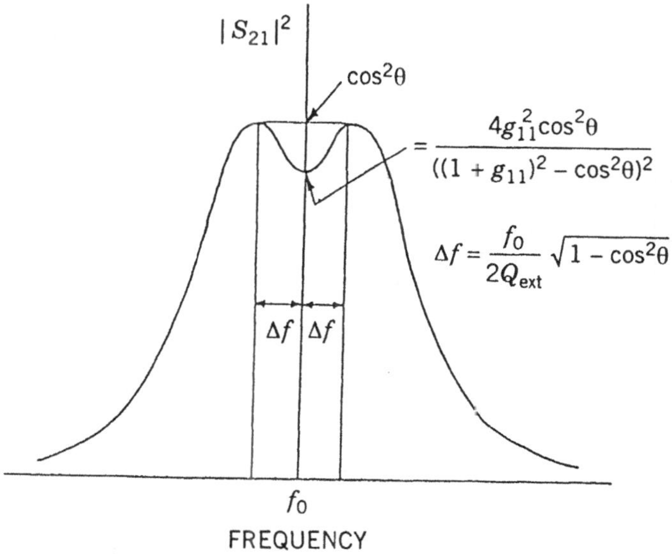

FIGURE 3.11 Power transmission of an asymmetric coupled annular ring resonator.

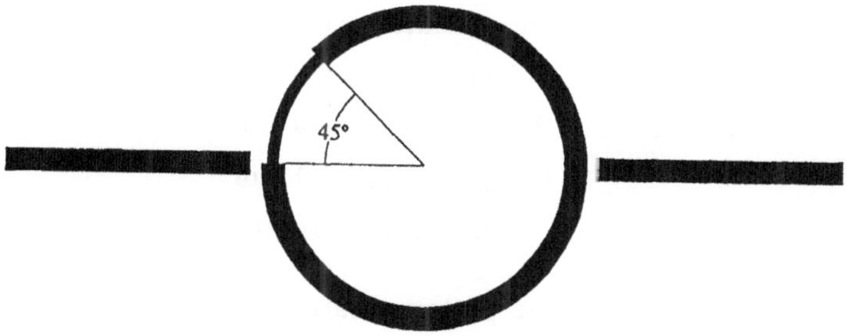

FIGURE 3.12 Layout of the symmetric coupled annular circuit with 45° LRS.

resonant modes with mode number $n = 4m$, where $m = 1, 2, 3$, and so on, have integer multiple of half guided-wavelength inside the perturbed sector. This means that these resonant modes can build up a local resonance and maintain the continuity of the standing-wave pattern inside the perturbed region. The other resonant modes that cannot meet the local resonant condition will suffer energy loss due to scattering inside the perturbed sector. According to the analysis of the standing-wave pattern, it is expected that only the fourth mode will maintain the resonant condition and the other modes will split. The theoretical and experimental results illustrated in Figure 3.13 agree very well. The test circuit was built on a RT/Duroid 6010.5 substrate with the following dimensions:

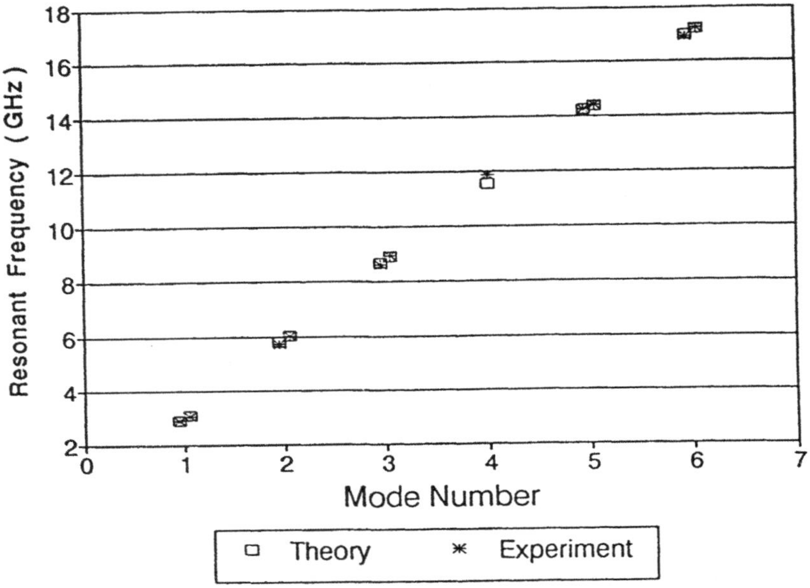

FIGURE 3.13 Resonant frequency vs. mode number for 45° LRS.

Substrate thickness = 0.635 mm
Line width = 0.6 mm
LRS line width = 0.4 mm
Coupling gap = 0.1 mm
Ring radius = 6 mm

Following the standing-wave pattern analysis, the mode phenomenon for the 45° LRS is found to be the same as that of the 135° LRS. The theoretical and experimental results for the 135° LRS is shown in Figure 3.14. They agree with the prediction of the standing-wave pattern analysis. The same results occur between the 60° and 120° LRS. Therefore the period of the annular degree for the LRS is 180°.

From the preceding discussion a general design rule for the use of local resonant split modes is concluded in the following:

> Given an annular degree $\phi = \theta_{LR}$ of the LRS, the resonant modes that have integer mode number $n = m \cdot 180°/|\theta_{LR}|$, for $-90° \leq \theta_{LR} \leq 90°$, or $n = m \cdot 180°/|\theta_{LR} - 180°|$, $90° \leq \theta_{LR} \leq 270°$, where $m = 1, 2, 3$, and so on, will not split.

3.4.3 Notch Perturbation Split Modes

Notch perturbation, as shown in Figure 3.10*c*, uses a small perturbation area with a high impedance line width on the coupled annular circuit [7]. If the

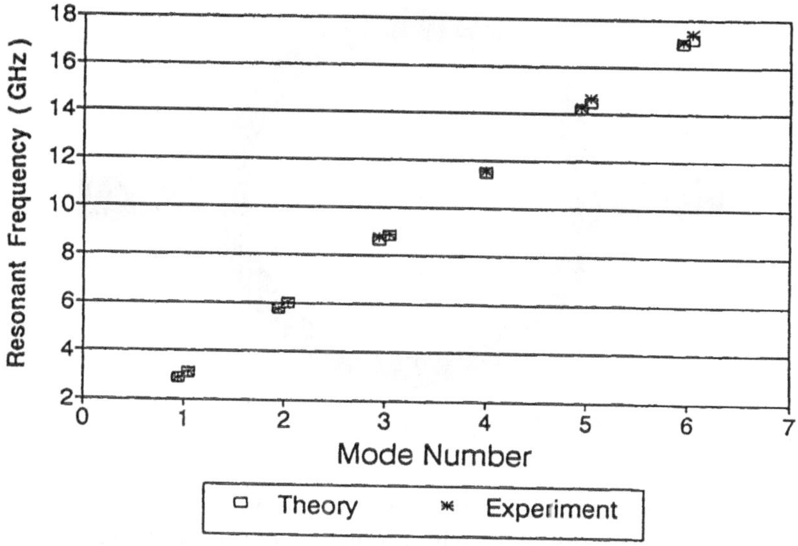

FIGURE 3.14 Resonant frequency vs. mode number for 135° LRS.

disturbed area is located at the position of the maximum or the minimum electric field for some resonant modes, then these resonant modes will not split [2, 6]. A general design rule for the *notch perturbation split mode* is concluded in the following:

> Given an annular degree $\phi = \theta_{no}$ of the notch perturbation, the resonant modes with integer mode number $n = m \cdot 90°/|\theta_{no}|$, for $-90° \leq \theta_{no} \leq 90°$, or $n = m \cdot 90°/|\theta_{no} - 180°|$, for $90° \leq \theta_{no} \leq 270°$, where $m = 1, 2, 3$, and so on, will not split. If the notch perturbation area is at 0° or 180° of the annular angle, then all the resonant modes will not split.

3.4.4 Patch Perturbation Split Modes

Patch perturbation utilizes a small perturbation area with low-impedance line width, as shown in Figure 3.10*d*. The design rule and analysis method is the same as for the notch perturbation. The advantage of using patch perturbation is the flexibility of the line width. A larger splitting range can be obtained by increasing the line width. The splitting range of the notch perturbation, on the other hand, is limited by a maximum line width [7]. As mentioned in the previous notch perturbation design rule, if the patch perturbation area is at 0° or 180° of the annular angle, then all the resonant modes will not split.

3.5 FURTHER STUDY OF NOTCH PERTURBATIONS

A ring-resonator circuit is said to be asymmetric, if when bisected one-half is not a mirror image of the other. Asymmetries are usually introduced either by

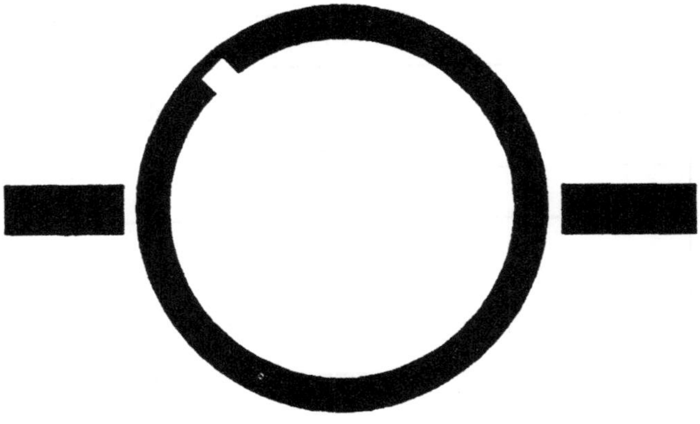

FIGURE 3.15 Layout of a notched ring resonator.

skewing one of the feed lines with respect to the other, or by introduction of a notch [2, 6]. A ring resonator with a notch is shown in Figure 3.15. Asymmetries perturb the resonant fields of the ring and split its usually degenerate resonant modes. Wolff [7] first reported resonance splitting in ring resonators by both introduction of a notch and by skewing one of the feed lines. To study the effect of such asymmetries, it is worthwhile to first consider the fields of a symmetric microstrip ring resonator. The magnetic-wall model solution [9] to the fields of a symmetric ring resonator are

$$E_z = \{AJ_n(kr) + BN_n(kr)\}\cos(n\phi) \tag{3.3a}$$

$$H_r = \frac{n}{j\omega\mu_0 r}\{AJ_n(kr) + BN_n(kr)\}\sin(n\phi) \tag{3.3b}$$

$$H_\phi = \frac{k}{j\omega\mu_0}\{AJ_n'(kr) + BN_n'(kr)\}\cos(n\phi) \tag{3.3c}$$

where A and B are constants; $J_n(kr)$ is the Bessel function of the first kind of order n; $N_n(kr)$ is the Bessel function of the second kind of order n; and k is the wave number; the other symbols have their usual meaning. A close scrutiny of the solution would indicate that another set of degenerate fields, one that also satisfy the same boundary conditions, is also valid. These fields are given by

$$E_z = \{AJ_n(kr) + BN_n(kr)\}\sin(n\phi) \tag{3.4a}$$

$$H_r = \frac{-n}{j\omega\mu_0 r}\{AJ_n(kr) + BN_n(kr)\}\cos(n\phi) \tag{3.4b}$$

$$H_\phi = \frac{k}{j\omega\mu_0}\{AJ_n'(kr) + BN_n'(kr)\}\sin(n\phi) \tag{3.4c}$$

These two solutions could be interpreted as two waves, one traveling clockwise, and the other anticlockwise. If the paths traversed by these waves before extraction are of equal lengths, then the waves are orthogonal, and no resonance splitting occurs. However, if the path lengths are different, then the normally degenerate modes split. Path-length differences and hence resonance splitting can be caused by disturbing the symmetry of the ring resonator. This can be done by placement of a notch along the ring. However, resonance splitting has a strong functional dependence on the position of the notch, and on the mode numbers of the resonant peaks. For very narrow notches, if the notch is located at azimuthal angles of $\phi = 0°, 90°, 180°$, or $270°$, then one of the two degenerate solutions goes to zero and only one solution exists. This is based on the assumption that a narrow notch does not perturb the fields of the symmetric ring appreciably, since the fields are at their maximum at these locations. However, if $\phi = 45°, 135°, 225°$, or $315°$, then for odd n both solutions exist and the resonances split because the symmetry of the ring is disturbed; for even n, one of the solutions goes to zero as discussed earlier, and hence the resonances do not split. For other angles, the splitting is dependent on whether or not solutions exist. Although the preceding equations can be used to predict resonance splitting, it is very difficult to estimate the degree of splitting, as it is dependent on the mode number, the width of the notch, and the depth of the notch. Using the distributed transmission-line model reported in the previous chapter, the degree of resonance splitting can be accurately predicted. The notch was modeled as a distributed transmission line with step discontinuities at the edges. The modes that split, the degree of splitting, and the insertion loss were all estimated using this model. To compare with experiments, circuits were designed to operate at a fundamental frequency of approximately 2.5 GHz. These designs were delineated on a RT/Duroid 6010 ($\varepsilon_r = 10.5$) substrate with the following dimensions:

$$\text{Substrate thickness} = 0.635\,\text{mm}$$

$$\text{Line width} = 0.573\,\text{mm}$$

$$\text{Coupling gap} = 0.25\,\text{mm}$$

$$\text{Mean radius of the ring} = 7.213\,\text{mm}$$

$$\text{Notch depth} = 0.3\,\text{mm}$$

$$\text{Notch width} = 2.0\,\text{mm}$$

Figures 3.16 and 3.17 show the experimental results for notches located at $\phi = 0°$ and $135°$, respectively. When $\phi = 0°$, there is no resonance splitting. When $\phi = 135°$, the odd modes split. Figure 3.18 shows a comparison of theory and experiment for the degree of resonance splitting of odd modes. The good agreement demonstrates that not only can the modes that split be predicted, but so can the degree of splitting.

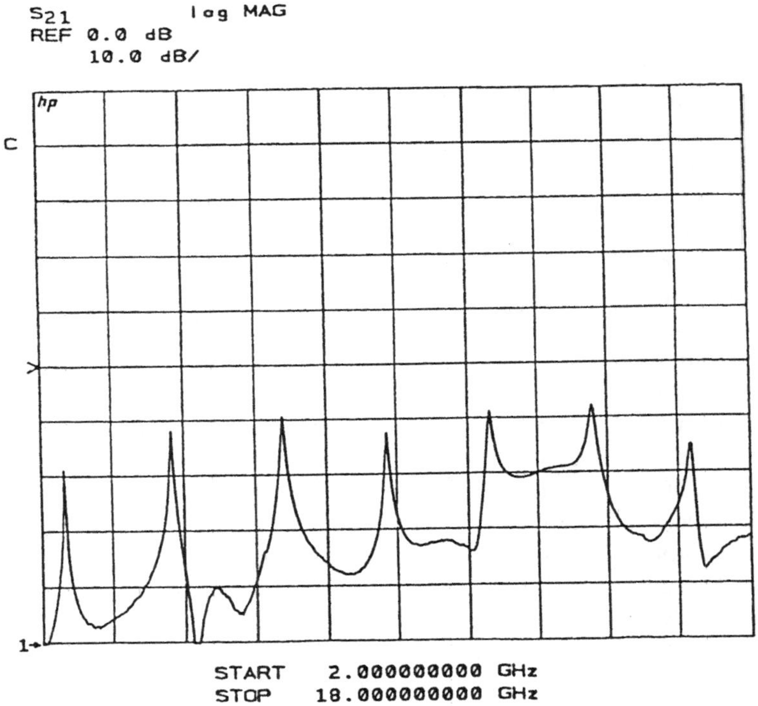

FIGURE 3.16 $|S_{21}|$ vs. frequency for notch at $\phi = 0°$ [6]. (Permission from *Electronics Letters.*)

Resonance splitting can also be obtained by skewing one feed line with respect to the other. However, the degree of resonance splitting is very small because the asymmetry is not directly located in the path of the fields. In this case, resonance splitting occurs because the loading effect of the skewed feed line is different for the counterclockwise fields as compared to the clockwise fields, or vice versa.

3.6 SLIT (GAP) PERTURBATIONS

The attractive characteristics exhibited by the microstrip ring resonator have elevated it from the state of being a mere characterization tool to one with other practical applications; practical circuits require integration of devices such as varactor and PIN diodes. Toward this end, slits have to be made in the ring resonator, to facilitate device integration. Concomitantly, there exists the problem of field perturbation to be contended with [2, 10]. Fortunately, this problem can be alleviated by strategically locating these slits. The introduction of slits will excite the forced resonant modes.

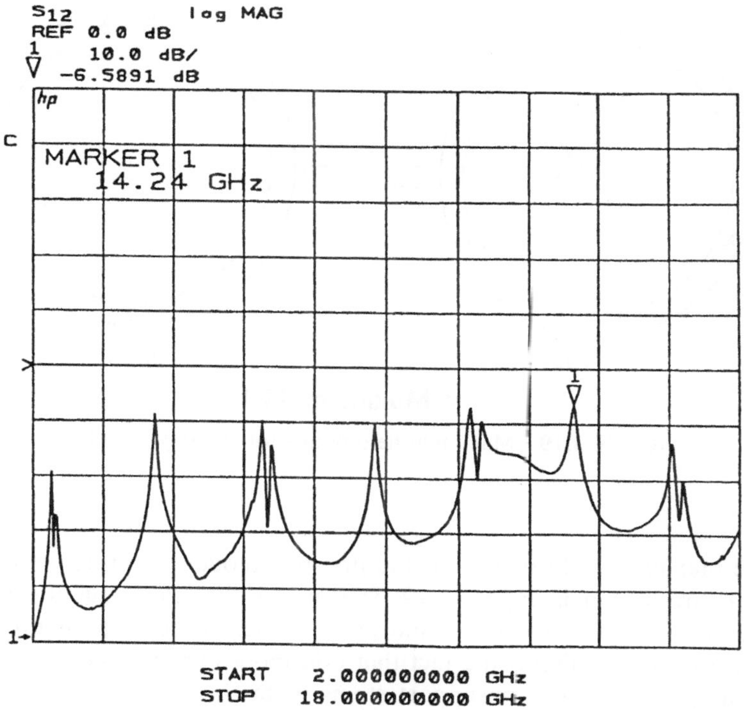

FIGURE 3.17 $|S_{21}|$ vs. frequency for notch at $\phi = 135°$ [6]. (Permission from *Electronics Letters.*)

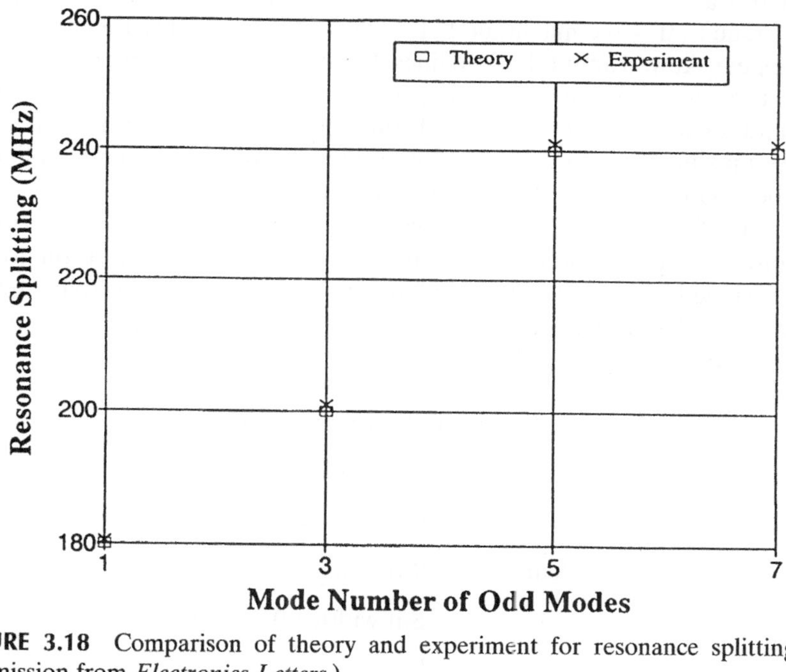

FIGURE 3.18 Comparison of theory and experiment for resonance splitting [6]. (Permission from *Electronics Letters.*)

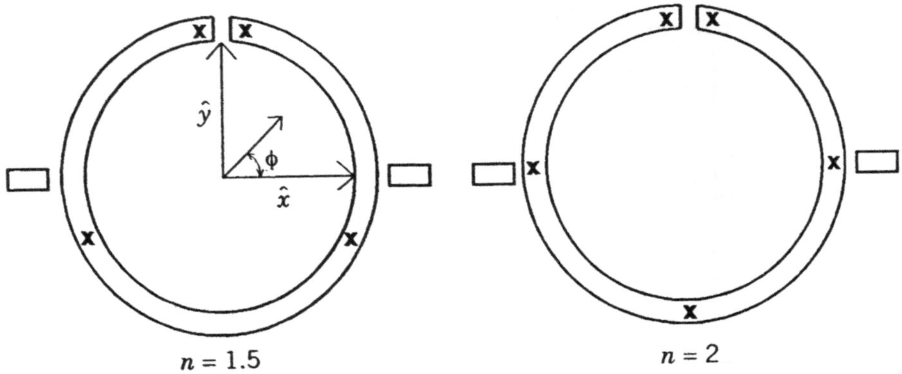

FIGURE 3.19 Maximum field points for slit at $\phi = 90°$ [10].

The maximum field points for the first two modes of a ring with a slit at $\phi = 90°$ are shown in Figure 3.19. The modes that this structure supports are the $n = 1.5, 2, 2.5, 3.5, 4, \ldots$, and so on, modes of the basic ring resonator. Also worth mentioning is the fact that odd modes are not supported in this slit configuration. This nonsupport stems from the contradictory boundary condition requirements of an odd mode in a closed ring (field minimum at $\phi = \pm 90°$), and the slit (field maximum at slit). As can be seen from Figure 3.19, however, half-modes are supported. In the presence of slits, the fields in the resonator are altered so that the corresponding boundary conditions are satisfied. Due to this, the maximum field points of some modes are not collinear, but appear skewed about the feed lines. To efficiently extract microwave power from a given mode, the extracting feeding line has to be in line with the maximum field point of that mode. If this condition is not satisfied, the modes whose maximum field points are not in line with the extracting feed line will not be coupled efficiently to the feed line as compared to those whose maximum field points do line up with the feed line. In order to verify this proposition experimentally, slits were etched into a plain ring resonator that was designed to operate at a fundamental frequency of approximately 2.5 GHz. These designs were delineated on a RT/Duroid 6010 ($\varepsilon_r = 10.5$) substrate with the following dimensions:

Substrate thickness = 0.635 mm

Line width = 0.573 mm

Coupling gap = 0.25 mm

Mean radius of the ring = 7.213 mm

Slit width = 0.25 mm

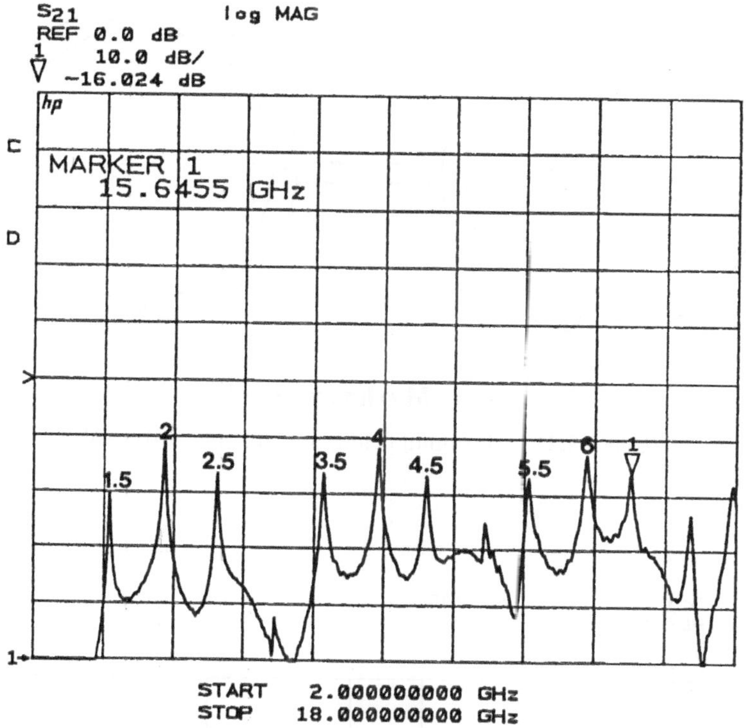

FIGURE 3.20 $|S_{21}|$ vs. frequency for a slit at $\phi = 90°$ [10].

The measured results are shown in Figure 3.20. As can be seen, the first resonant peak occurs at approximately 3.75 GHz, which corresponds to the $n = 1.5$ half-mode; the even modes centered between the half-modes can also be seen. The half-modes are partially supressed as compared to the even modes, because their maximum field points are not in line with the extraction feed line. The $n = 1.5$ mode is approximately 10 dB down as compared to the $n = 2$ mode. The distributed transmission-line model was applied to the circuit just given, and the aforementioned observations were verified.

To further the preceding study, a ring resonator with two slits located at $\phi = \pm 90°$ was considered. The maximum field points for the first two modes supported by this structure are shown in Figure 3.21. The modes that this structure supports are the $n = 2, 4, 6, \ldots,$ and so on, modes of the basic ring resonator; all odd modes are suppressed, and there are no half-modes. The measurement corresponding to this device is shown in Figure 3.22. As can be seen, the first resonance occurs at approximately 5 GHz ($n = 2$), the second at 10 ($n = 4$), and so on. Resonance splitting in this figure is attributed to the differences in path lengths of the normally orthogonal modes of the ring resonator. This difference stems from the few degrees of error in slit placement that occurred during mask design.

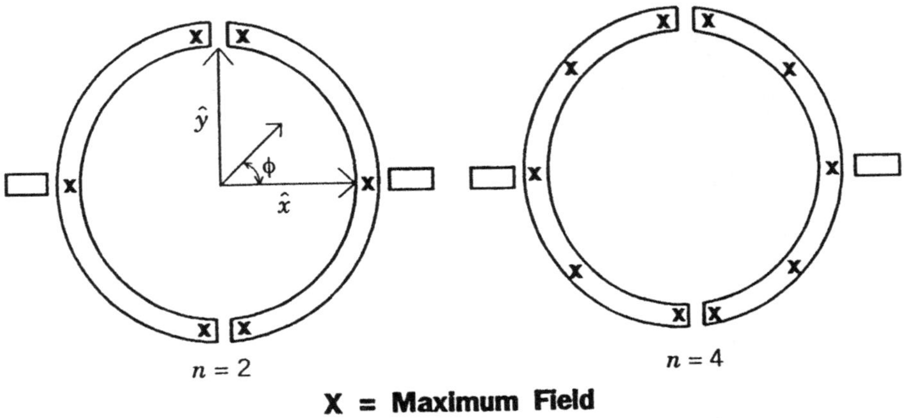

FIGURE 3.21 Maximum field points for slits at $\phi = \pm 90°$.

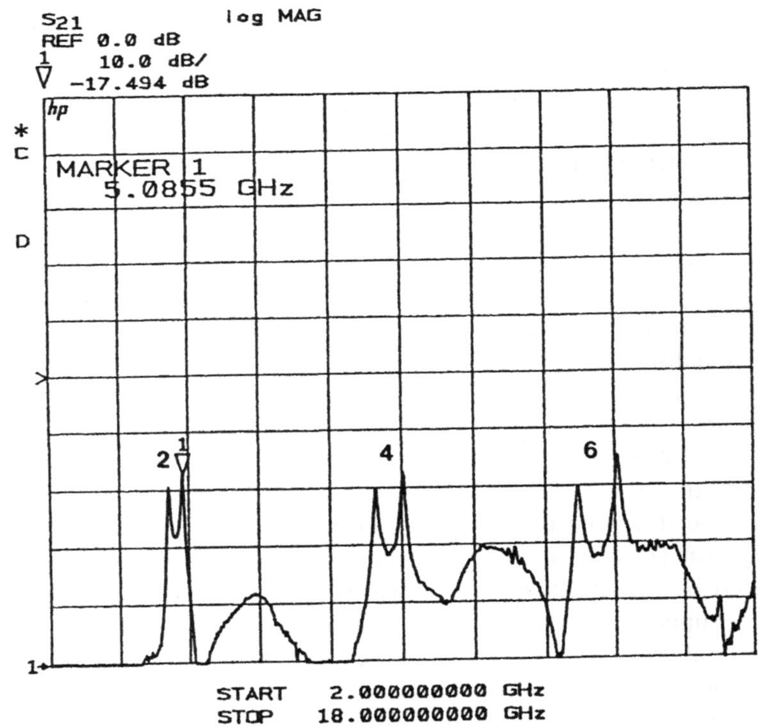

FIGURE 3.22 $|S_{21}|$ vs. frequency for slits at $\phi = \pm 90°$.

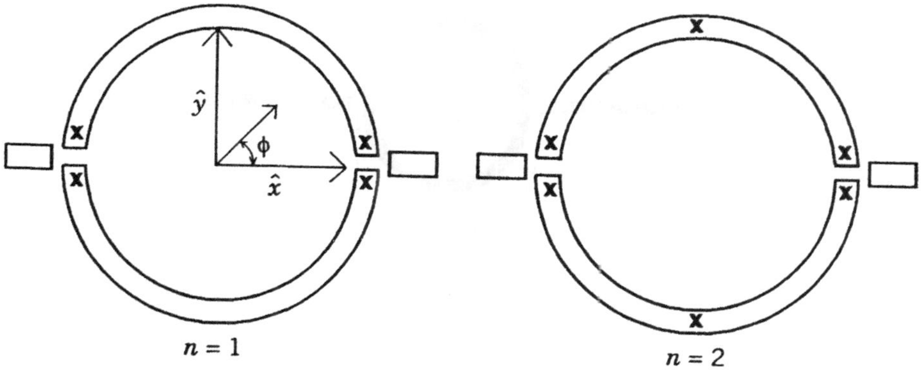

FIGURE 3.23 Maximum field points for slits at $\phi = 0°$ and 180° [10].

The mode configuration of the structure least susceptible to slit-related field perturbation is shown in Figure 3.23. These modes are identical to those shown in Figure 3.1 for the basic ring resonator. To experimentally verify this, a circuit with two slits, one at $\phi = 0°$ and the other at $\phi = 180°$ was fabricated; the circuit dimensions were the same as those mentioned previously. On measurement, the results obtained were identical to that of Figure 3.2 (corresponding to the basic ring), and hence are not shown separately. Thus, it has been clearly demonstrated that by strategically locating discontinuities such as notches and slits, a variety of modes can be obtained.

3.7 COUPLING METHODS FOR MICROSTRIP RING RESONATORS

Coupling efficiency between the microstrip feedlines and the annular microstrip ring element will affect the resonant frequency and the Q-factor of the circuit. Choosing the right coupling for the proper application circuit is important [2, 4]. According to the different coupling peripheries, the coupling schemes can be classified into the following [4]: (1) loose coupling [9] or matched loose coupling [11], (2) enhanced coupling [2, 12], (3) annular coupling, (4) direct connection, and (5) side coupling [13]. These five types of coupling schemes are shown in Figure 3.24*a–f*.

The *loose-coupling* scheme shown in Figure 3.24*a* results in the least disturbed type of coupling. The high-Q resonator application uses the loose coupling. Unfortunately the loose coupling suffers from the highest insertion loss because of its small effective coupling area [2, 12]. There is one variety of loose coupling that was developed to increase the coupling energy by using a matched coupling stub. Figure 3.24*b* shows this type of matched loose coupling [11].

The enhanced-coupling scheme shown in Figure 3.24*c* is designed by punching the feed lines into the annular ring element. This type of coupling is used

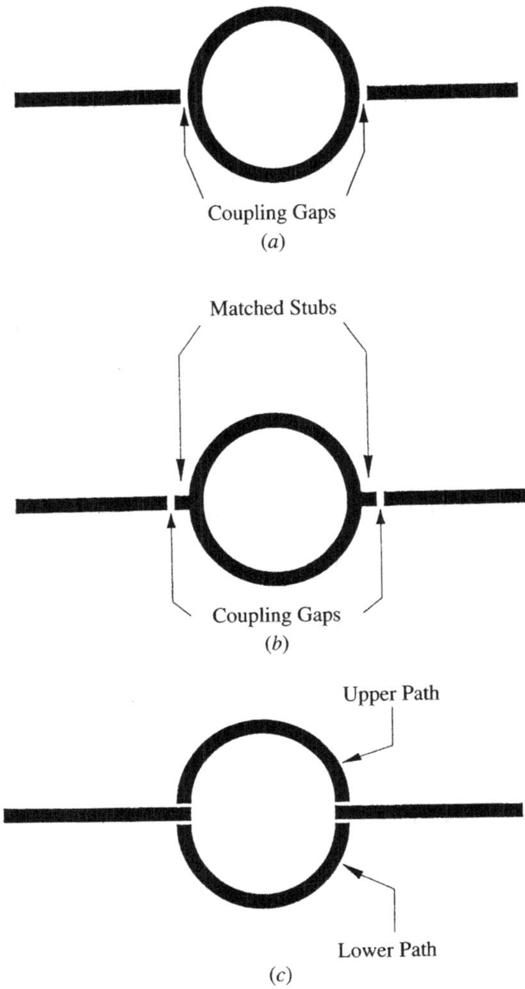

FIGURE 3.24 Coupling methods of annular ring element: (*a*) loose coupling; (*b*) matched loose coupling; (*c*) enhanced coupling; (*d*) annular coupling; (*e*) direct connection; (*f*) side coupling.

to increase the coupling periphery, but it slightly degrades the Q-factor of the resonator [2, 12]. By breaking the unity of the annular element, two parallel linear resonators that have a certain amount of curvature are formed. This type of coupling is also called *quasi-linear coupling*.

The third type of coupling as illustrated in Figure 3.24*d* is called annular *coupling*. This type of coupling scheme is developed to achieve the highest energy coupling. The coupling length is designed in terms of two annular angles, that is, θ_{in} and θ_{out}. By increasing the coupling length, higher coupling energy will be achieved. This type of coupling is used for a circuit design that

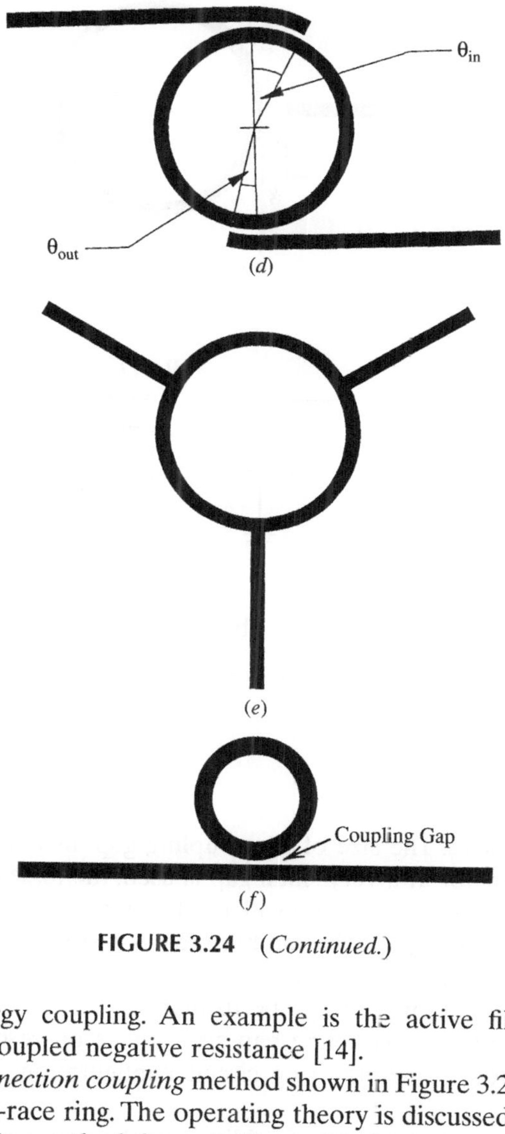

FIGURE 3.24 (*Continued.*)

needs large energy coupling. An example is the active filter design that requires a large coupled negative resistance [14].

This *direct-connection coupling* method shown in Figure 3.24*e* is used in the hybrid ring or rat-race ring. The operating theory is discussed in Chapter 8.

The *side-coupling* method shown in Figure 3.24*f* was reported in [13]. It was found that two distinctive but very close resonant peaks exist due to odd- and even-mode coupling. Introducing proper breaks in the ring will maintain the resonance characteristics of one mode while shifting the other peak away from the region of interest [13].

3.8 EFFECTS OF COUPLING GAPS

The coupling gap is an important part of the ring resonator. It is the separation of the feed lines from the ring that allows the structure to only support

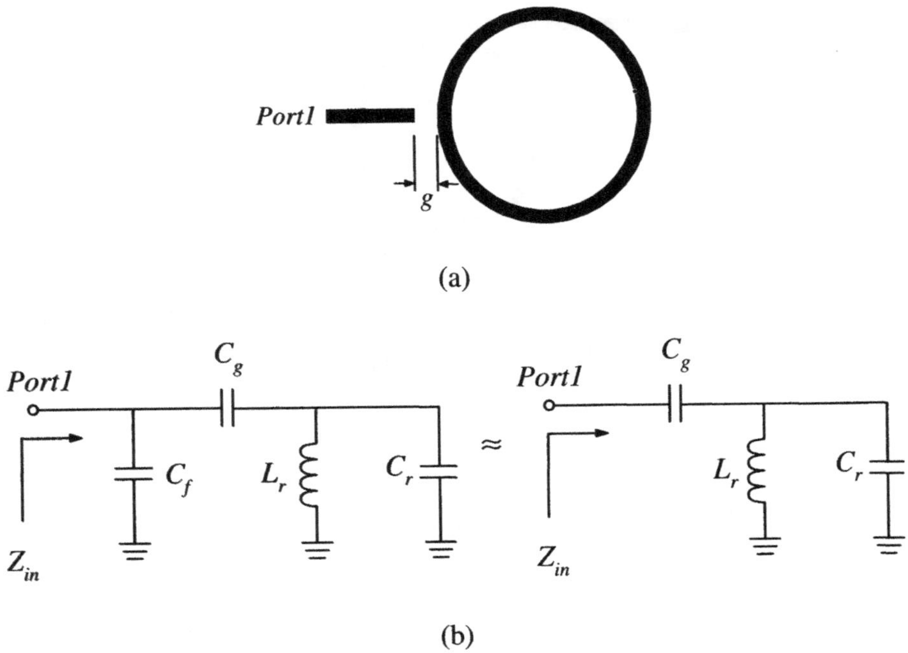

Figure 3.25 One-port ring circuit (a) configuration and (b) equivalent circuit.

selective frequencies. The size of the coupling gap also affects the performance of the resonator. If a very small gap is used, the losses will be lower but the fields in the resonant structure will also be more greatly affected. A larger gap results in less field perturbation but greater losses. It is intuitive that the larger the percentage of the ring circumference the coupling region occupies, the greater the effect on the ring's performance.

First, considering the coupling gap size effects on resonant frequencies, Figure 3.25 shows a one-port ring circuit configuration and its equivalent circuit.

The coupling gap between the feed line and the ring is represented by a L-network capacitance C_g and C_f [15]. The lossless ring resonator is expressed by a shunt circuit of L_r and C_r. In addition, comparing C_g and C_f, the coupling gap is significantly dominated by C_g. To simplify the calculation of the input impedance, the fringe capacitance C_f is neglected as shown on the right of Figure 3.25b. The total input impedance obtained from the simple equivalent circuit is given by

$$Z_{in} = \frac{j[\omega^2 L_r(C_r + C_g) - 1]}{\omega C_g(1 - \omega^2 L_r C_r)} \tag{3.5}$$

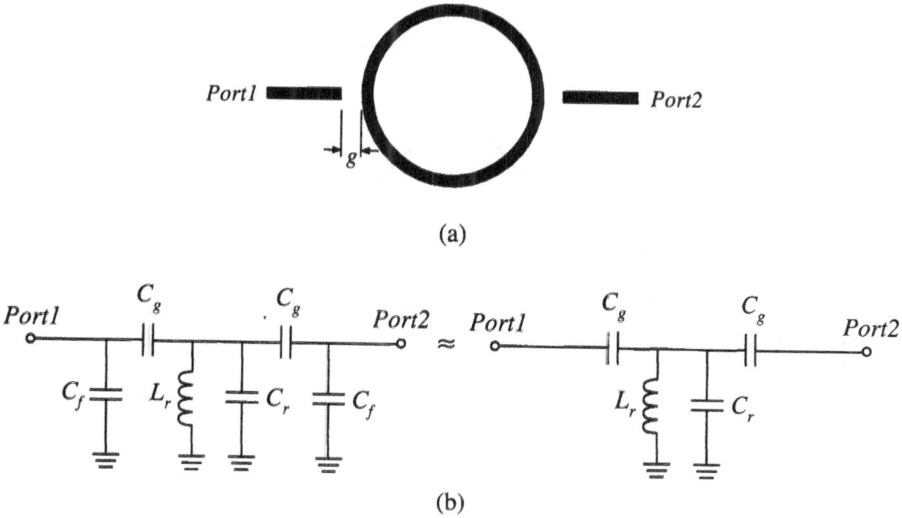

Figure 3.26 Two-port ring circuit (a) configuration and (b) equivalent circuit.

where ω is the angular frequency. At resonance, $Z_{in} = 0$ and the resonant angular frequency can be found as

$$\omega_o = \frac{1}{\sqrt{L_r(C_r + C_g)}} \tag{3.6}$$

Inspecting Equation (3.6), if the coupling gap size g is decreased (C_g increases), and therefore, the resonant frequencies move to lower locations. This equation shows the smaller size of coupling gap the lower resonant frequency.

The coupling gap size effect on the insertion loss can be observed from the two-port ring circuit in Figure 3.26.

S_{21} of the simplified equivalent circuit on the right of Figure 3.26b is given by

$$S_{21}\big|_{\omega=\omega_o} = \frac{2}{2(1+Z_gY)+Z_g(2+Z_gY)/Z_o+YZ_o} \tag{3.7}$$

where $Z_g = \dfrac{1}{j\omega_o C_g}$, $Y = \dfrac{j(\omega_o^2 L_r C_r - 1)}{\omega_o L_r}$, and Z_o is the characteristic impedance.

Inspecting Equation (3.7), when the coupling gap size g is decreased (increased), C_g and S_{21} increases (decreases). To verify above observations in Equations (3.6) and (3.7), a two-port ring circuit designed at a fundamental frequency of 2 GHz is simulated using IE3D [16].

In Figure 3.27, it can be found that a smaller (larger) gap size g has a lower (higher) insertion loss and more (less) significant effect on resonant frequency.

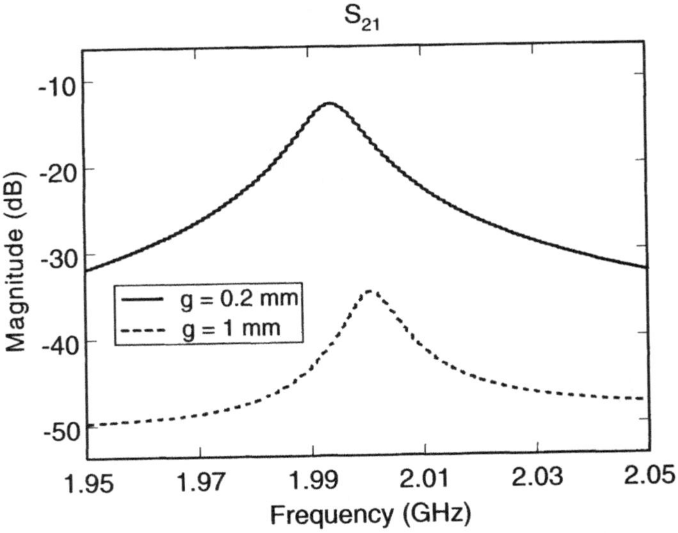

Figure 3.27 Simulated results for two different coupling gap size g = 0.2 mm and 1 mm.

Also, as the gap size g is increased (decreased), the loaded Q-factor decreases (increases) as expected.

In many of the ring's applications, the resonant frequency is measured in order to determine another quantity. For example, the resonant frequency is used to determine the effective permittivity (ε_{eff}) of a substrate and its dispersion characteristics. It is important in this measurement that the coupling gap not affect the resonant frequency of the ring and introduce errors in the calculation of ε_{eff}. Troughton realized this and took steps to minimize any error that was introduced [17]. He would initially use a small gap. The resonant frequency was measured and then the gap was etched back. Through repeated etching and frequency measurements the point was determined at which the feed lines were not seriously disturbing the fields of the resonator. This is a very tedious and time-consuming process. It would be very useful if a method could be developed that would enable the effects of the coupling gap on the resonant frequency to be determined.

The transmission-line method [18, 19] has the ability to predict the effects of the gap on the resonant frequency. It has been verified that the proposed equivalent circuit does give acceptable accuracy, but it should be pointed out that if the circuit does have a weakness it is the model used to represent the coupling gap. To verify the ability of the model to predict the gap dependence of the resonant frequency, experimental data were compiled and compared to the theoretical predictions.

Another method to predict the coupling gaps was proposed by Zhu and Wu [20]. They presented a joint field/circuit mode for coupling gaps of a ring circuit. The equivalent circuit model was derived from field theory and expressed in terms of a circuit network.

3.9 ENHANCED COUPLING

Although the loose-coupling method shown in Figure 3.24*a* is the most commonly used of the six types discussed earlier, it suffers from high insertion loss. To improve high insertion loss caused by loose couplings, many new configurations were introduced [21–25]. The philosophy underlying the design of these schemes is to increase the coupling strength (C_g) between feed lines and ring resonators. This has been discussed in Section 3.8. The enhanced coupling ring circuit with minimum perturbation shown in Figure 3.28 is designed to improve the insertion loss [2, 12].

As was mentioned in Section 3.5, the fields of the ring are least perturbed if discontinuities are present at points of field maximum (i.e., at $\phi = 0°$ and

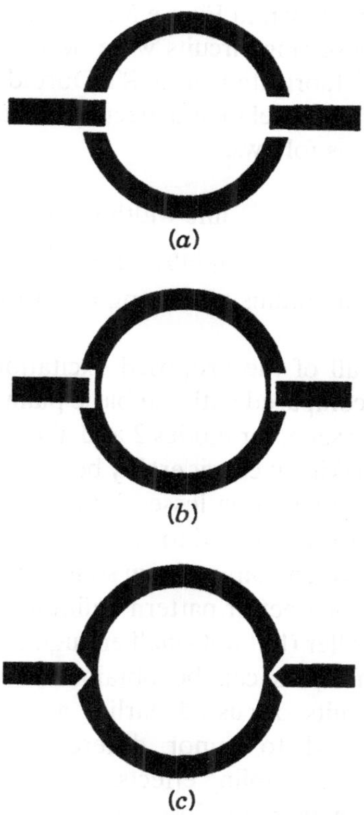

Figure 3.28 Three novel excitation schemes with much lower insertion losses: *a*, *b*, and *c* [12]. (Permission from *Electronics Letters.*)

TABLE 3.1 Comparison of Resonant Frequencies of Different Modes

	Resonant Frequency (GHz)			
Mode Number	Plain Ring	Scheme A	Scheme B	Scheme C
1	2.48	2.5	2.48	2.46
2	4.88	4.96	4.91	4.88
3	7.36	7.48	7.39	7.34
4	9.76	9.92	9.76	9.7
5	12.08	12.3	12	12
6	14.4	14.68	14.44	14.36
7	16.64	16.96	16.62	16.56

$\phi = 180°$). Hence, by increasing the coupling periphery at these points, the insertion loss of the ring can be reduced with minimal field perturbation. The measured results of the ring resonator shown in Figure 3.28*a* were given in Figure 3.29 and Table 3.1 for the first seven resonant frequencies. The measured data for resonators shown in Figure 3.28*b* and *c* are also given in Figure 3.29*b* and Table 3.1. These ring circuits were designed at a fundamental frequency of 2.5 GHz and fabricated on a RT/Duroid 6010.5 substrate with a thickness $h = 0.635$ mm and a relative dielectric constant $\varepsilon_r = 10.5$. The dimensions of the circuits are as follows:

$$\text{line width} = 0.573\,\text{mm}$$
$$\text{coupling gap} = 0.25\,\text{mm}$$
$$\text{mean radius of the ring} = 7.213\,\text{mm}$$

Inspecting the results, all of the proposed excitation schemes have a much lower insertion loss as compared with the basic plain ring. Also, superiority of scheme C can be clearly seen; for modes 2 and above the insertion loss of this scheme is about 5 dB, making it considerably better than the other circuits. The inconsistent trends in the insertion losses for the basic ring and the ring corresponding to scheme B, is attributed to variations associated with the process of circuit etching. However, if conventional solid-state photolithographic techniques are used, then much better pattern definition can be obtained. Also, if the gap size is made smaller (but not small enough to cause an RF short), then even smaller insertion losses can be obtained. In Table 3.1 the resonant frequencies of the circuits discussed earlier are compared; the frequency differences are attributed to minor differences in the lengths of the resonating section and the coupling effects.

To obtain a better low insertion loss, a ring resonator with more coupling periphery is shown in Figure 3.30. This configuration is usually designed for a filter application.

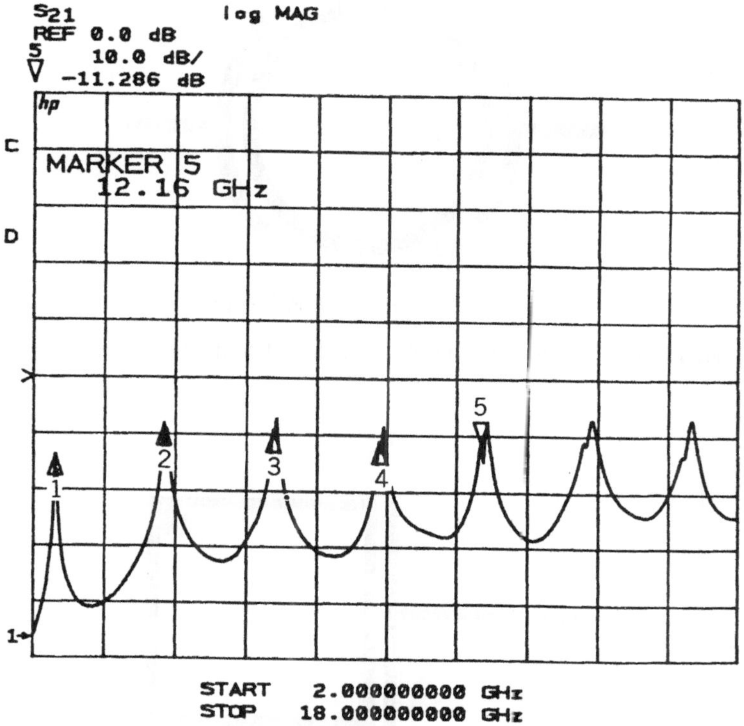

(a)

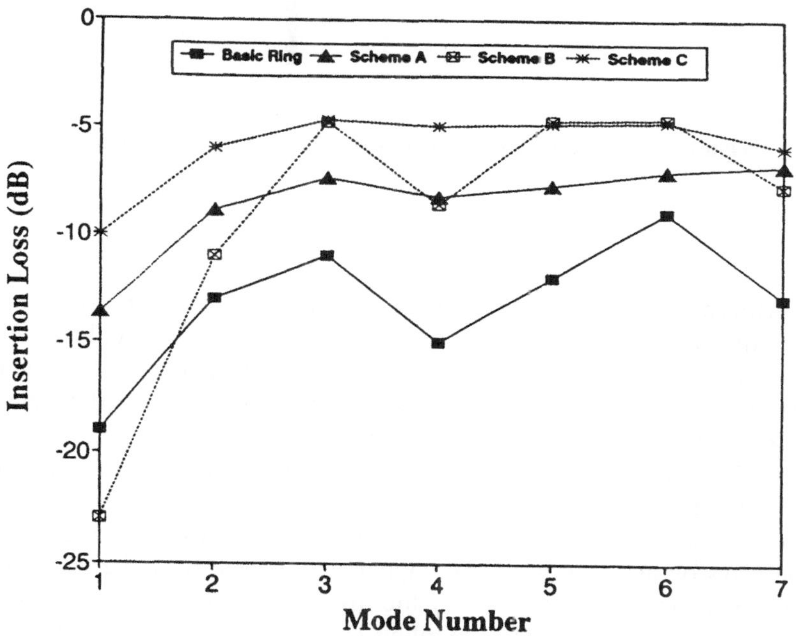

(b)

Figure 3.29 (a) $|S_{21}|$ vs. frequency for scheme A and (b) insertion loss vs. mode number for different ring resonators.

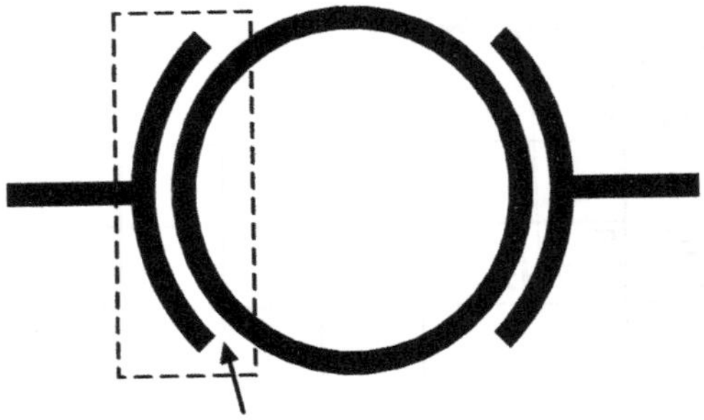

Figure 3.30 Configuration of an enhanced coupling ring resonator.

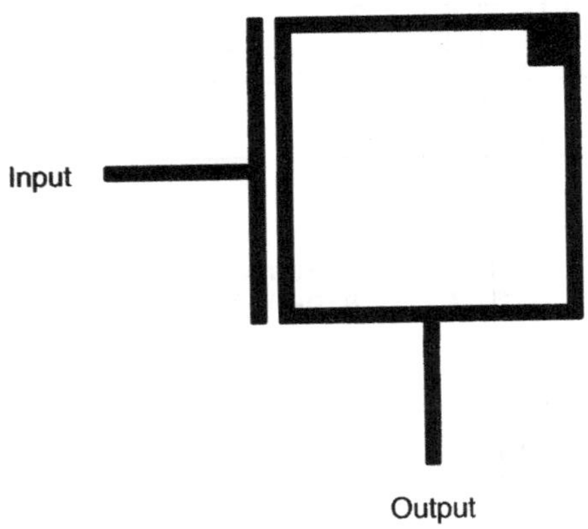

Figure 3.31 A dual-mode ring circuit with one single coupling gap.

Furthermore, to reduce the coupling gap effect on insertion loss and resonant frequencies, Figure 3.31 shows a ring circuit with one coupling gap [26]. Observing this circuit, with one coupling gap, the ring resonator has less effect on resonant frequencies and a low insertion loss can be reduced because one of two coupling gaps has been eliminated.

Another method to increase the coupling and lower the insertion loss is to use the dielectrically shielded ring resonator [18, 27] or dielectric overlay on top of the gaps [28]. Insertion loss of less than 1 dB can be achieved in these ways by using an insulated copper tape placed over the gap [28]. The coupling capacitance is formed by the insulation material between the tape and the microstrip line. This coupling capacitance corresponds to a much smaller gap.

3.10 UNIPLANAR RING RESONATORS AND COUPLING METHODS

Although the microstrip is the most mature and widely used planar transmission line, other forms of transmission lines are available for flexibility in ring circuit design [1, 29, 30]. These uniplanar transmission lines include coplanar waveguide (CPW), slotline, and coplanar strips (CPS). The characteristics of these transmission lines are listed in [31, p. 299].

The coplanar waveguide can be an alternative to the microstrip in hybird microwave integrated circuits (MIC) and monolithic microwave integrated circuits (MMIC). The center conductor and ground planes are on the same side of the substrate to allow easy series and shunt connections of passive and active solid-state devices. Use of CPW also circumvents the need for via holes to connect the center conductor to ground and helps to reduce processing complexity in monolithic implementations.

The slotline ring resonator was first proposed by Kawano and Tomimuro [32] for measuring the dispersion characteristics of slotline. The theorectical and experimental results agree well within 0.5% in their measurement. In 1983 Stephan et al. [33] developed a quasi-optical polarization-duplexed balanced mixer using a slotline ring antenna. The technique reported in [33] used the dual-mode feature of the slotline ring antenna. Slotline rings have also been implemented in a frequency-selective surface [34–36] and a tunable resonator [30, 37]. As a frequency-selective surface, the ring array has a reflection bandwidth of about 26% and a transmission/reflection-band ratio of 3:1. the varactor-tuned slotline ring resonator in [37] has a tuning bandwidth of over 23% from 3.03 GHz to 3.83 GHz.

The slotline ring resonator has been analyzed with equivalent transmission-line model [33], distributed transmission-line model [30, 37], spectral domain analysis [38], and Babinet's equivalent circular locp [39, 40]. The distributed transmission-line method provides a simple and straight-forward solution.

Coupling between the external feed lines and slotline ring can be classified into the following three types: (1) microstrip coupling, (2) CPW coupling, and (3) slotline coupling. Figure 3.32 shows these three possible coupling schemes.

As shown in Figure 3.32, the microstrip coupling that utilizes the microstrip-slotline transition [31, 41] is a capacitive coupling. The lengths of input and output microstrip coupling stubs can be adjusted to optimize the loaded-Q values. However, less coupling may effect the coupling efficiency and cause higher insertion loss. The trade-off between the loaded-Q and coupling loss depends on the application. Figure 3.33 shows the measured and calculated frequency responses of insertion loss for the microstrip-coupled slotline ring resonator. The test circuit was built on a RT/Duroid 6010.5 substrate with the following dimensions: substrate thickness h = 0.635 mm, characteristic impedance of the input/output microstrip feed lines $Z_{m0} = 50\,\Omega$, input/output microstrip feed lines with line width W_{m0} = 0.57 mm, characteristic impedance of the slotline ring $Z_S = 70.7\,\Omega$, slotline ring line width W_S = 0.2 mm, and slotline ring mean radius r = 18.21 mm. The S-parameters were measured using

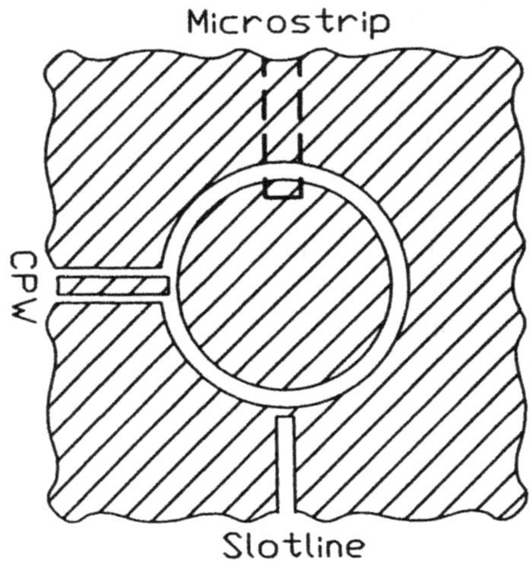

Figure 3.32 Three possible feed configurations for the slotline ring resonators [29]. (Permission from IEEE.)

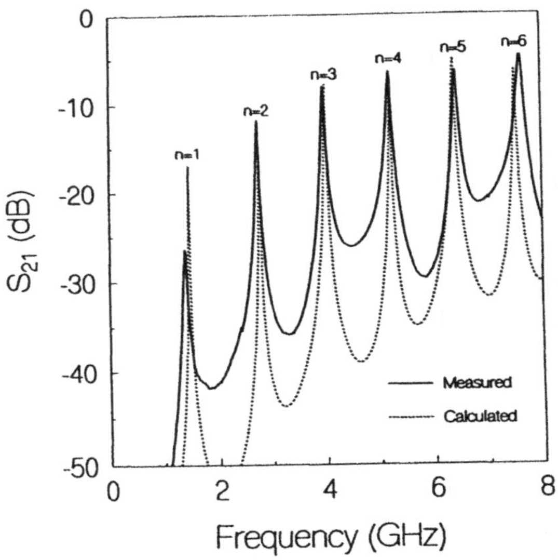

Figure 3.33 Measured and calculated frequency responses of insertion loss for a microstrip-coupled slotline ring resonator from 2 GHz to 8 GHz [29]. (Permission from IEEE.)

standard SMA connectors with an HP-8510 network analyzer. The calculated results were obtained from the distributed transmission-line model.

The CPW-coupled slotline ring resonator using CPW-slotline transition is also a capacitively coupled ring resonator. The CPW coupling is formed by a

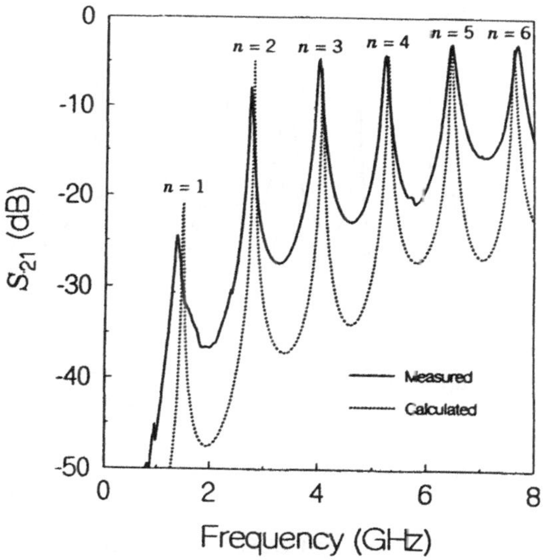

Figure 3.34 Measured and calculated frequency responses of insertion loss for a CPW-coupled slotline ring resonator from 2 GHz to 8 GHz.

small coupling gap between the external CPW feed lines and the slotline ring. The loaded-Q value and insertion loss are dependent on the gap size. The smaller gap size will cause a lower loaded-Q and smaller insertion loss. This type of slotline ring resonator is truly planar and also allows easy series and shunt device mounting. Figure 3.34 shows the measured and calculated frequency responses of insertion loss for the CPW-coupled slotline ring resonator. The test circuit was built on a RT/Duroid 6010.5 substrate with the following dimensions: substrate thickness $h = 0.635$ mm, characteristic impedance of the input/output CPW feed lines $Z_{C0} = 50\,\Omega$, input/output CPW feed lines gap size $G_{C0} = 0.56$ mm, input/output CPW feed lines center conductor width $S_{C0} = 1.5$ mm, characteristic impedance of the slotline ring $Z_S = 70.7\,\Omega$, slotline line width $W_S = 0.2$ mm, slotline ring mean radius $r = 18.21$ mm, and coupling gap size $g = 0.2$ mm.

The slotline ring coupled to slotline feeds is an inductively coupled ring resonator. The metal gaps between the slotline ring and external slotline feeds are for the coupling of magnetic field energy. Therefore, the maximum electric field points of this resonator are opposite to those of the capacitively coupled slotline ring resonators. Figure 3.35 shows the measured and calculated results of insertion loss for the slotline ring resonator with slotline feeds. The test circuit was built on a Duroid/RT 6010.5 substrate with the following dimensions: substrate thickness $h = 0.635$ mm, characteristic impedance of the input/output slotline feed lines $Z_{S0} = 56.37\,\Omega$, slotline feeds line width $W_{S0} = 0.1$ mm, characteristic impedance of the slotline ring $Z_S = 70.7\,\Omega$, slotline ring

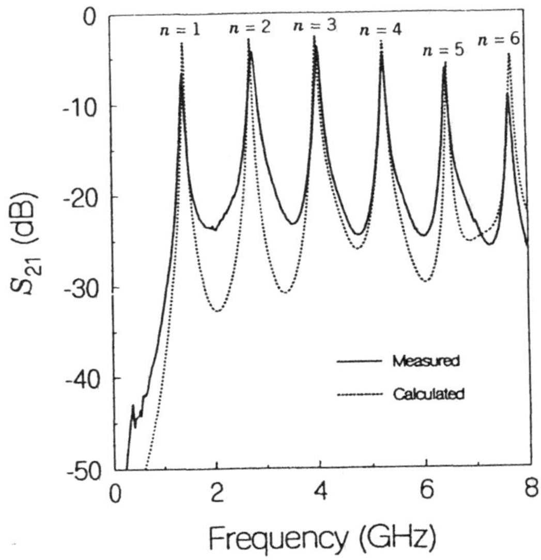

Figure 3.35 Measured frequency responses of insertion loss for a slotline ring resonator with slotline feeds.

line width $W_S = 0.2$ mm, slotline ring mean radius $r = 18.21$ mm, and coupling gap $g = 0.2$ mm.

As mentioned previously, the inductively slotline ring is the dual of the capacitively coupled slotline ring. The coupling of the capacitively coupled slotline ring resonators, as shown in Figure 3.33 and 3.34, becomes more efficient at higher frequencies. However, the coupling of the inductively coupled slotline ring with slotline feeds is less efficient at higher frequencies as shown in Figure 3.35. The reason for this phenomenon is the difference between the capacitive coupling and inductive coupling.

A uniplanar CPW ring resonator can also be constructed [30]. Figure 3.36 shows such a circuit. The circuit can be analyzed using a distributed transmission-line model similar to that described for the microstrip ring resonator in Chapter 2.

To demonstrate the performance of a CPW ring resonator, a ring was built with a mean diameter of 21 mm using 0.5-mm slotlines spaced 1.035 mm apart on 0.635-mm Duroid/RT Duroid 6010.5.

Figure 3.37 shows that the performance of the CPW ring is corrupted by the propagation of even-coupled slotline modes along the ring. To suppress these unwanted modes, the center disk of the ring must be maintained at ground potential. Wire bonding can be used at the input and output of the ring and along the ring itself to maintain the center disk ground potential but may prove to be labor intensive. A cover maintains the center disk at ground

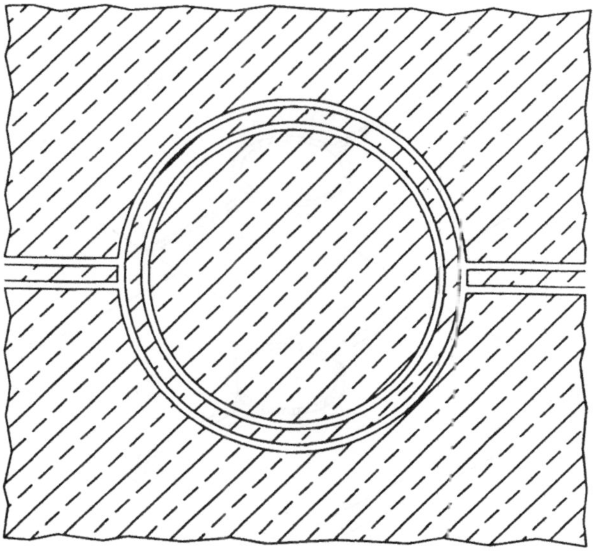

Figure 3.36 CPW ring resonator fed by CPW transmission lines.

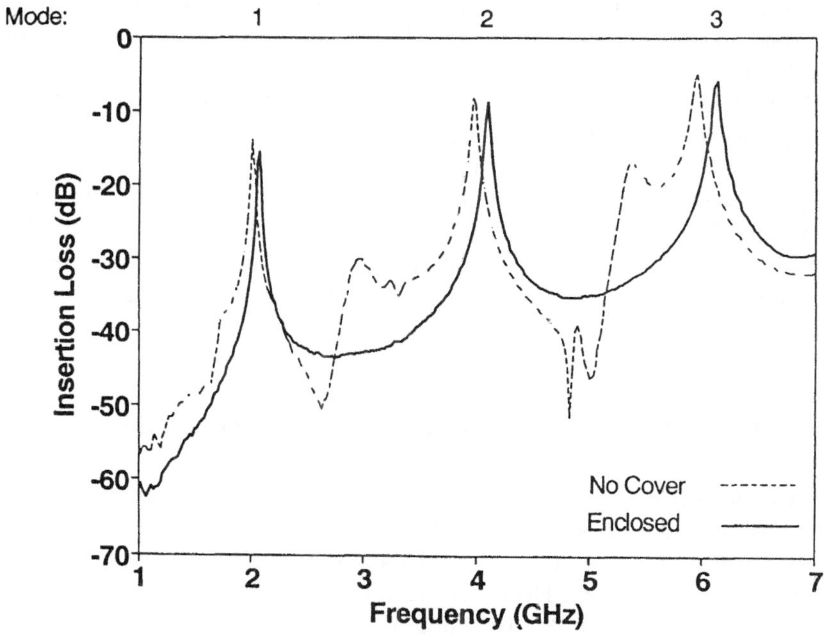

Figure 3.37 Insertion loss of a CPW ring with even and odd modes propagating [30]. (Permission from IEEE.)

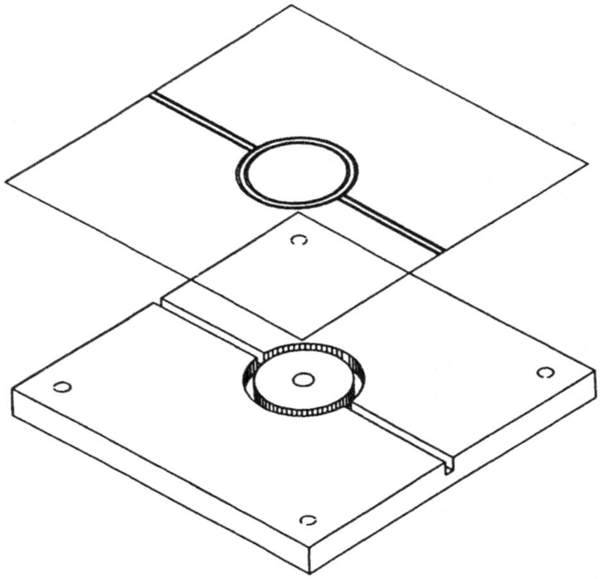

Figure 3.38 The enclosure for the CPW ring assembly [30]. (Permission from IEEE.)

potential all along the circumference of the ring; it also as seals and protects the circuit. The enclosure suppresses all even-mode propagation and reduces its inductive effect on the CPW odd mode. The enclosure and assembly shown in Figure 3.38 avoids wire bonding and soldering but requires alignment and good pressure contact with the ring. The height and width of the enclosure do not require high-tolerance machining.

Figure 3.39 shows the theoretical and measured results for the enclosed CPW ring. The theoretical results were obtained based on the distributed transmission-line equivalent circuit. The transmission-line parameters were determined based on formulas in [31, p. 275]. The gap capacitances were determined empirically. The agreement is within 2.91%.

3.11 PERTURBATIONS IN UNIPLANAR RING RESONATORS

As the microstrip ring resonator, the uniplanar ring structure could support both cosine and sine solutions. For feed lines located at 0° and 180°, the maximum E-field points are at 0° and 180°, and only the cosine mode satisfies the boundary conditions. However, perturbations will excite the other mode and cause mode split [1, 29]. By using a microstrip perturbation on the backside of the slotline ring, the regular resonances will become split resonant

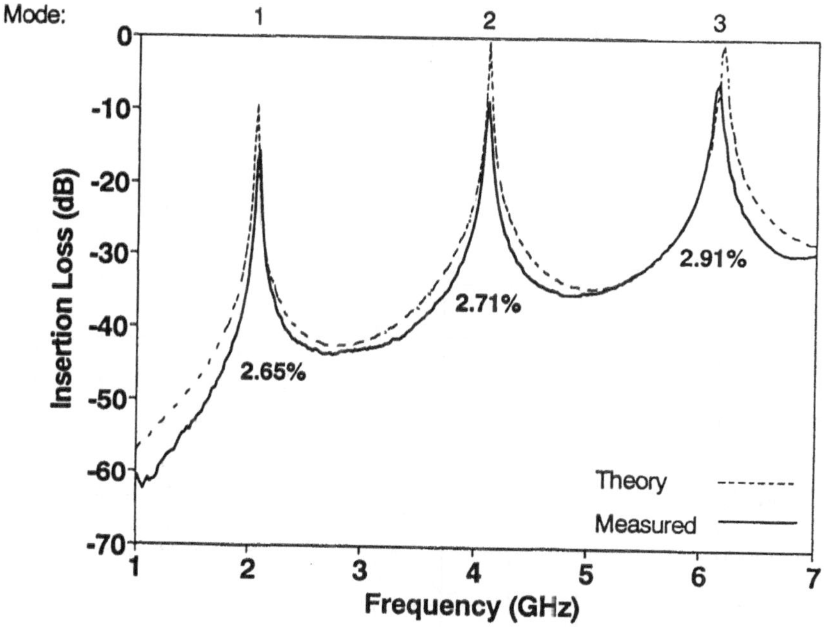

Figure 3.39 Theorectical vs. measured insertion loss and resonant frequencies of a CPW ring resonator [30]. (Permission from IEEE.)

modes. Figure 3.40 shows the measured frequency response of insertion loss for a slotline ring resonator with a microstrip perturbation at 45°. As shown in the figure, the resonant modes with mode numbers

$$n = m\frac{180°}{\theta} \tag{3.8}$$

where $\theta = 45°$ and $m = 1, 2, 3, \ldots$, will not split. According to the E-field mode chart, the 45° location of the microstrip perturbation is just at the maximum E-field point of the fourth resonant mode. The maximum E-field point corresponds to a magnetic-wall point that will not be disturbed by the microstrip perturbation. Other measured results for the perturbed slotline ring resonators with microstrip perturbation at 60° and 36° on the rear also agree with the general design rule of Equation 3.38.

The slotline type of varactor-tuned resonator deals the forced resonant modes of ring structure. The forced resonant modes of ring structure are excited by either short or open boundary conditions with respect to the electric or magnetic field. Figure 3.41 shows the measured frequency response of

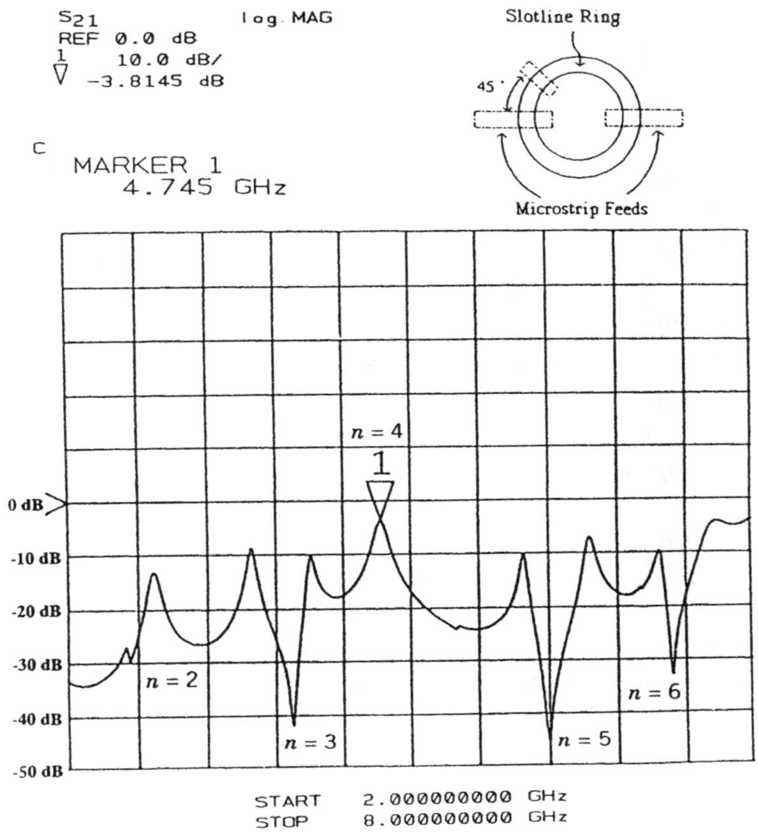

Figure 3.40 Measured frequency response of insertion loss for a slotline ring resonator with backside microstrip perturbation at 45°.

insertion loss for the slotline ring resonator with a short plane at 90°. According to the E-field mode chart, the maximum E-field points of the even resonant modes are located at the short boundary point. This means that the even resonant modes cannot exist in this perturbed ring structure, whereas the half-wavelength resonant modes will be excited due to the short boundary condition at 90°. The mode numbers of forced resonant modes shown in Figure 3.41 are given by

$$n = m/2 \tag{3.9a}$$

for half-wavelength resonant modes, where $m = 1, 3, 5, \ldots$, or

$$n = 2m - 1 \tag{3.9b}$$

for full-wavelength resonant modes were $m = 1, 2, 3, \ldots$.

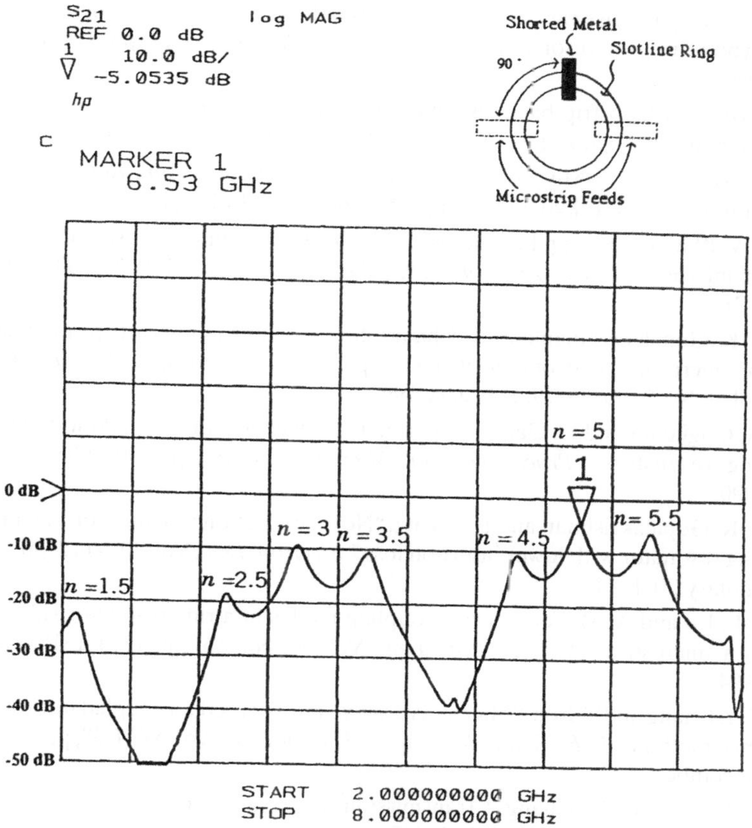

Figure 3.41 Measured frequency response of insertion loss for the forced resonant modes of a slotline ring resonator with an *E*-field short plane at 90°.

REFERENCES

[1] C. Ho, "Slotline, CPW ring circuits and waveguide ring cavities for coupler and filter applications," Ph.D. dissertation, Texas A&M University, College Station, May 1994.

[2] G. K. Gopalakrishnan, "Microwave and optoelectronic performance of hybrid and monolithic microstrip ring resonator circuits," Ph.D. dissertation, Texas A&M University, College Station, May 1991.

[3] R. P. Owens, "Curvature effect in microstrip ring resonator," *Electron. Lett.*, Vol. 12, No. 14, pp. 356–357, July 8, 1976.

[4] C. Ho and K. Chang, "Mode phenomenons of the perturbed annular ring elements," Texas A&M University Report, College Station, September 1991.

[5] I. Wolff and V. K. Tripathi, "The microstrip open-ring resonator," *IEEE Trans. Microwave Theory Tech.*, Vol. MTT-32, pp. 102–106, January 1984.

[6] G. K. Gopalakrishnan and K. Chang, "Bandpass characteristics of split-modes in asymmetric ring resonators," *Electron. Lett.*, Vol. 26, No. 12, pp. 774–775, June 7, 1990.

[7] I. Wolff, "Microstrip bandpass filter using degenerate modes of a microstrip ring resonator," *Electron. Lett.*, Vol. 8, No. 12, pp. 302–303, June 15, 1972.

[8] T. Okoshi and T. Miyoshi, "Analysis of planar circuit," *Ann. Rep. Eng. Res. Int.*, Univ. of Tokyo, Tokyo, Vol. 30, pp. 153–168, 1971 (In English).

[9] I. Wolff and N. Knoppik, "Microstrip ring resonator and dispersion measurements on microstrip lines," *Electron. Lett.*, Vol. 7, No. 26, pp. 779–781, December 30, 1971.

[10] G. K. Gopalakrishnan and K. Chang, "Study of slits in microstrip ring resonators for microwave and optoelectronic application," *Microwave Opt. Technol. Lett.*, Vol. 5, No. 2, pp. 76–79, February 1992.

[11] M. Guglielmi and G. Gatti, "Experimental investigation of dual-mode microstrip ring resonators," *Proc. 20th Eur. Microwave Conf.*, pp. 901–906, September 1990.

[12] G. K. Gopalakrishnan and K. Chang, "Novel excitation schemes for the microstrip ring resonator with lower insertion loss," *Electron. Lett.*, Vol. 30, No. 2, pp. 148–149, January 20, 1994.

[13] S. L. Lu and A. M. Ferendeci, "Coupling modes of a ring resonator side coupling to a microstrip line," *Electron. Lett.*, Vol. 30, No. 16, pp. 1314–1315, August 4, 1994.

[14] C. Y. Chang and T. Itoh, "Microwave active filters based on coupled negative resistance method," *IEEE Trans. Microwave Theory Tech.*, Vol. MTT-38, pp. 1879–1884, December 1990.

[15] C. C. Yu and K. Chang, "Transmission-line analysis of a capacitively coupled microstrip-ring resonator," *IEEE Trans. Microwave Theory Tech.*, Vol. 45, No. 11, pp. 2018–2024, November 1997.

[16] IE3D Version 8.0, Zeland Software Inc., Fremont, CA, January 2001.

[17] P. Troughton, "Measurement techniques in microstrip," *Electron Lett.*, Vol. 5, No. 2, pp. 25–26, January 23, 1969.

[18] K. Chang, T. S. Martin, F. Wang, and J. L. Klein, "On the study of microstrip ring and varactor-tuned ring circuits," *IEEE Trans. Microwave Theory Tech.*, Vol. MTT-35, pp. 1288–1295, December 1987.

[19] T. S. Martin, "A study of the microstrip ring resonator and its applications," M.S. thesis, Texas A&M University, College Station, December 1987.

[20] L. Zhu and K. Wu, "A joint field/circuit model of line-to-ring coupling structures and its application to the design of microstrip dual-mode filters and ring resonator circuits," *IEEE Trans. Microwave Theory Tech.*, Vol. 47, No. 10, pp. 1938–1948, October 1999.

[21] G. K. Gopalakrishnan and K. Chang, "Novel excitation schemes for the microstrip ring resonator with lower insertion loss," *Electron. Lett.*, Vol. 30, No. 2, pp. 148–149, January 1994.

[22] J. S. Hong and M. J. Lancaster, "Bandpass characteristics of new dual-mode microstrip square loop resonators," *Electron. Lett.*, Vol. 31, No. 11, pp. 891–892, May 1995.

[23] J. Y. Park and J. C. Lee, "A new enhanced coupling structure of microstrip ring resonator with two coupled lines and a slit," *IEEE MTT-S Int. Microwave Symp. Dig.*, pp. 805–808, 1998.

[24] W. C. Jung, H. J. Park and J. C. Lee, "Microstrip ring bandpass filters with new interdigital side-coupling structure," *Asia Pacific Microwave Conf.*, Vol. 3, pp. 678–681, 1999.

[25] C. E. Saavedra, "Microstrip ring resonator using quarter-wave couplers," *Electron. Lett.*, Vol. 37 No. 11, pp. 694–695, May 2001.

[26] L.-H. Hsieh and K. Chang, "Compact dual-mode elliptic-function bandpass filter using a single ring resonator with one coupling gap," *Electron. Lett.*, Vol. 36, No. 19, pp. 1626–1627, September 2000.

[27] K. Chang and J. Klein, "Dielectrically shielded microstrip (DSM) lines," *Electron. Lett.*, Vol. 23, No. 10, pp. 535–537, May 7, 1987.

[28] T. S. Martin, F. Wang, and K. Chang, "Theoretical and experimental investigation of novel varactor-tuned switchable microstrip ring resonator circuits," *IEEE Trans. Microwave Theory Tech.*, MTT-36, pp. 1733–1739, December 1988.

[29] C. Ho, L. Fan, and K. Chang, "Slotline annular ring elements and their applications to resonator, filter, and coupler design," *IEEE Trans. Microwave Theory Tech.*, Vol. MTT-41, pp. 1648–1650, September 1993.

[30] J. A. Navarro and K. Chang, "Varactor-tunable uniplanar ring resonators," *IEEE Trans. Microwave Theory Tech.*, Vol. MTT-41, pp. 760–766, May 1993.

[31] K. C. Gupta, R. Garg, and I. J. Bahl, *Microstrip Lines and Slotlines*, Artech House, Dedham, Mass, 1979.

[32] K. Kawano and H. Tomimuro, "Slot ring resonator and dispersion measurement on slot lines," *Electron. Lett.*, Vol. 17, No. 24, pp. 916–917, November 26, 1981.

[33] K. D. Stephan, N. Camilleri, and T. Itoh, "A quasi-optical polarization-duplexed balanced mixer for millimeter-wave applications," *IEEE Trans. Microwave Theory Tech.*, vol. MTT-31, pp. 164–170, February 1983.

[34] E. A. Parker and S. M. A. Hamdy, "Rings as elements for frequency selective surfaces," *Electron. Lett.*, Vol. 17, No. 17, pp. 612–614, August 20, 1981.

[35] E. A. Parker, S. M. A. Hamdy, and R. J. Langley, "Arrays of concentric rings as frequency selective surfaces," *Electron. Lett.*, Vol. 17, No. 22, pp. 880–881, October 10, 1981.

[36] R. Cahill and E. A. Parker, "Concentric ring and Jerusalem cross arrays as frequency selective surfaces for a 45° incidence diplexer," *Electron. Lett.*, Vol. 18, No. 2, pp. 313–314, April 15, 1982.

[37] J. A. Navarro and K. Chang, "Varactor-tunable uniplanar ring resonators," in *1992 IEEE MTT-S Int. Microwave Symp. Dig.*, pp. 951–954, June 1992.

[38] K. Kawano and H. Tomimuro, "Spectral domain analysis of an open slot ring resonator," *IEEE Trans. Microwave Theory Tech.*, Vol. MTT-30, pp. 1184–1187, August 1982.

[39] G. Dubost, "Theoretical radiation resistance of an isolated slot ring resonator," *Electron. Lett.*, Vol. 23, No. 18, pp. 928–930, August 27, 1987.

[40] G. Dubost, "Large bandwidth circular slot at resonance with directional radiation," *Electron. Lett.*, Vol. 24, No. 23, pp. 1410–1411, November 10, 1988.

[41] R. N. Simons, S. R. Taub, R. Q. Lee, and P. G. Young, "Microwave characterization of slot line and coplanar strip line on high-resistivity silicon for a slot antenna feed network," *Microwave Opt. Technol. Lett.*, Vol. 7, No. 11, pp. 489–494, August 5, 1994.

CHAPTER FOUR

Electronically Tunable Ring Resonators

4.1 INTRODUCTION

In this chapter the varactor tuned resonator is studied [1]. The varactor is a two-terminal solid-state device that utilizes the voltage variable capacitance of the *PN* junction. When a varactor is mounted in series in the transmission line of the ring, the variable capacitance is used to change the resonant frequency of the structure. The arrangement of the varactor-tuned ring is given in Figure 4.1. The resonant frequency of the ring is normally determined by its physical dimensions. The addition of the capacitance in the ring is equivalent to the addition of a given length of transmission line to the ring's circumference. A larger capacitance will result in a larger circumference, and thus lower the resonant frequency. As the capacitance is decreased, the resonant frequency will increase. Because the capacitance in the *PN* structure of the varactor diode is voltage dependent, the resonant frequencies of the varactor-tuned resonator can be tuned electronically.

Before the varactor-tuned resonator can become useful in microwave circuits, it is important that the effects of the introduced capacitance on the resonant frequency be predictable by analysis. If the resonant frequency cannot be accurately determined by a theoretical method, then the design of the varactor—ring circuits will be the result of a trial-and-error process. Trial-and-error design methods are time-consuming and would make the varactor-tuned ring less likely to be used.

Microwave Ring Circuits and Related Structures, Second Edition,
by Kai Chang and Lung-Hwa Hsieh
ISBN 0-471-44474-X

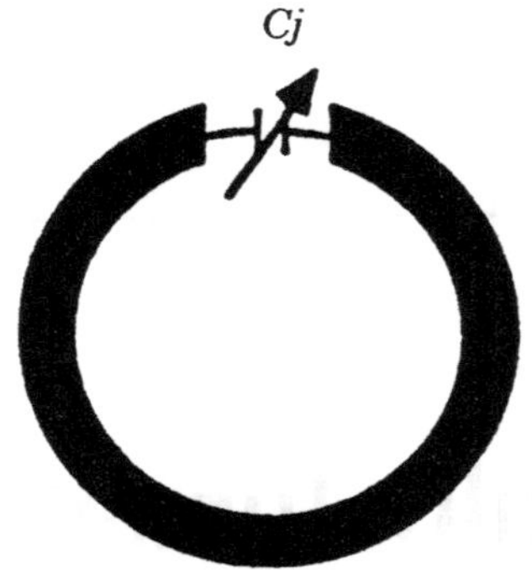

FIGURE 4.1 A varactor-tuned ring resonator.

The magnetic-wall model is the most accurate ring analysis technique. What is gained in accuracy is in turn lost in flexibility. The magnetic-wall model cannot easily be altered to include the effect of a varactor. This is the primary reason that the transmission-line method has been developed. It is shown in Chapter 2 that the transmission-line method can accurately determine the resonant frequency of a simple microstrip ring. This analysis can easily be altered to include the varactor. To include the varactor in the analysis would require that an equivalent circuit for the varactor diode be incorporated into the already proposed equivalent circuit for the ring. In this chapter an equivalent circuit for the varactor is proposed and the varactor-tuned ring is analyzed by the transmission-line method outlined in Chapter 2, although for better accuracy, the distributed transmission-line method, also given in Chapter 2, can be used.

4.2 SIMPLE ANALYSIS

The varactor-tuned ring was first introduced in 1986 in a paper by Makimoto and Sagawa [2]. Conventional varactor-tuned filters have been quarter- or half-wavelength linear resonators. The disadvantages of linear resonators are discussed in Chapter 2. The authors realized the disadvantage of conventional filters and proposed the varactor-tuned ring because of its increased stability and steeper attenuation slope in the stopband.

For the resonator circuit shown in Fig. 4.2, the input admittance of the circuit could be given as

$$Y_i - jY_o \frac{Y_o \sin\theta_T - 2\omega C_T(1-\cos\theta_T)}{Y_o \cos\theta_T - \omega C_T \sin\theta_T} \tag{4.1}$$

where Y_o is the characteristic admittance of the line; θ_T is the total electrical length of the line; and C_T is the tuning capacitance. The steps taken to obtain

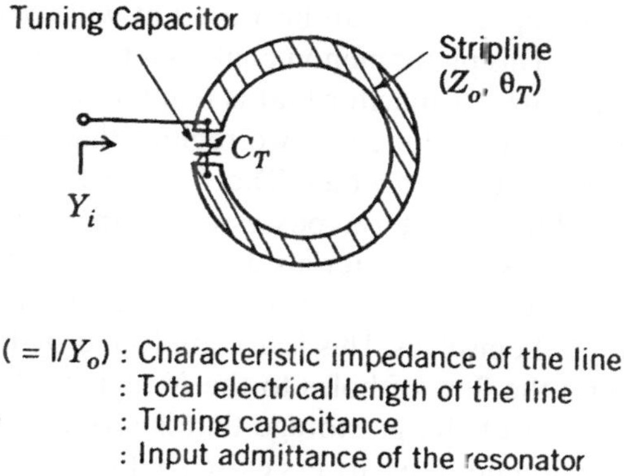

FIGURE 4.2 Structure of ring resonator for analysis [2]. (Permission from IEEE.)

Equation (4.1) are not explained in the reference, but it does take a form similar to the transmission-line equation

$$Y(\theta) = jY_o \frac{Y_o \sin\theta - jY_L \cos\theta}{Y_o \cos\theta - jY_L \sin\theta} \tag{4.2}$$

where Y_L is the load admittance. Using Equation (4.1), the tuning range of the varactor-tuned circuit was predicted. Experimental results showed that the varactor could be used to tune the resonant frequency of the ring, but the authors note that (4.1) could not accurately predict the response. The resonant frequencies were lower than expected and the tuning range was smaller. Errors were attributed to stray inductance that was not included in the analysis. This method is also unable to include the effect of the coupling gap in the circuit. More accuracy might have been obtained if the parasitics of the varactor had been included in the admittance equation.

4.3 VARACTOR EQUIVALENT CIRCUIT

The varactor is a solid-state diode whose capacitance is a result of the *PN* junction. Every semiconductor diode has some internal junction capacitance. Usually, however, this internal capacitance is insignificant because it is intentionally kept as small as possible so that it will not degrade normal diode operation. Basically, the varactor is a special-purpose junction diode. But it differs from other diodes in one important respect: it is designed and fabricated specifically to make its junction capacitance useful [1, 3]. This design is such that the varactor has a usable internal capacitance, high parallel resistance, and low series resistance. Thus, capacitance, which is an unavoidable nuisance in conventional diodes, is deliberately cultivated in the varactor. It is both novel and useful that this capacitance can be varied at will by varying the voltage applied to the diode. This phenomenon enables a tiny varactor to do the work of a conventional variable capacitor that is many times larger.

In the formation of a *PN* junction two regions that possess opposite types of conductivity are brought together. The *P* material possesses holes as its majority carrier, and the *N* material possesses electrons as its majority carrier. The Fermi levels of these two materials differ as a result of the different conductivity. When the two materials are brought together in contact, the Fermi levels work to align themselves. This is accomplished by the flow of electrons to the *P* region and the flow of holes to the *N* region. The Fermi level alignment results in a layer of charge storage that is termed the depletion region. In the depletion region no free carriers exist, so it is effectively called an insulator. The diode then has the appearance of a positive region (*P*) separated from a negative region (*N*) by an intrinsic or insulating region. This structure is identical to two flat plates separated by a dielectric. This arrangement describes any two-plate capacitor. In this type of capacitor the capacitance is directly proportional to the effective area of the plates and the dielectric constant of the dielectric, and is inversely proportional to the separation of the plates. The junction capacitance can then be expressed as

$$C_j = \frac{\varepsilon A}{d} \tag{4.3}$$

where ε is the dielectric constant; A is the plate area; and d is the plate separation. By analogy, in the semiconductor junction, A is the area of the N and P regions that face each other across the junction; d is the thickness of the depletion region; and ε is the dielectric constant of the depletion layer semiconductor.

For a given diode both A and ε will be constant, but d can be varied depending on the applied voltage. If the diode is forward biased, the depletion region is decreased and the internal capacitance increases. If the diode is reverse biased, the depletion region increases and the internal capacitance decreases. Reverse biasing results in a small reverse current. Figure 4.3 shows the actual plot of a varactor capacitance as a function of voltage obtained with a *C-V* meter and *x-y* plotter for an M/A COM varactor diode (Model MA-46600). In Figure 4.3*a* the forward-biased varactor is shown, and in Figure 4.3*b* the reverse-biased varactor capacitor is shown. The varactor is usually operated as reverse biased since forward-biased voltage results in a large leakage current and low Q.

The junction capacitance can be expressed as [4]

$$C_j(V) = C_{j0}\left(1 - \frac{V}{V_{bi}}\right)^{-\gamma} \tag{4.4}$$

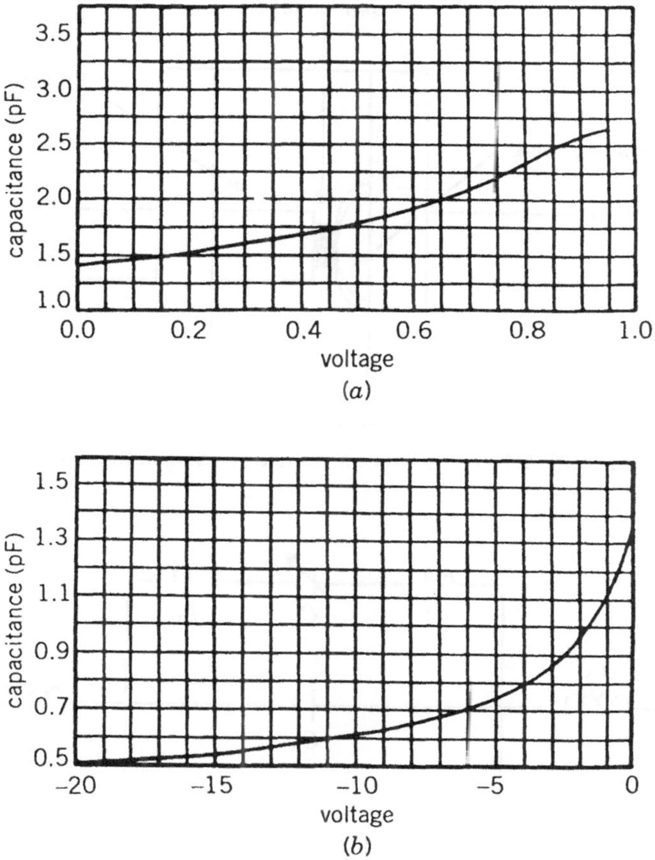

FIGURE 4.3 *C-V* traces for the (*a*) forward- and (*b*) reverse-biased varactor diode.

where C_{j0} is the capacitance at zero bias voltage, V_{bi} is the built-in potential of 1.3 volts for GaAs, and γ is a parameter depending on the *PN* junction doping profile.

Any varactor used in a circuit will also introduce parasitic components from the packaging in addition to the resistance and capacitance of the semiconductor. A typical cross-sectional view of a packaged diode is shown in Fig. 4.4. It is seen that the package consists of an insulating casing separating two metallic end pieces sealed in such a manner as to provide a hermitic encapsulation for the semiconductor within. Within the package the semiconductor is usually mounted on a post or pedestal with a suitable strap making contact to the opposite end of the diode. Both the metallic ends and insulating ceramic parts contribute inductance and capacitance between the contacts of the actual semiconductor element and the connections to the external diode housing or package. From the consideration of the package an equivalent circuit can be

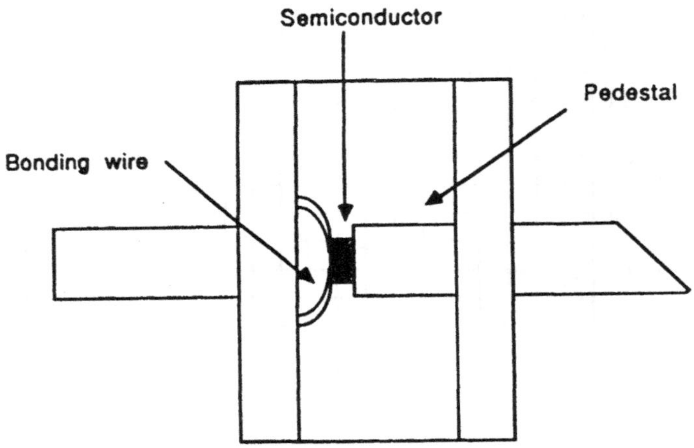

FIGURE 4.4 Diagram of a varactor package cross section.

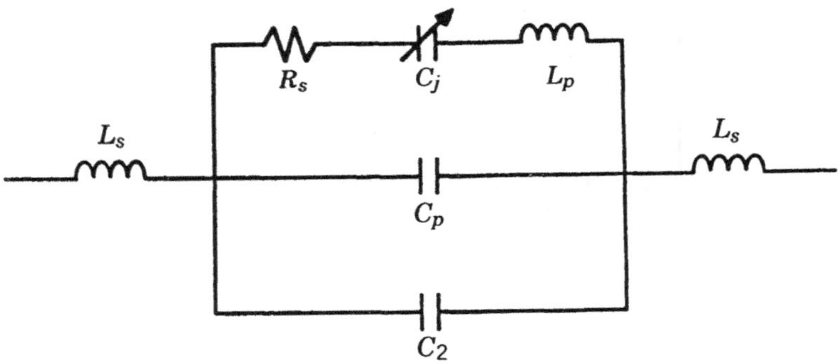

FIGURE 4.5 Equivalent circuit of a packaged varactor.

proposed. The equivalent circuit given in Figure 4.5 can also be used for diodes other than varactors. The only difference will be the value of the parameters.

In Figure 4.5 C_j is obviously the capacitance that arises from the semiconductor junction. It is this value in which we are most interested; all the others are undesirable but unavoidable. The value R_s is the series resistance due primarily to the bulk resistance of the semiconductor. Minimizing R_s increases the Q of the varactor (here, $Q = 1/\omega R_s C_j$), reducing power losses in the circuit and increasing the overall circuit Q. Typically higher Q-values are obtainable with hyperabrupt junction varactors because of the lower bulk resistance.

The parameters C_p, L_p, and L_s are the parasitics introduced by the package. The capacitance C_p, which appears in shunt, is a combination of the capacitance that exists between the upper contact and the metallic mount of the semiconductor and the insulating housing. Because of the close spacing required in microwave frequency circuits, particularly for small elements that possess small junction capacitances, the capacitance contribution can become quite significant. The capacitance C_2 is also included in Figure 4.5. Here C_2 is the capacitance that arises from the gap in the transmission line across which the diode will be mounted. This is the same gap capacitance discussed in Chapter 2. The gap shunt capacitance, C_1, is omitted because its effects are considered to be negligible.

In addition to the capacitances, all metallic portions of the package will introduce inductance. The inductance is divided into two components L_s and L_p. The inductance L_p appears in series with the junction capacitance. The most significant contributions of the inductance come from the metallic contacting strap and the post upon which the semiconductor element is mounted. The contributions are significant because of the very small cross-sectional dimensions of the parts with lengths that are comparable to the dimensions of the package. The inductance L_s represents the series inductance of the outer end parts to the external contacting points. This can become very large if long leads are required for bonding to the circuit.

The equivalent circuit does to some extent actually represent the physical contributions of the typical packaged diode structure and can be useful over a wide range of frequencies. Values for the equivalent circuit will vary for each diode type and package style. Because the packaged-diode equivalent circuit is widely recognized, manufacturers usually supply typical parameter values for each package style and diode type.

4.4 INPUT IMPEDANCE AND FREQUENCY RESPONSE OF THE VARACTOR-TUNED MICROSTRIP RING CIRCUIT

Now that the equivalent circuit for the varactor has been proposed, the input impedance of the circuit can be determined [1, 3]. In Chapter 2 it was verified that the transmission-line method could be used to accurately determine the resonant frequency of the microstrip ring resonator. The equivalent circuit of Figure 2.12 should then adequately represent the ring and coupling gaps. The varactor-tuned ring will differ only slightly from the plain ring resonator.

To mount the varactor in the circuit, the ring will be cut at two points and the varactor placed across one of the cuts, while a dc block capacitor is mounted across the other cut. The dc block capacitor is chosen to have a large value. The capacitor is required so that a dc bias voltage can be applied across the cathode and anode of the varactor. At microwave frequencies the capacitance will appear as a short and have very little effect. For low frequency, however, the capacitance appears as an open circuit and allows the varactor to be biased. To apply the voltage to the device, bias lines connect to the ring. The bias lines are high impedance lines. The bias lines act as RF chokes, preventing the leakage of RF power, while at the same time allowing the applied dc bias voltage to appear across the terminals of the device. The layout for the varactor-tuned ring is given in Figure 4.6.

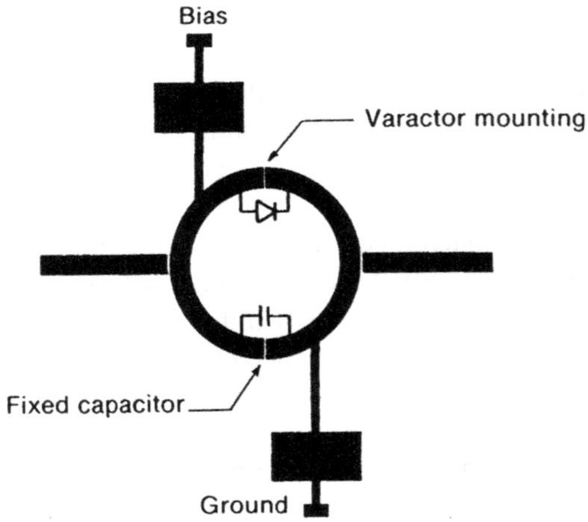

FIGURE 4.6 Diagram of varactor-tuned ring resonator [3]. (Permission from IEEE.)

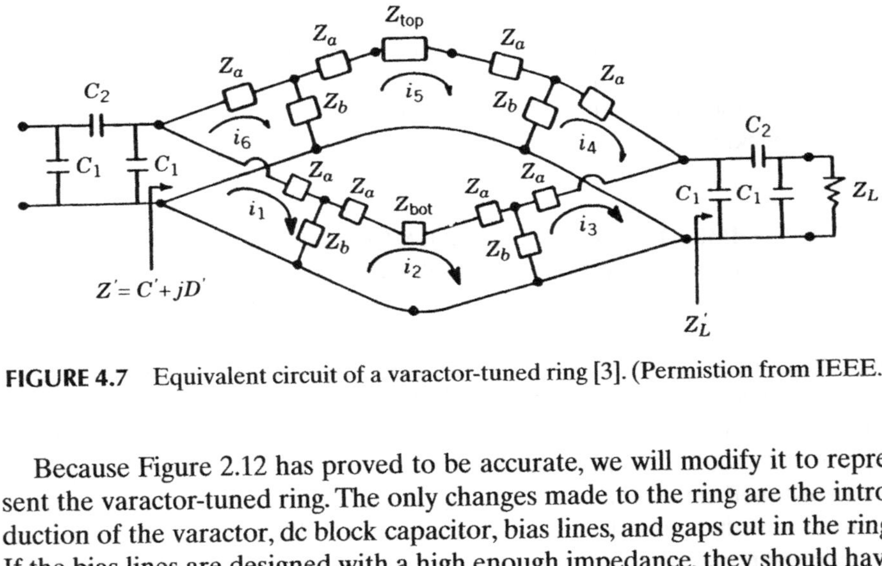

FIGURE 4.7 Equivalent circuit of a varactor-tuned ring [3]. (Permistion from IEEE.)

Because Figure 2.12 has proved to be accurate, we will modify it to represent the varactor-tuned ring. The only changes made to the ring are the introduction of the varactor, dc block capacitor, bias lines, and gaps cut in the ring. If the bias lines are designed with a high enough impedance, they should have little effect on the circuit and will be neglected in the analysis. The proposed equivalent circuit for the varactor-tuned ring is given in Figure 4.7. The parameters C_1 and C_2 are discussed in Chapter 2 and are used to model the input and output coupling gaps. The parameters Z_a and Z_b are from the T-model for the transmission line of the ring, also discussed in Chapter 2. The impedance Z_{bot} represents the bypass capacitor. Because the bypass capacitor wilt be large (usually 10 pF or larger), the capacitance of the gap across which the dc block is mounted can be neglected. In fact, because the bypass capacitor is large, it acts as a very low impedance (short circuit) at microwave frequencies. Thus, for this application the dc block capacitor could be neglected, but it can be included to make the input impedance equations more flexible for other applications. The impedance Z_{top} represents the varactor mounted in the ring. The equivalent circuit for the varactor was given in Figure 4.5.

The load seen by the ring at the output coupling gap is given as Z_L' where

$$Z_L' = A + jB \tag{4.5}$$

and A and B are defined in Chapter 2. The ring structure is not symmetrical and therefore cannot be reduced through combinations of series and parallel impedances. A unit voltage is applied to the circuit and six loop currents are visualized. From the six loop currents, a system of six equations and six unknowns is formed. The input impedance looking into the gap, Z', can be calculated by solving the sixth-order system of equations for the currents due to a unit source. The system to be solved is

$$\mathbf{V} = \mathbf{IZ} \tag{4.6}$$

where

$$\mathbf{V} = \begin{pmatrix} V_{\text{unit}} \\ 0 \\ 0 \\ 0 \\ 0 \\ V_{\text{unit}} \end{pmatrix}$$

$$\mathbf{I} = \begin{pmatrix} i_1 \\ i_2 \\ i_3 \\ i_4 \\ i_5 \\ i_6 \end{pmatrix}$$

and

$$\mathbf{Z} = \begin{pmatrix} Z_a + Z_b & -Z_b & 0 & 0 & 0 & 0 \\ -Z_b & 2Z_a + 2Z_b + Z_{\text{bot}} & -Z_b & 0 & 0 & 0 \\ 0 & -Z_b & Z_a + Z_b + Z_L' & Z_L' & 0 & 0 \\ 0 & 0 & Z_L' & Z_a + Z_b + Z_L' & -Z_b & 0 \\ 0 & 0 & 0 & -Z_b & 2Z_a + 2Z_b + Z_{\text{top}} & -Z_b \\ 0 & 0 & 0 & 0 & -Z_b & Z_a + Z_b \end{pmatrix}$$

Once the currents are known, then

$$Z' + C' + jD' = \frac{V_{\text{unit}}}{i_1 + i_6} \tag{4.7}$$

The input impedance of the circuit, Z_{in}, can be found by replacing C and D of Chapter 2 by C' and D', respectively.

To facilitate the solution of (4.7) the IMSL subroutine LEQ2C was used [6]. The IMSL library is a collection of mathematical and statistical subroutines written in Fortran. The subroutine LEQ2C is used to solve a system of complex equations.

The resonant frequency of Figure 4.7 can be determined in two ways. The first method was discussed in Chapter 2, the bisectional method. Using the bisection algorithm the frequency can be determined at which $X_{in} = 0$. The second method uses the S-parameters of the circuit. The ratio of the reflected power over the incident power can be determined from

$$S_{11} = \frac{Z_{in} - Z_o}{Z_{in} + Z_o} \tag{4.8}$$

where Z_{in} is the input impedance of the circuit and Z_o is the characteristic impedance. From S_{11}, the ratio of transmitted power over the incident power, for a lossless circuit, can be determined from

$$S_{21} = \sqrt{1 - S_{11}^2} \tag{4.9}$$

The resonant frequency is the point at which S_{12} reaches a maximum, resulting in maximum power transfer. The condition S_{12} = max and $X_{in} = 0$ occur at the same frequency, and it is equally correct to apply either condition. The S-parameter method will become more important later when the attenuation at some frequency is desired.

Using (4.8) and (4.9) the frequency response of a typical varactor-tuned ring can be compared to a plain ring resonator of similar dimensions. Figure 4.8*a* shows the frequency response of a typical ring resonator. Figure 4.8*b* shows the frequency response of a typical varactor-tuned ring. A few interesting things can be seen in the comparison of Figure 4.8*a* and Figure 4.8*b*. The odd modes in the varactor-tuned ring disappear while the even modes remain unaffected and coincide exactly with the even modes of the plain ring. Introduced in the varactor-tuned ring are what can be called "half-modes." If the varactor is removed from the circuit, but the ring is still cut, the half-modes will lie exactly between the even and missing odd modes.

Figure 4.9 is used to explain the mode phenomena. This figure displays the positive maximum and negative maximum electric field distribution on a ring with a gap in it. The boundary condition at the gap requires that there be either a positive maximum or negative maximum at that point. In the even modes ($n = 2$ and $n = 4$), this condition is satisfied with or without the gap and the fields are not disturbed. In the odd modes ($n = 1$ and $n = 3$), the boundary conditions cannot be satisfied and therefore the modes cannot exist. Because the

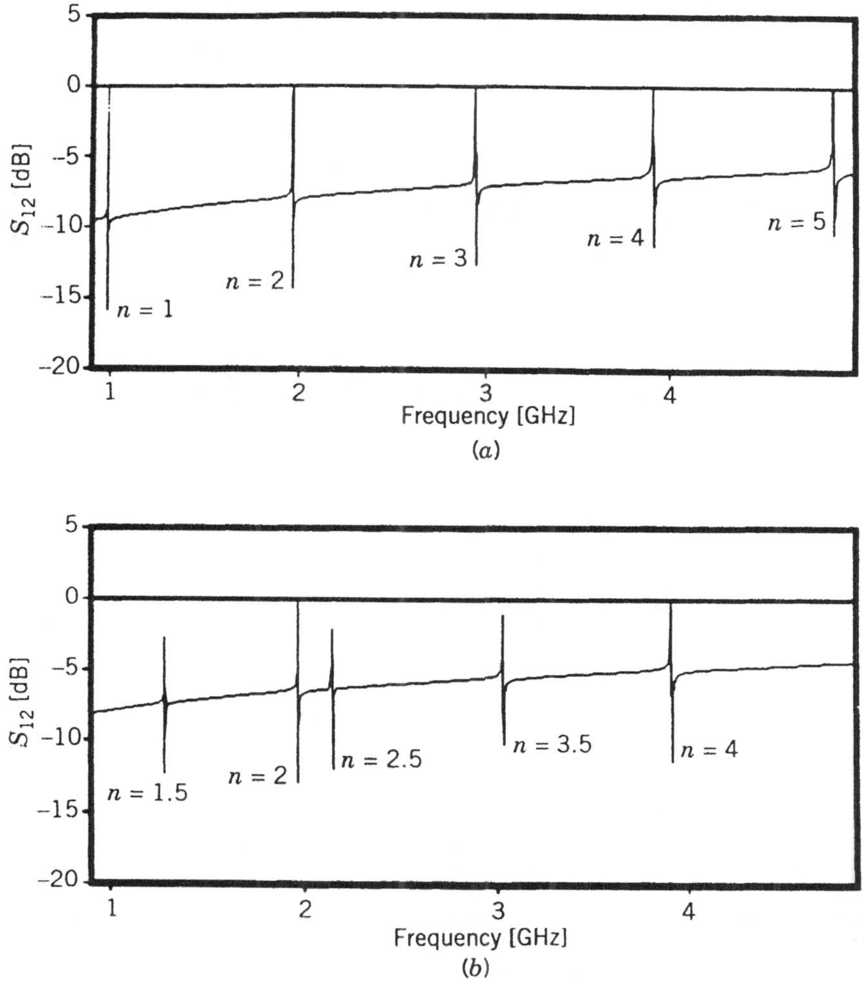

FIGURE 4.8 Typical frequency response of (*a*) a ring and (*b*) a varactor-tuned ring.

potential across the gap does not have to be continuous (of the same sign), the new half-modes, which satisfy the boundary conditions, are formed.

When the varactor is mounted across the gap in the ring, it is similar to an open circuit when the diode is operated as reverse biased. It would be safe to assume that the even modes would not be affected and the odd modes would disappear. The half-modes should also appear. We now have only the even and half-modes present. Figure 4.10 shows the excitation at the varactor for the even modes. For any amount of impedance change of the varactor the overall circuit impedance remains unchanged. Figure 4.11 shows the excitation of the varactor for the half-modes. An impedance change on the varactor will result in a change of the overall impedance and therefore change the resonant frequency. From these arguments it can be expected that for the varactor-tuned

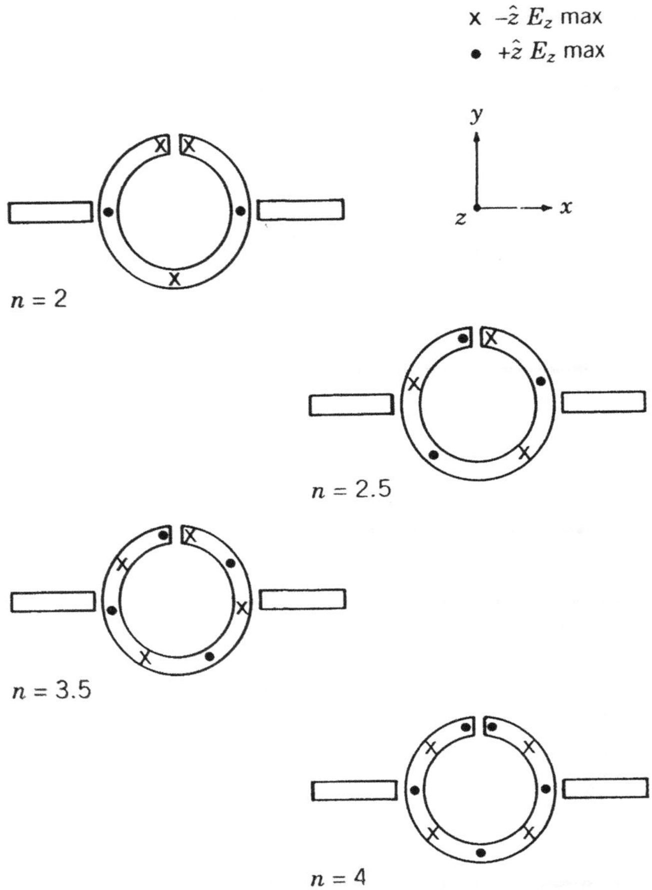

FIGURE 4.9 Mode chart for a varactor-tuned ring [3]. (Permission from IEEE.)

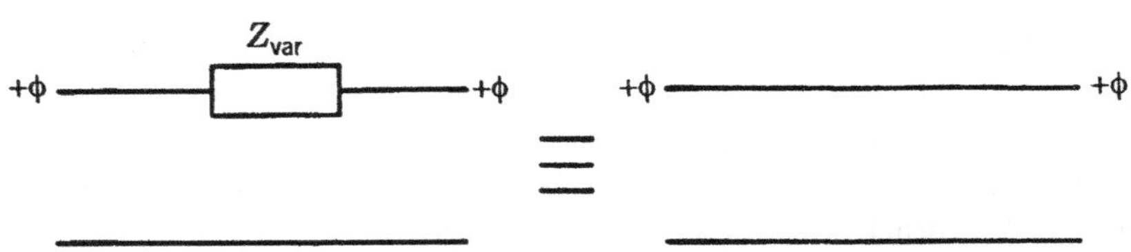

FIGURE 4.10 Excitation of the varactor for the even modes.

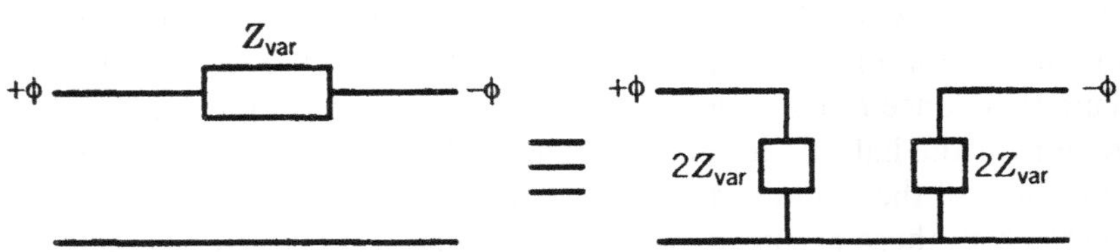

FIGURE 4.11 Excitation of the varactor for the half-modes.

ring the newly introduced half-modes will be tuned, the even modes will remain unchanged, and the odd modes will disappear.

4.5 EFFECTS OF THE PACKAGE PARASITICS ON THE RESONANT FREQUENCY

It is important that the effects of the package parasitics on the resonant frequency are understood [1]. A figure of merit for the varactor-tuned ring will be its tuning range. The package parameters could greatly affect this tuning range. It would be useful to know which parasitics degrade the tuning performance so that devices that minimize the parasitics can be used. Likewise it would be useful to known if any of the parameters enhance the tuning range so that they can be maximized in the varactor being used. The parasitics that we are concerned with are those in Figure 4.5, L_s, L_p, C_p, and R_s. The bulk resistance of the semiconductor, R_s, and L_p and C_p are due to the varactor packaging. Typical values for R_s, L_p, and C_p are given by manufacturers in their databooks for a given device and package style. The parameter L_s is the inherent inductance introduced in the circuit due to the package leads and bonding. This value may become quite large if long lead lines are used. The size of L_s depends on the application.

The resonant frequency as a function of varactor capacitance has been plotted for various parameters in Figures 4.12 through 4.15. The ranges for the parameters are typical values that can be expected for a packaged varactor. In Figure 4.12 the effect of the package capacitance on the resonant frequency is displayed. The package capacitance C_p is in parallel with the tuning capacitance, C_j. Because capacitances in parallel are added, the effective varactor

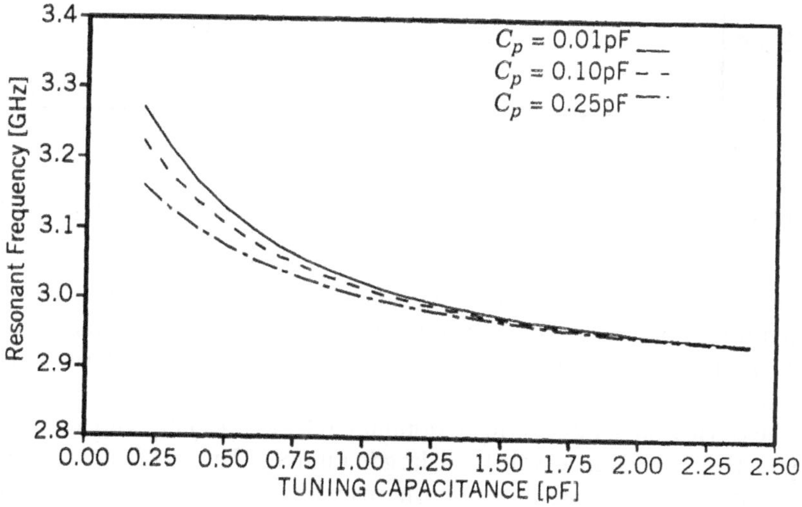

FIGURE 4.12 Effect of C_p on the resonant frequency as it is varied from 0.01 to 0.25 pF.

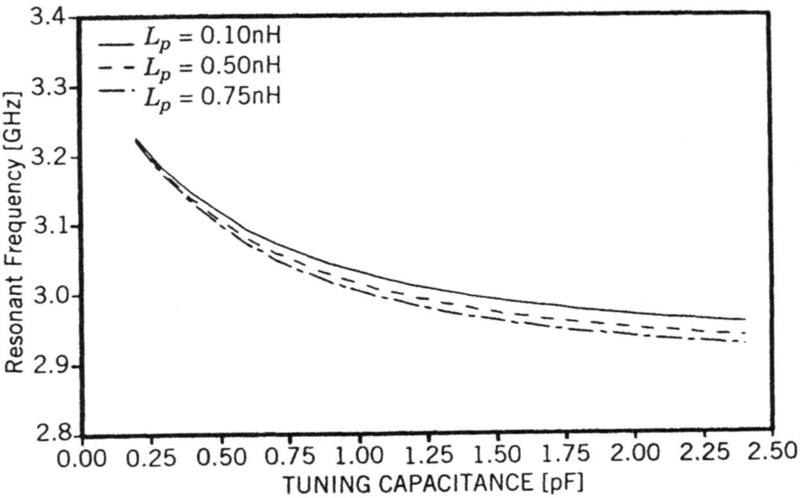

FIGURE 4.13 Effect of L_p on the resonant frequency as it is varied from 0.10 to 0.75 nH.

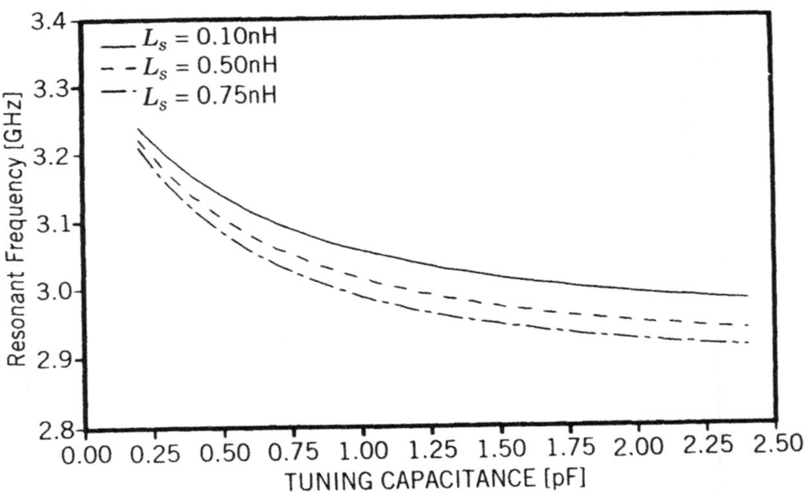

FIGURE 4.14 Effect of L_s on the resonant frequency as it is varied from 0.10 to 0.75 nH.

capacitance (neglecting C_2) can be written as $C_p + C_j$. From Figure 4.12 we can see that for a small varactor junction capacitance the package capacitance can result in a large change in the resonant frequency, while for a large junction capacitance, the effect is small. If a package with a large capacitance is used, then the device capacitance will be dominated by the package capacitance and the effective capacitance will be a larger number. The small device capacitances will have less of an effect on the resonant frequency, the result being a smaller tuning range. This is shown in Figure 4.12. As the package capacitance

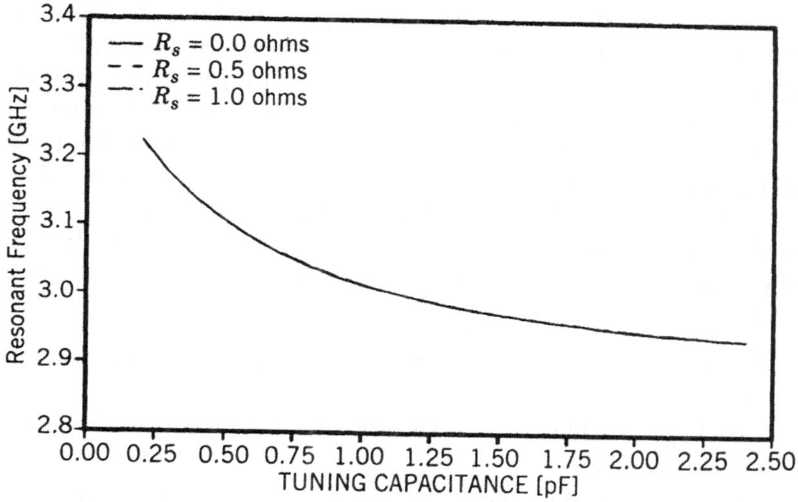

FIGURE 4.15 Effect of R_s on the resonant frequency as it is varied from 0.0 to 1.0 Ω.

is increased while all other parameters remain constant, the frequency tuning range for a given capacitance range is smaller. To ensure the maximum tuning range possible, it is important that a package with a small capacitance be chosen.

The inductance L_p is also introduced in the device package. Figure 4.13 shows the effects of the package inductances on the resonant frequency. As the inductance is increased, the tuning range is also slightly increased. The inductance does not degrade the performance of the circuit but seems to enhance it. This is both novel and convenient. It is generally conceived that all package parasitics should be minimized in order to maximize the performance of any circuit, but this is not the case for this application. Many package styles offer relatively high inductances (as high as 2.0 nH). In this varactor application the package inductance does not degrade the performance of the circuit and thus if given a choice, a package with a large inductance should be chosen.

The bonding inductance is not actually a package parasitic in the strictest sense because it does not lie within the package itself. The inductance L_s arises from the embedding of the varactor into the circuit. The leads from the device to the circuit and the bonding of the leads gives rise to L_s. Information on this inductance cannot be supplied by the vendor because it varies for each application. The effect of L_s on the resonant frequency is given in Figure 4.14. The range of L_s is arbitrarily chosen, but one would expect L_s to be at least comparable to L_p because of the physical dimensions involved. As can be seen in Figure 4.14, the inductance L_s does not degrade the frequency tuning range and may actually improve it slightly. As the inductance is increased, the whole tuning curve is lowered. This gives the same effect as increasing the mean circumference of the ring. Longer bonding wires give rise to a larger inductance and a longer single path. The longer signal path increases the mean circumference of the ring, and as would be expected, lowers the resonant frequency.

The effect of the bulk resistance on the resonant frequency is shown in Figure 4.15. As can be seen, the resistance does not affect the resonant frequency of the circuit. It should be noted that it is important to minimize R_s so that the circuit Q will be as high as possible and the insertion loss kept as low as possible.

The effect of the package parasitics on the turning range is now known. From this information a device and package can be chosen so that the frequency tuning range is maximized. The following is a guideline for choosing a varactor:

1. The tuning capacitance, C_j, should span a large range of junction capacitance values.
2. A package should be chosen such that the package capacitance, C_p, is as small as possible.
3. A package should be chosen such that the package inductance, L_p, is as large as possible.
4. The bonding wires will not degrade the tuning range, but should be kept as short as possible so that L_s will be more predictable.
5. The bulk resistance should be as small as possible.

4.6 EXPERIMENTAL RESULTS FOR VARACTOR-TUNED MICROSTRIP RING RESONATORS

The operation of the varactor-tuned ring resonator has been explained using transmission-line analysis. An equivalent circuit for the varactor was formed from considerations of the actual packaged device and incorporated into the total equivalent circuit for the ring resonator that was verified in Chapter 2. From this expanded equivalent circuit the frequency response of the varactor-tuned circuit was observed using the S-parameters. It was shown that the odd modes should disappear, the even modes remain unaffected, and the newly introduced half-modes should be tuned by varying the varactor capacitance. The effects of the package parasitics on the frequency tuning range were also examined. This allowed the development of guidelines to be followed when choosing a varactor so the maximum tuning range can be obtained. It is important that the theoretical results be verified with experimental data [1, 3].

The first step to verify the theoretical results is to choose a varactor. The varactors chosen for the circuit were from the MA-46600 series from M/A-COM. The MA-46600 series comprises gallium arsenide microwave tuning varactors with an abrupt junction, and feature Q-factors in excess of 4000. A variety of capacitance ranges are available, which run from 0.5 pF to 3.0 pF. Case style 137, which is specifically made for stripline implementation, was chosen as a package for the varactor. It has leads that may be attached to the circuit using silver epoxy or solder. Case style 137 is also acceptable from the guidelines outlined in the previous section. The typical capacitance, C_p, is quoted as 0.05 pF. A value for the package inductance, L_p, is not quoted, but similar packages have typical values of 1.0 nH. Thus we can summarize the advantages of package style 137 as low package capacitance, high package inductance, and leads that are easily attached to the microstrip ring.

Various circuits were designed and tested to verify the results using the varactors discussed. The results for each circuit tested were consistent, and thus only one will be presented here. The parameters for one of circuits tested are as follows:

$$\begin{aligned} \text{Substrate} &= \text{Rogers RT/Duroid 6010.5} \\ \varepsilon_r &= 10.5 \\ \text{Thickness} &= 0.645\,\text{mm}\,(25\,\text{mil}) \\ \text{Width} &= 0.538\,\text{mm}\,(21.2\,\text{mil}) \\ \text{Coupling gap} &= 0.079\,\text{mm}\,(3.1\,\text{mil}) \\ \text{Device gap} &= 0.132\,\text{mm}\,(5.2\,\text{mil}) \\ \text{dc block capacitor} &= 10\,\text{pF} \\ \text{Mean radius} &= 1.837\,\text{cm} \end{aligned}$$

The actual mask used to manufacture this circuit is given in Figure 4.16. Note the "bow-tie" configuration on the dc bias lines. The bow tie acts as a bandstop filter to minimize RF leakage at the designed frequency. The coupling and device gaps may not be distinguishable because they are very small.

The circuit was manufactured and the device gap and dc block capacitance gap were both filled with a conductive silver epoxy. This gave the effect of a simple ring resonator. As was expected all modes were present and spaced approximately an equal distance apart. The silver epoxy was then removed from the dc block capacitance gap and a 10-pF chip capacitor was soldered across the gap. Again as would be expected, the odd modes disappeared, the

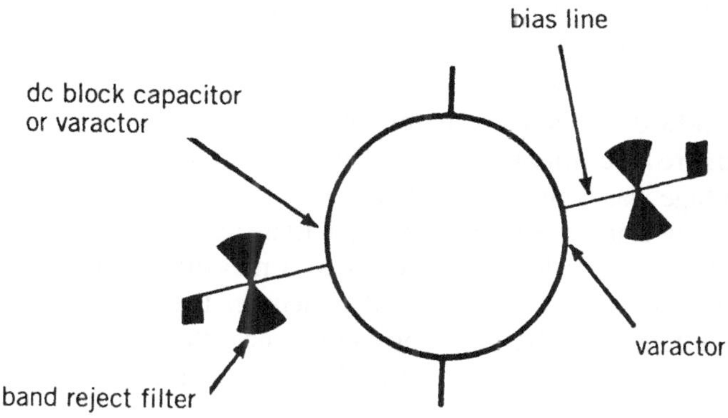

FIGURE 4.16 Mask of the experimental varactor-tuned ring.

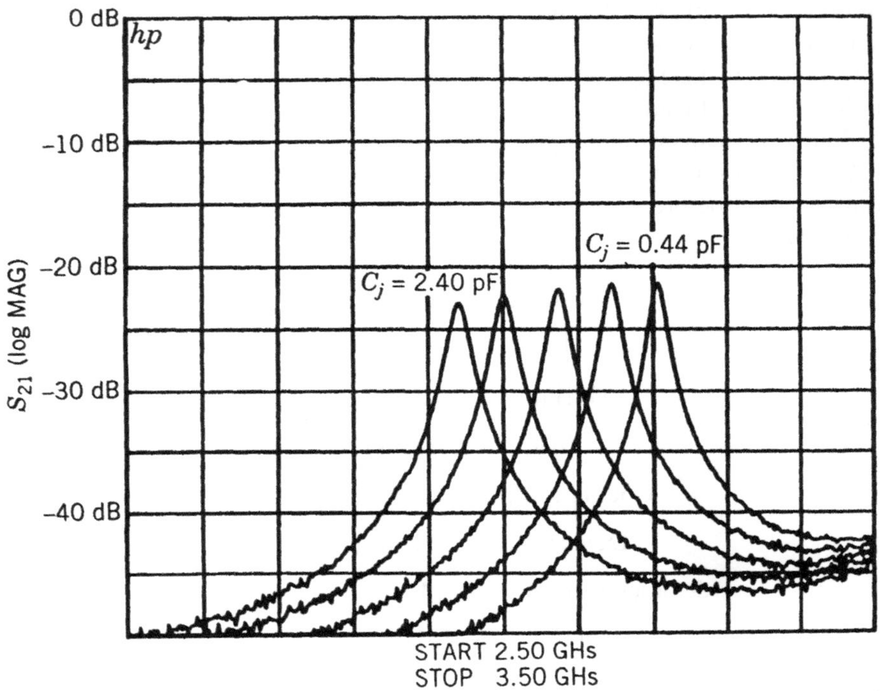

FIGURE 4.17 Frequency response of the varactor-tuned ring for a bias voltage ranging, from +0.85 to −30.0.

even modes were unaffected, and the half-modes appeared exactly between the even and missing odd modes. The silver epoxy was then removed from the device gap and the varactor was put into place. The even modes remained in place, and the half-modes shifted slightly lower.

The circuit was then ready for an applied voltage across the varactor. The voltage on the varactor was varied from +0.85 to −30.0 volts. When the voltage was varied and thus the capacitance in the circuit changed, the resonant frequency of the half-modes could be controlled. An example of the frequency response for various applied voltages is given in Figure 4.17. The resonant frequency varies from 2.94 GHz at +0.85 volts to 3.20 GHz at −30.0 volts. This is a tuning bandwidth of approximately 9%.

To compare the theoretical predictions and experimental results, the applied voltage was converted to its corresponding varactor capacitance. This can be done by using the *x*-*y* plot of capacitance versus voltage in Figure 4.3*a* and *b*. Each voltage value corresponds to a measured varactor capacitance. The measured varactor capacitance also includes the parallel package capacitance. To obtain only the varactor capacitance, the package capacitance is subtracted from the measured values. Table 4.1 is formed using Figure 4.3*a* and *b* and the experimental applied voltage.

TABLE 4.1 Varactor Capacitance Values for the Applied Voltages for the Single Varactor-Tuned Circuit

Applied Voltage (V)	Resonant Frequency (GHz)	Capacitance (pF)
+0.85	2.940	2.40
0.0	3.000	1.35
−2.5	3.075	0.85
−9.0	3.145	0.58
−30.0	3.208	0.44

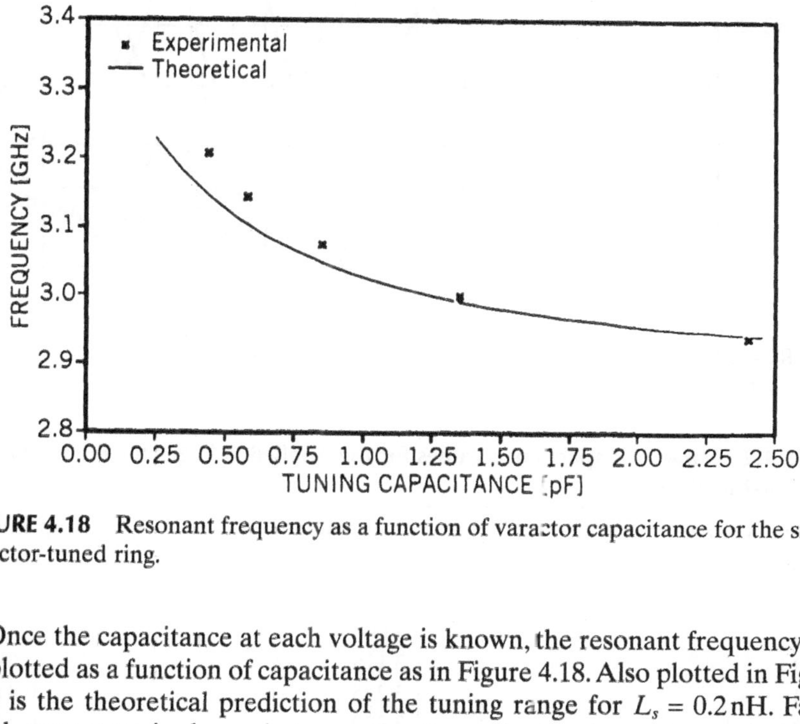

FIGURE 4.18 Resonant frequency as a function of varactor capacitance for the single varactor-tuned ring.

Once the capacitance at each voltage is known, the resonant frequency can be plotted as a function of capacitance as in Figure 4.18. Also plotted in Figure 4.18 is the theoretical prediction of the tuning range for $L_s = 0.2\,\text{nH}$. Fairly good agreement is shown between the theoretical and experimental results, From Figure 4.18 it would seem that the measured capacitance values are approximately 0.20 pF larger than the actual values of the varactor. This error was possibly introduced in the capacitance measurement. Any parallel capacitance, such as the capacitance from the leads of the *C-V* meter, will increase the overall measured capacitance.

4.7 DOUBLE VARACTOR-TUNED MICROSTRIP RING RESONATOR

The single varactor-tuned ring resonator offers a 9% tuning bandwidth. To increase the tuning bandwidth the two-varactor ring resonator is proposed

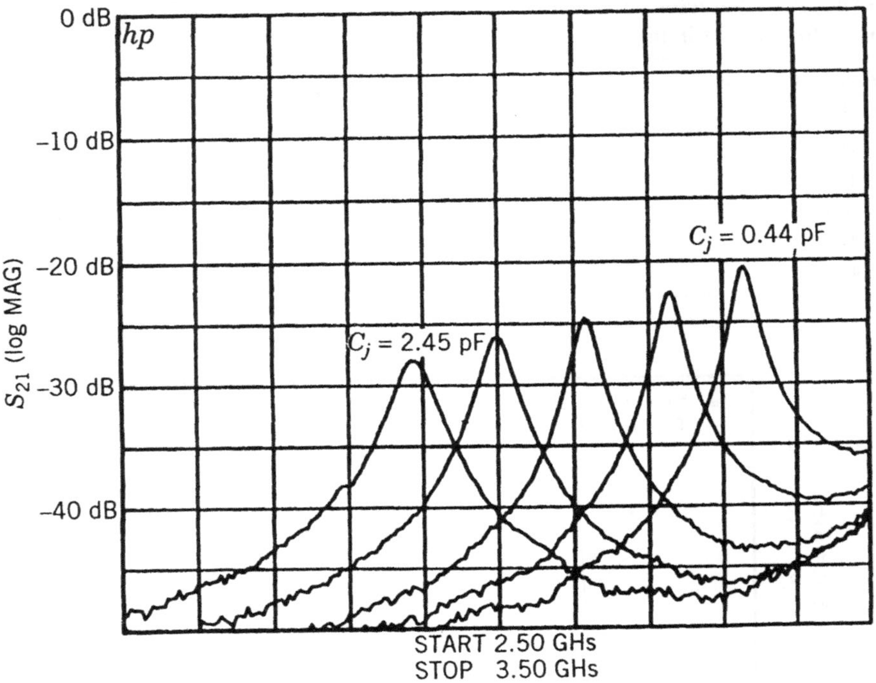

FIGURE 4.19 Frequency response of the double varactor-tuned ring for a bias voltage ranging from +0.90 to −30.0.

TABLE 4.2 Varactor Capacitance Values for the Applied Voltages for the Double Varactor-Tuned Circuit

Applied Voltage (V)	Resonant Frequency (GHz)	Capacitance (pF)
+0.9	2.885	2.45
0.0	3.000	1.35
−5.0	3.115	0.69
−15.0	3.225	0.49
−30.0	3.330	0.44

[1, 3]. The same circuit that is used for the single varactor can be used for two varactors. The dc block capacitor is replaced by another varactor. Correct biasing can still be achieved and an increase in the tuning bandwidth is offered. The frequency response of a two-varactor circuit is presented in Figure 4.19. Close comparison with the single-varactor response (Figure 4.17) does indeed show an increased tuning range. The tuning bandwidth is increased to approximately 15%. To compare the theoretical and experimental results it was assumed that the two varactors are identical and then the same procedure can be followed as in the single varactor case. The experimental results are summarized in Table 4.2.

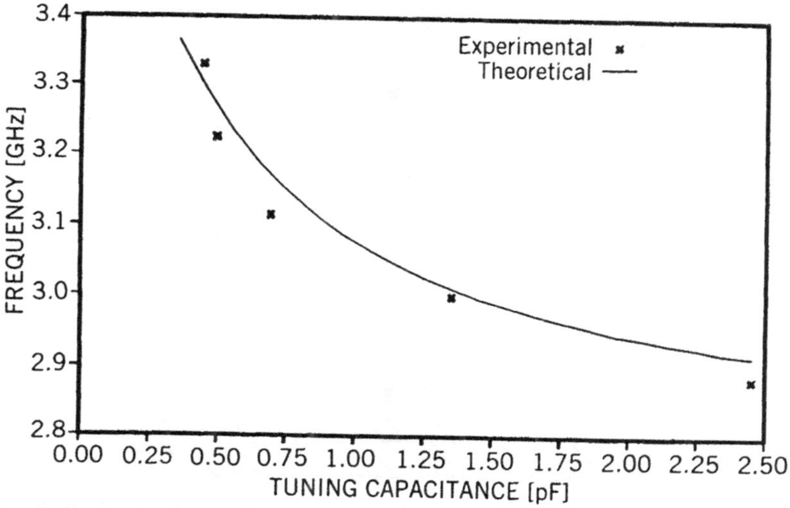

FIGURE 4.20 Resonant frequency as a function of a varactor capacitance for the double varactor-tuned ring.

The resonant frequency as a function of tuning capacitance is presented in Figure 4.20. The agreement of the experimental results and theoretical predictions is quite good, especially when one considers that the two varactors were considered to be identical.

4.8 VARACTOR-TUNED UNIPLANAR RING RESONATORS

Varactor diodes can be incorporated into the uniplanar ring resonators to make the resonant frequencies electronically tunable [7]. Examples are given here for both slotline and coplanar waveguide ring resonators.

Figure 4.21 shows the CPW-fed slotline ring configuration. A distributed transmission-line model was used to analyze the slotline ring. A 50-Ω CPW line feeds an 85-Ω slotline ring through a series gap. The gap can be represented by a capacitor that controls the coupling efficiency into the slotline ring and is inversely proportional to the gap spacing. The effect of the size of the coupling gap is shown in Figure 4.22 for two gap sizes of approximately 0.50 and 0.05 mm. The 0.05-mm gap reduces the insertion loss by increasing the coupling into and out of the resonator. The ring has a mean radius of 11.26 mm and uses a 0.50-mm slotline on a 0.63-mm-thick RT/Duroid 6010 substrate. The relative dielectric constant is 10.5.

The circuit was first tested without the varactor diodes. Figure 4.23*a* shows the theoretical and experimental insertion loss for a 0.095-mm gap. The theo-

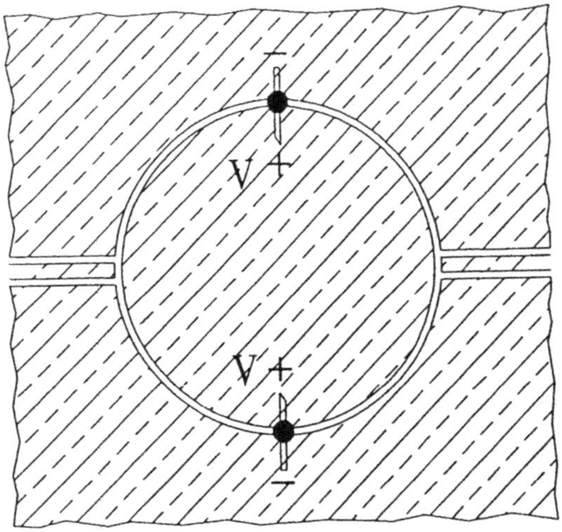

FIGURE 4.21 The varactor-tunable slotline ring configuration [7]. (Permission from IEEE.)

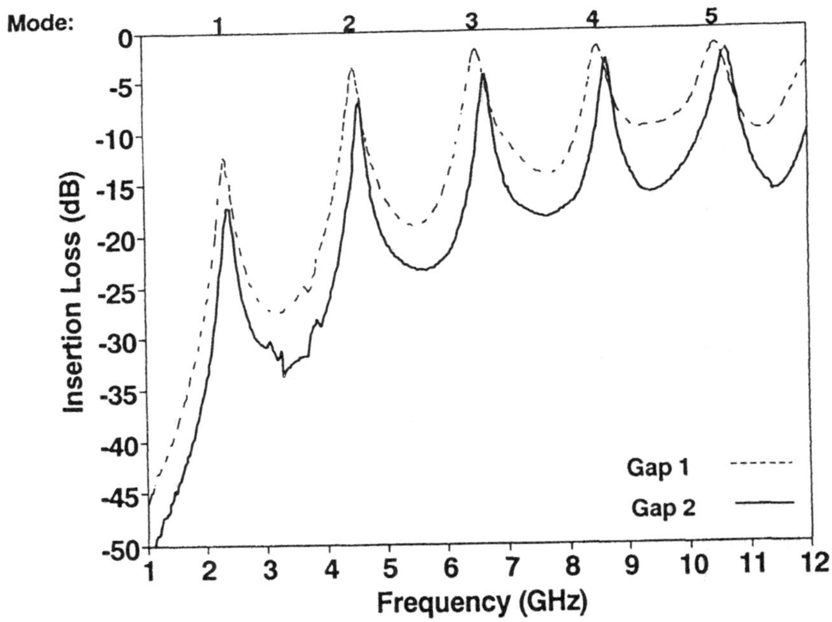

FIGURE 4.22 Effect of gap spacing on input/output coupling to slotline ring. Gap 1 is 0.05 mm, and gap 2 is 0.50 mm [7]. (Permission from IEEE.)

retical results were obtained based on the distributed transmission-line model discussed in Chapter 2. The slotline ring is formed by cascading many small sections of slotlines together. The input coupling gap is approximated using a small series capacitor. The transmission-line parameters were determined

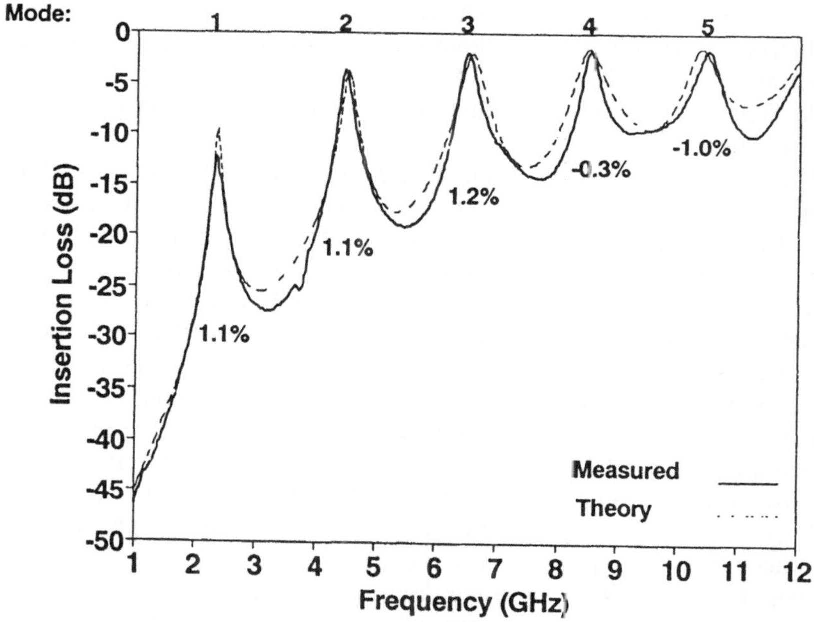

(*a*)

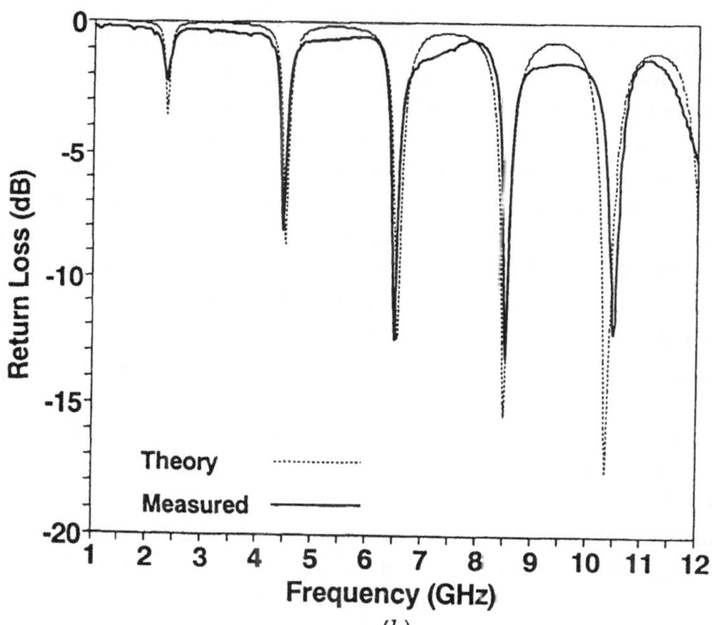

(*b*)

FIGURE 4.23 Theoretical vs. measured insertion loss and resonant frequencies of a slotline ring resonator: (*a*) insertion loss; (*b*) return loss [7]. (Permission from IEEE.)

based on formulas in [8, p. 215]. The gap capacitances were determined empirically from measurements. The theoretical results agree fairly well with measurement over a wide bandwidth. The errors for resonant frequencies are within 1.2%. Figure 4.23*b* shows the return loss that indicates the typical input matching condition.

The varactors located at 90 and 270 degrees along the ring tune the even modes of the resonator and allow a second mode electronic tuning bandwidth of 940 MHz from 3.13 to 4.07 GHz for varactor voltages of 1.35 to 30 volts. Figure 4.24*a* shows the experimental results. The first peak is for the first mode, which is stationary during the electronic tuning. A return loss of 6.4, 7.7, and 8.5 dB was achieved for varactor voltages of 5, 10, and 30 volts, respectively. Improved return loss could be achieved using matching elements at the coupling points. Figure 4.24*b* shows a comparison between the theoretical and the actual tuning range with reasonable agreement. The increase in loss as the frequency is lowered is due, in part, to a reduction in input/output coupling. The loss increases linearly from 6 dB at 4.07 GHz to 11 dB at 3.13 GHz.

In order to reduce the insertion loss, a 3 × 3 × 0.3-mm capacitive overlay [9] placed over the input and output of the slotline ring was used to increase the coupling and reduce the discontinuity radiation. This overlay reduced the loss and slightly lowered the frequencies of operation due to greater capacitive loading. The tuning bandwidth becomes 3.03 to 3.83 GHz. The 800-MHz tuning range centered at 3.4 GHz is shown in Figure 4.25. As shown, the overlay helps to improve the insertion loss of the tunable resonator, The 23% tuning range from 3.03 to 3.83 GHz has an insertion loss of 4.5 dB ± 1.5 dB for varactor voltages of 1.35 to 30 volts. As shown in Figure 4.25, the varactors have little effect on the first mode of the slotline ring resonator while capacitively tuning the second mode. The 3-dB points on the passband vary from 4.85% at 3.03 GHz to 5.17% at 3.83 GHz. The insertion loss at ±10% away from the second mode resonant frequency is about ≥15 dB. The increase in insertion roll-off for the lower frequency end of the tuning range is due to the stationary third mode. As the varactor bias level is lowered further, the second mode continues to approach the stationary first mode.

The CPW-fed varactor-tuned CPW ring configuration is shown in Figure 4.26. The CPW ring is divided into many sections and the distributed transmission-line model is used for analysis. Two 50-Ω CPW lines feed the CPW ring via a series gap. The ring has a mean diameter of 21 mm and uses 0.5-mm slotlines spaced 1.035 mm apart on a 0.635-mm RT/Duroid 6010 substrate with a relative dielectric constant of 10.5.

Advantages of the CPW ring over the slotline ring are that both series and shunt devices can be mounted easily along the ring and two shunt varactors can be placed at each circuit point to increase the tuning range and reduce the diode real resistance. A varactor and PIN diode can be placed at a single node to obtain switching and tuning with the same ring resonator.

The varactors located at 90 and 270 degrees along the ring tune the even modes of the resonator and allow a second resonant mode electronic tuning

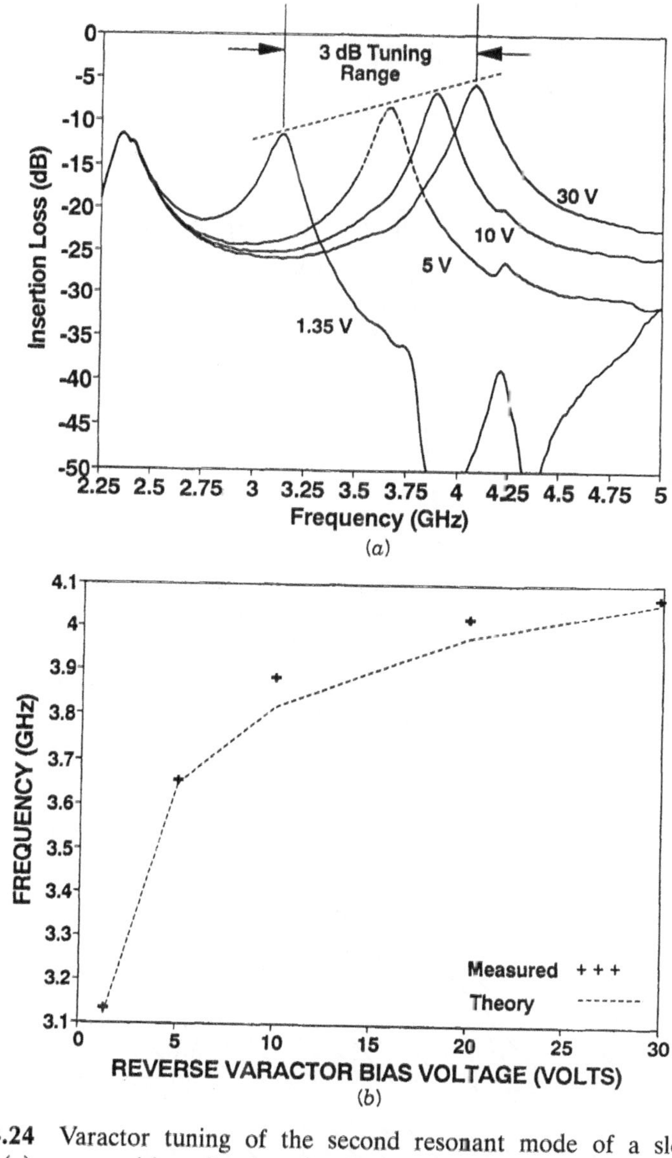

FIGURE 4.24 Varactor tuning of the second resonant mode of a slotline ring resonator: (*a*) measured insertion loss for different varactor voltages; (*b*) theoretical vs. measured second resonant mode frequency as a function of varactor voltage [7]. (Permission from IEEE.)

bandwidths of 710 MHz from 2.88 to 3.59 GHz for varactor voltages of 0 to 30 volts, Figure 4.27*a* shows the experimental results, and Figure 4.27*b* shows a comparison of theoretical and measured resonant frequency at different varactor bias levels. The increase in loss as the frequency is lowered is due, in

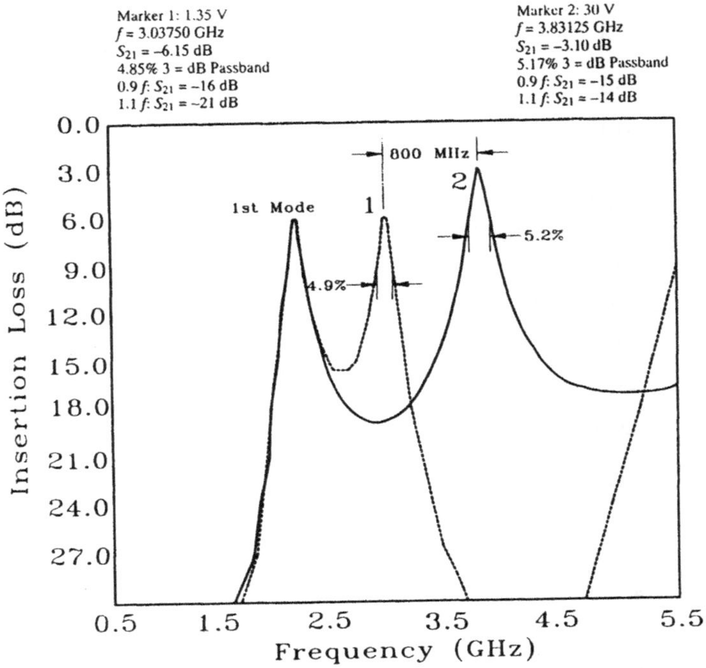

FIGURE 4.25 Measured varactor tuning range of a slotline ring with dielectric overlays over the coupling gaps [7]. (Permission from IEEE.)

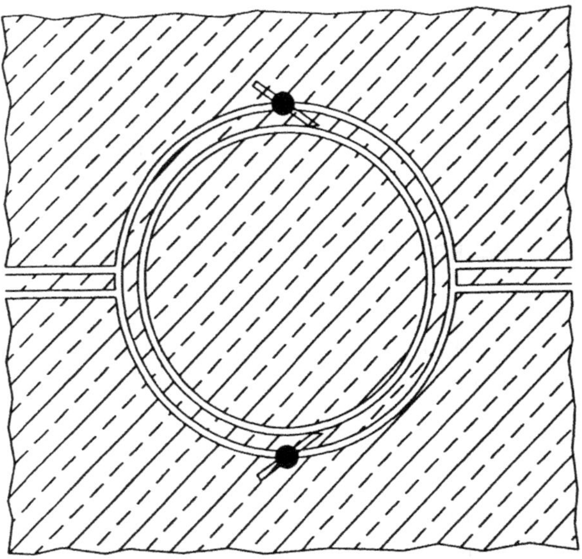

FIGURE 4.26 The varactor-tunable CPW ring configuration [7]. (Permission from IEEE.)

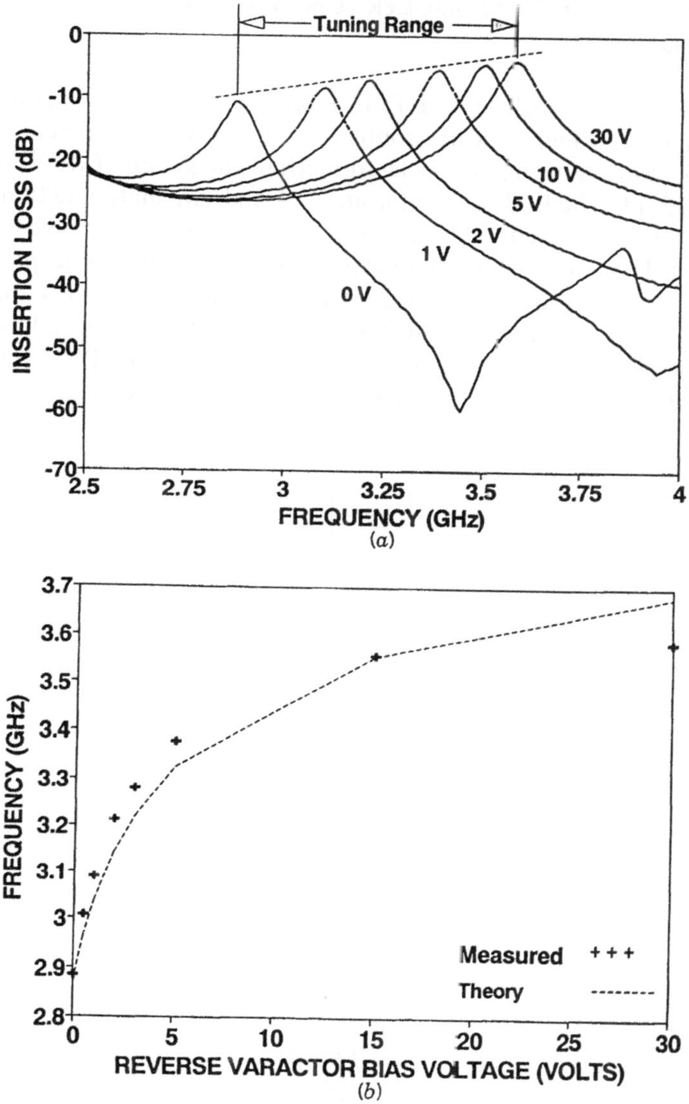

FIGURE 4.27 Varactor tuning of the second resonant mode of a CPW ring resonator: (*a*) measured insertion loss for different varactor voltages; (*b*) theoretical vs. measured second resonant mode frequency as a function of varactor voltage [7]. (Permission from IEEE.)

part, to a reduction in input/output coupling. The loss increases linearly from 4 dB at 3.59 GHz to 10.5 dB at 2.88 GHz. Although two varactors can be used at either point on the ring, only one was used for this investigation. The insertion loss of the CPW ring could be reduced by using a similar dielectric overlay at the input and output as was used in the slotline ring.

4.9 PIEZOELECTRIC TRANSDUCER TUNED MICROSTRIP RING RESONATOR

The piezoelectric transducer (PET) is an electromechanical operation using electric signals to mechanically control or tune microwave circuits [10–12]. Similar operations using microelectromechanical system (MEMS) can be found in [13, 14]. The PET is a composition of lead, zirconate, and titanate [15]. The PET shown in Figure 4.28 consists of two piezoelectric layers and one shim layer. The center shim laminated between the two same polarization piezoelectric layers adds mechanical strength and stiffness. Also, the shim is connected to one polarity of a DC voltage to deflect the PET and move it up or down vertically. A dielectric perturber is attached to the PET. This motion makes it possible to change the effective dielectric constant of the ring

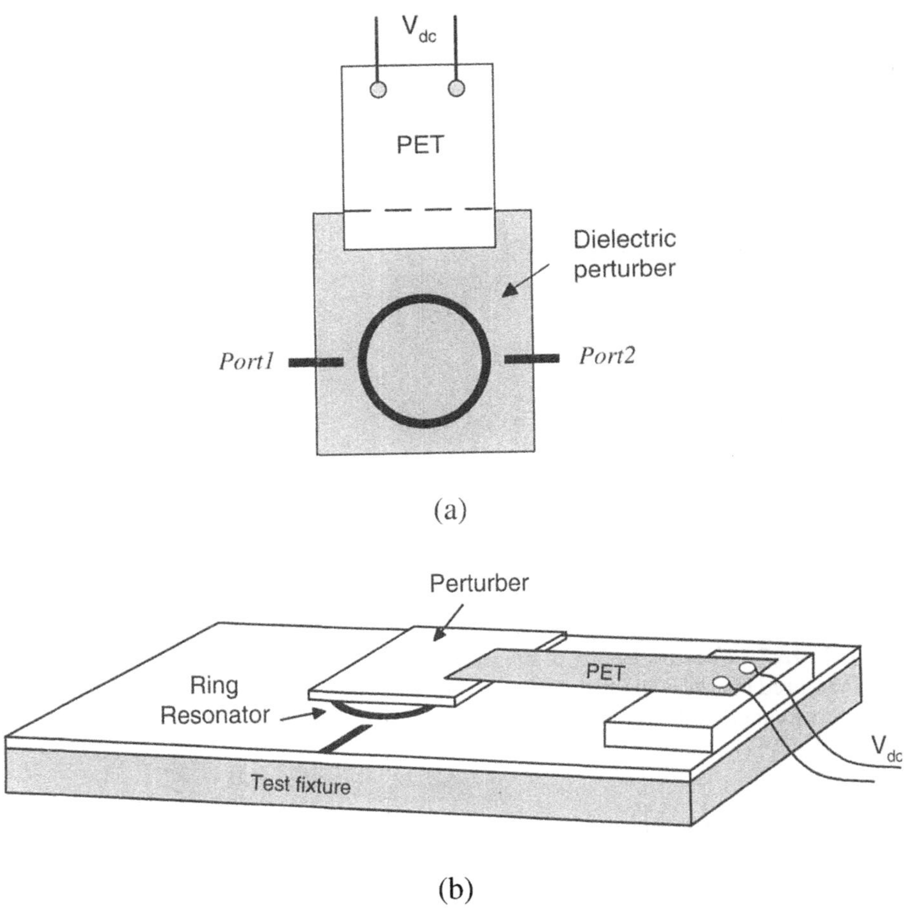

FIGURE 4.28 Configuration of the tunable ring resonator using a PET: (a) top view and (b) 3-dimensional view.

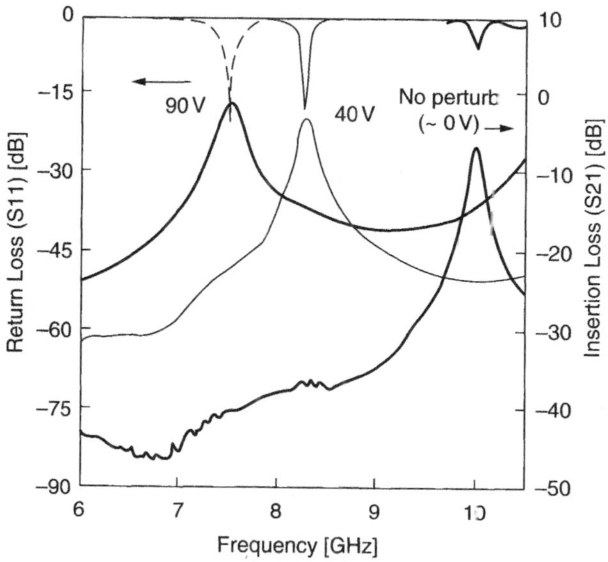

FIGURE 4.29 Measured results for the PET tuned ring resonator [10]. (Permission from IEEE.)

resonator, thus varying resonant frequency of the ring resonator. The PET can be deflected over +/– 1.325 mm at +/– 90 V with an applied current of 1 μA.

The ring resonator shown in Figure 4.28 is designed at fundamental of 5 GHz and fabricated at RT/Duroid 5870 with substrate 20 mil and dielectric constant $\varepsilon_r = 2.33$. The dielectric constant of the perturber is $\varepsilon_r = 10.8$. The measured results shown in Figure 4.29 are for the second mode near 10 GHz, and the tuning range is 2.5 GHz with a 28.5% about 10 GHz. Inspecting the results, the radiation and mismatch loss are gradually reduced by overlaying the dielectric perturber [16].

REFERENCES

[1] T. S. Martin, "A study of the microstrip ring resonator and its applications," M.S. thesis, Texas A&M University, College Station, December 1987.

[2] M. Makimoto and M. Sagawa, "Varactor tuned bandpass filters using microstripline resonators," in *1986 IEEE MTT-S Int. Microwave Symp. Dig.*, pp. 411–414, June 1986.

[3] K. Chang, T. S. Martin, F. Wang, and J. L. Klein, "On the study of microstrip ring and varactor-tuned ring circuits," *IEEE Trans. Microwave Theory Tech.*, Vol. MTT-35, pp. 1288–1295, December 1987.

[4] K. Chang, *Microwave Solid-State Circuits and Applications*, Wiley, New York, 1994, Chap. 5.

[5] K. E. Mortenson, *Variable Capacitance Diodes*, Artech House, Dedham, Mass., 1974.

[6] IMSL Library Reference Manual, Houston, Texas.

[7] J. A. Navarro and K. Chang, "Varactor-tunable uniplanar ring resonators," *IEEE Trans. Microwave Theory Tech.*, Vol. 41, No. 5, pp. 760–766, May 1993.

[8] K. C. Gupta, R. Garg, and I. J. Bahl, *Microstrip Lines and Slotlines*, Artech House, Dedham, Mass., 1979.

[9] D. F. Williams and S. E. Schwarz, "Design and performance of coplanar waveguide bandpass filters," *IEEE Trans. Microwave Theory Tech.*, Vol. MTT., No. 7, pp. 558–566, July 1983.

[10] T. –Y. Yun and K. Chang, "Piezoelectric-transducer-controlled tunable microwave circuits," *IEEE Trans. Microwave Theory Tech.*, Vol. 50, No. 5, pp. 1303–1310, May 2002.

[11] L. –H. Hsieh and K. Chang, "Piezoelectric transducer tuned bandstop filter," *Electron. Lett.*, Vol. 38, No. 17, pp. 970–971, August 2002.

[12] L. –H. Hsieh and K. Chang, "Tunable microstrip bandpass filters with two transmission zeros," *IEEE Trans. Microwave Theory Tech.*, Vol. 51, No. 2, pp. 520–525, February 2003.

[13] S. Barker and G. M. Rebeiz, "Distributed MEMS true-time delay phase shifters and wideband switches," *IEEE Trans. Microwave Theory Tech.*, Vol. 46, No. 11, pp. 1881–1890, November 1998.

[14] A. Borgioli, Y. Liu, A. S. Nagra, and R. A. York, "Low-low distributed MEMS phase shifter," *IEEE Microwave Guided Wave Lett.*, Vol. 10, pp. 7–9, January 2000.

[15] R. C. Buchanan, Editor, *Ceramic Material for Electronics*, Marcel Dekker, New York, Chap. 3 1986.

[16] K. Chang and K. Klein, "Dielectrically shielded microstrip (DSM) lines," *Electron. Lett.*, Vol. 23, No. 10, pp. 535–537, May 1987.

CHAPTER FIVE

Electronically Switchable Ring Resonators

5.1 INTRODUCTION

It has already been explained that the ring resonator exhibits a bandpass frequency response. The modes (or frequencies) that pass through the circuit are only those whose guided wavelength is an integral multiple of the mean circumference. The number of wavelengths present on the ring at resonance defines the mode numbers. There are infinitely many resonant frequencies and therefore infinitely many mode numbers. From Chapter 3, it is seen that these modes are not all equally affected when the ring circuit is changed. An example of this would be when the ring was cut at 90° radially from the feed point. To satisfy the new boundary conditions the odd-numbered modes disappeared and the new half-modes appeared. If it was possible to repair the cut so that the ring was complete again, the half-modes would disappear and the odd modes would reappear again. This idea can be used to develop a switch/filter [1].

The rind resonator alone acts naturally as a bandpass filter. PIN diodes can be mounted in the ring to facilitate the mode switching. The result is an electronic switch/filter. Like other diodes the PIN diode acts as an open circuit when reverse biased and a short circuit when forward biased. If the diodes are mounted in the ring resonator across the gaps at $\phi = 90°$ and 270°, the odd modes can be switched off and on at will by varying the bias on the diode. When the diode is forward biased it is as if there are no gaps in the ring and all integer-numbered (even and odd) modes are passed. When the diode is reverse biased, the boundary conditions will not allow the odd-numbered

Microwave Ring Circuits and Related Structures, Second Edition,
by Kai Chang and Lung-Hwa Hsieh
ISBN 0-471-44474-X

modes to propagate, and they will have a high attenuation. So by changing the diodes from forward to reverse bias the odd modes will disappear. And by forward biasing the diodes, the odd modes will again appear.

A similar circuit can also be used to switch the half-modes on and off. For this purpose only one PIN diode is placed at $\phi = 90°$ and a dc block capacitor placed at $\phi = 270°$. Because the procedure is the opposite of the odd-mode switching, this could become very confusing, but as will be seen in a later application, it is quite important. It may be helpful to refer to the mode chart of Figure 4.9. When the diode is forward biased, only the even-numbered modes are present (no half-modes). When the diode is reverse biased, the half-modes appear due to the boundary conditions (and the odd-numbered modes disappear).

As before with the varactor diode, when a PIN diode is mounted in the ring the resonant frequencies will shift. This shift is due to the impedance introduced by the diode in the circuit. Even when the diode is forward biased (short circuit) the resonant frequency will be affected by the impedance of the device package. It is important that the resonant frequency of the circuit for the forward- and reverse-bias conditions be predictable by some type of analysis. The most obvious analysis is the transmission-line method. The transmission-line method has accurately predicted the response of a similar circuit, the varactor-tuned ring. The varactor-tuned ring and the switch/filter ring circuit are identical, except that the varactor and dc block capacitor are replaced by the PIN diodes. All that is necessary to analyze the switch/filter is to replace the varactor equivalent circuit by the PIN equivalent and follow the same procedure described earlier in Chapter 4. Of particular interest will be the resonant frequency of the circuit and the isolation between the on and off states.

5.2 PIN DIODE EQUIVALENT CIRCUIT

To analyze the switchable ring resonators, one will need an equivalent circuit for the PIN diode [1,2]. In Chapter 4 it was explained how the depletion region arises in a *PN* structure. The depletion region is a result of the Fermi-level alignment of the *P* and *N* regions. With no bias applied the width of the depletion region (so called because it is depleted of carriers or is intrinsic in nature) depends on the doping of the *P* and *N* regions that it separates. It was also explained how the structure actually represented a two-plate capacitor whose capacitance was expressed in Equation 4.3. As the width of the depletion region is increased, the capacitance per unit area of the junction decreases. The PIN-diode structure is a *PN* junction separated by an (almost) intrinsic region. Thus comes the name PIN; *P*-type Intrinsic *N*-type. The PIN diode is merely an extension of the *PN* diode. When a forward bias is applied to the PIN, the result is a short circuit as in the *PN* diode. The difference occurs when reverse biased. If the PIN diode is reverse biased, the depletion region will increase and the junction capacitance decreases. Because the PIN structure has an added intrinsic region, its depletion region will be larger than the *PN* structure. The increased depletion width results in a smaller than normal junction capacitance when reverse biased. The complex impedance of a capacitor is represented as

$$Z_{\text{cap}} = \frac{1}{j\omega C} \tag{5.1}$$

where ω is the angular frequency and C is the capacitance. Thus, a smaller capacitance will result in a larger impedance and a better approximation of an open circuit. So the purpose of the PIN diode is to better achieve an open circuit when reverse biased while still representing a short circuit when forward biased.

To develop an equivalent circuit for the PIN to be used in our analysis we can draw on the knowledge gained from the varactor in Chapter 4. Both the varactor and PIN are merely diodes. All diodes can be represented by the same equivalent circuit. The various parameters for the diodes are cultivated to improve the performance for a given application. The parameters of the varactor are cultivated to give a workable junction capacitance and a low series resistance. This allows the capacitance of the varactor to be more effectively used in frequency tuning applications. The PIN, unlike the varactor, is designed to work as either forward or reverse biased. When forward biased, the series resistance should be low so that the diode will better represent a short circuit. When reverse biased, the junction capacitance should be as small as possible and nearly constant over a wide range of reverse-biased voltages. This allows the PIN to more effectively represent two distinct states; open circuit (forward bias) and short circuit (reverse bias).

Because all diodes typically have the same equivalent circuit, Figure 4.5, the equivalent circuit for the varactor can be used for the PIN. Figure 4.5 represents a reverse-biased diode. Because the PIN also operates in the forward-biased condition, Figure 4.5 should be altered to include this state. The proposed equivalent circuit for the PIN is given in Figure 5.1 [1, 2]. Note that the circuit can be switched to its two distinct states. In Figure 5.1, R_f is the series resistance of the forward-biased diode. A typical value for R_f is $1\,\Omega$. When reverse-biased, C_j represents the junction capacitance and R_r represents the series resistance. A typical value for C_j is 0.1 pF and R_r can be expected to be approximately $1\,\Omega$. These values will vary, depending on the PIN application and amount of bias current applied. Because very little current passes when the diode is reverse biased, the value of R_r is not particularly important. More important is R_f, which is present when large currents may be present. The other parameters in Figure 5.1 are the package parasitics explained in Chapter 4. The package style for the PIN may differ from that used for the varactor, but the equivalent circuit remains the same. Only the parameters change for a new package style.

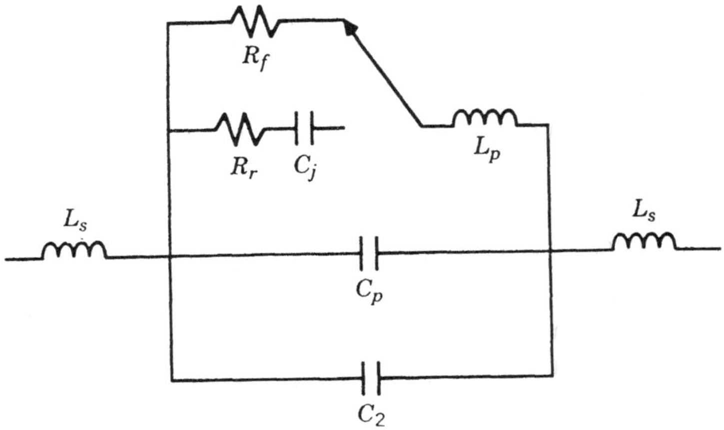

FIGURE 5.1 Equivalent circuit for the packaged PIN diode [3]. (Permission from IEEE.)

5.3 ANALYSIS FOR ELECTRONICALLY SWITCHABLE MICROSTRIP RING RESONATORS

The varactor analysis in Chapter 4 can be used to determine the resonant frequency of the PIN-diode ring resonator [1, 3]. In the varactor analysis, the first step was to propose a model for the circuit. Then an expression for the input impedance was obtained from the model. The resonant frequency could then be determined from the input impedance by two methods: solving for the frequency at which the imaginary part of the input impedance equals zero, and solving for the frequency at which S_{12} is a maximum. Both methods are equally correct. This method of analysis yielded reasonably good results for the varactor and can be altered slightly to apply to the PIN-diode circuit.

A circuit arrangement similar to the varactor-tuned ring shown in Figure 4.16 can be used for the PIN switch/filter circuit. The total equivalent circuit for the varactor-tuned ring is given in Figure 4.7. The impedance Z_{top} and Z_{bot} represent the varactor diodes for the double varactor-tuned ring. If we let Z_{top} and Z_{bot} represent the impedance of the PIN diodes, then the expression for the overall input impedance given in Equation (4.7) is still valid. An expression for Z_{top} and Z_{bot} can be derived from the equivalent circuit for the PIN diodes given in Figure 5.1. Typical values for the parameters L_p, C_p, C_j, R_f, and R_r, can be obtained from product databooks.

The resonant frequency can be determined using the same methods outlined in Chapter 4. Of particular importance for this circuit will be the parameter S_{12}. If the forward-biased condition is considered and an odd-numbered mode is observed, then S_{12} will reach a maximum at the resonant frequency. If the reverse-biased condition is considered, then the odd-numbered modes will have a much higher attenuation and there will be no resonance. The difference in S_{12} at the resonant frequency for the forward-biased condition and S_{12} at the same frequency for the reverse-biased condition is called *isolation*. Isolation is a figure of merit for switches. It is desirable to have the "on" signal level and the "off" signal level as isolated as possible.

As an example, the PIN diodes used are the MA-47047 from M/A COM Silicon Products. There diodes are medium-power diodes packaged in glass with case style 54. A value for C_j is quoted as 0.3 pF at −50 V [4]. A typical value for R_f is 1.3 Ω at 100 mA. A value for R_r is not quoted, but other similar diodes have a resistance of 2 Ω. The values for the package parameters can be determined from case style considerations. A value of 0.1 pF and 2.0 nH is quoted for C_p and L_p, respectively [5]. The value used for the bonding inductance L_s is an approximated parameter. The value used in Chapter 4 for the varactor circuit is 0.2 nH. This should also be a reasonable value for the PIN circuit.

5.4 EXPERIMENTAL AND THEORETICAL RESULTS FOR ELECTRONICALLY SWITCHABLE MICROSTRIP RING RESONATORS

To verify that the varactor analysis can indeed be applied to the PIN switch/filter, theoretical and experimental results were gathered [1, 3]. The circuit was designed for a RT/Duroid 5880 substrate, which has a relative permittivity of 2.2. The tested circuit is shown in Figure 5.2.

The circuit dimensions were as follows:

$$\text{Height} = 0.762\,\text{mm}$$

$$\text{Line width} = 2.310\,\text{mm}$$

$$\text{Coupling gap} = 0.100\,\text{mm}$$

$$\text{Device gap} = 0.250\,\text{mm}$$

$$\text{Radius} = 3.484\,\text{cm}$$

The PIN diodes used were the MA-47047 diodes discussed in the previous section.

The theoretical analysis is shown in Figure 5.3 for the forward- and reverse-biased diode. For the parameters given the circuit has a resonance at approximately 2.74 GHz when the diode is forward biased. When the diode is reverse biased, there is no resonance present. The isolation is predicted to be greater than 20 dB, which is acceptable for switch applications. The resonant frequency calculated by simply using the approximation $2\pi r = n\lambda_g$ is 3.0 GHz. This is approximately a 10% error. Because the forward-biased PIN can be represented primarily by its package parasitics, it becomes very obvious that the parasitics significantly affect the resonant frequency.

It should be noted that the theoretical isolation was very dependent on the forward-bias resistance value, R_f. An R_f larger than the 1 Ω used would

FIGURE 5.2 The PIN switch/filter circuit that was tested.

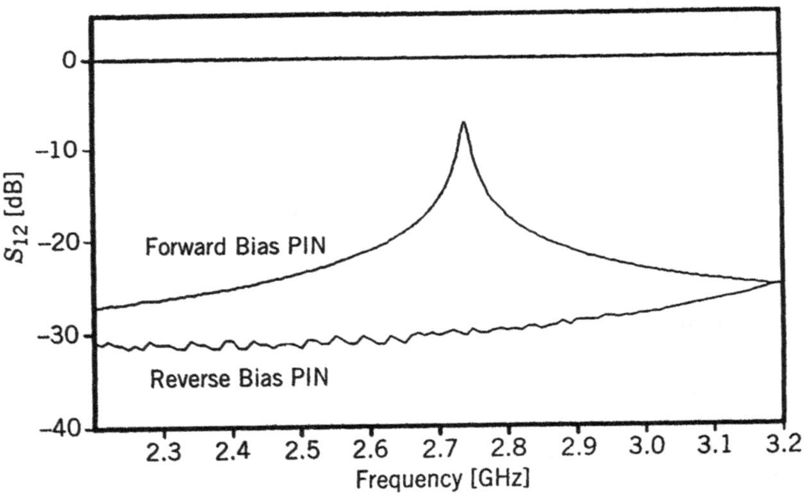

FIGURE 5.3 Theoretical results for the PIN switch/filter ring.

give less isolation because the resonant peak would not be as sharp. Thus it can be concluded that not only should a diode be chosen with a small reverse-bias junction capacitance, C_j, but a small forward resistance is also desirable.

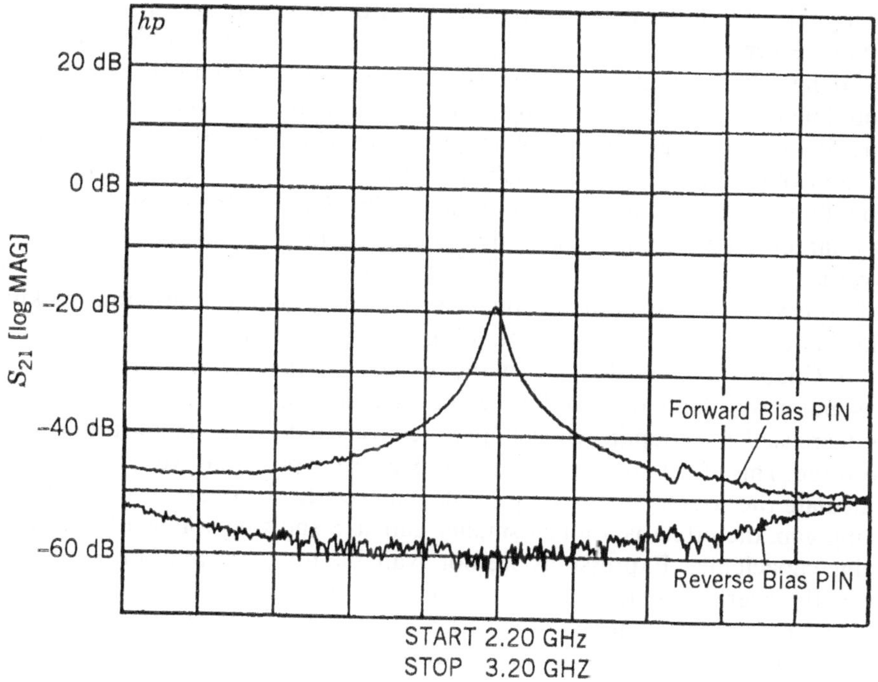

FIGURE 5.4 Experimental results for the PIN switch/filter ring [3]. (Permission from IEEE.)

Testing procedures were carried out on a Hewlett-Packard 8510 network analyzer. The experimental results are presented in Fig. 5.4. When the diodes are forward biased with a total current of 400 mA, the circuit has a resonance at 2.69 GHz. When the diodes are reversed biased to –50 V, the mode is turned off and there is no resonance. The signal isolation at 2.69 GHz when the circuit is reversed biased is better than 40 dB. The 40-dB isolation is excellent for a switch. The experimental isolation is quite a bit better than the theoretical predictions, which is probably due to the smaller than expected forward-bias resistance and smaller than expected reverse-bias capacitance. This is not at all unlikely because the predicted values are typical measurements taken at 100 MHz. The properties of the semiconductor are frequency dependent. As the frequency increases, the capacitance and resistance will actually decrease [4]. The resonant frequency was accurately predicted to within a respectable 2%. Errors in the resonant frequency estimation are more than likely due to the estimated value of L_s. If L_s were chosen to be 0.4 nH, the results would nearly agree exactly. Using a value of 0.4 nH for L_s would be completely justified because of the long wire leads extending from the glass package that are needed to bond the diode in the circuit.

From the experimental results in Fig. 5.4 it can be observed that the circuit has an insertion loss of 20 dB at resonance. The insertion loss in the circuit can be mainly attributed to radiation loss from the coupling regions of the resonator. Even for a very small coupling gap the loss associated with radiation from the coupling region is very substantial. It has been demonstrated that the coupling loss can be decreased by using an overlay to cover the radiating open circuits. By covering the coupling region with a dielectric of permittivity greater than that of the circuit substrate, the insertion loss resulting from the coupling region can be decreased dramatically [3, 6]. The result of the overlay is a larger coupling capacitance that is equivalent to a smaller gap size. The overlay also results in a lower loaded Q for the circuit. The lower Q results in a frequency response with a resonance peak that is not as sharp, thus making this undesirable for frequency measurement applications.

In Chapter 3 it was shown how the resonant frequency of a microstrip ring resonator was dependent on the size of the coupling gap. As the coupling gap is decreased, the coupling capacitance increases and the resonant frequency is lowered. This effect could be predicted because expressions were available to calculate the coupling capacitance. When an overlay is used to cover the coupling gap, the effect is a much smaller gap size and thus a much larger coupling capacitance. Expressions are not available for the capacitance resulting from the overlay, so the effect of the overlay on the resonant frequency cannot be theoretically determined.

To demonstrate the effect of an overlay, a switch/filter circuit was tested with the coupling gap covered by insulated copper tape. The circuit tested was the same circuit given in Figure 5.2. The results for the test are given in Figure 5.5. As can be seen, the insertion loss is decreased from 20 dB to less than 5 dB at resonance. The resonant frequency is also shifted from 2.74 GHz to 2.6 GHz as a result of the increased coupling capacitance. It can then be shown that for a frequency shift of this amount the coupling gap would have to be approximately 3 μm, which is not physically realizable.

5.5 VARACTOR-TUNED SWITCHABLE MICROSTRIP RING RESONATORS

A varactor when mounted in a ring resonator circuit was shown to tune the resonant frequency of what has come to be known as the half-modes [1, 3]. These half-modes arise because the varactor, which represents a high impedance when reverse biased, is almost equivalent to an open circuit at the mounting position. This almost open circuit forces boundary conditions on the ring that allow the half-modes to appear and odd-numbered modes to disappear.

It was also demonstrated in the previous sections that the half-modes could be turned off and on by correctly mounting a PIN diode in the circuit. The two states of the PIN diode, forward and reverse biased, present different boundary conditions on the ring to be satisfied. A forward-biased PIN diode represents a short circuit, and the circuit behaves as a normal ring resonator with all integer-numbered modes being present. When the PIN diode is reverse biased, it represents an open circuit and the frequency response of the circuit

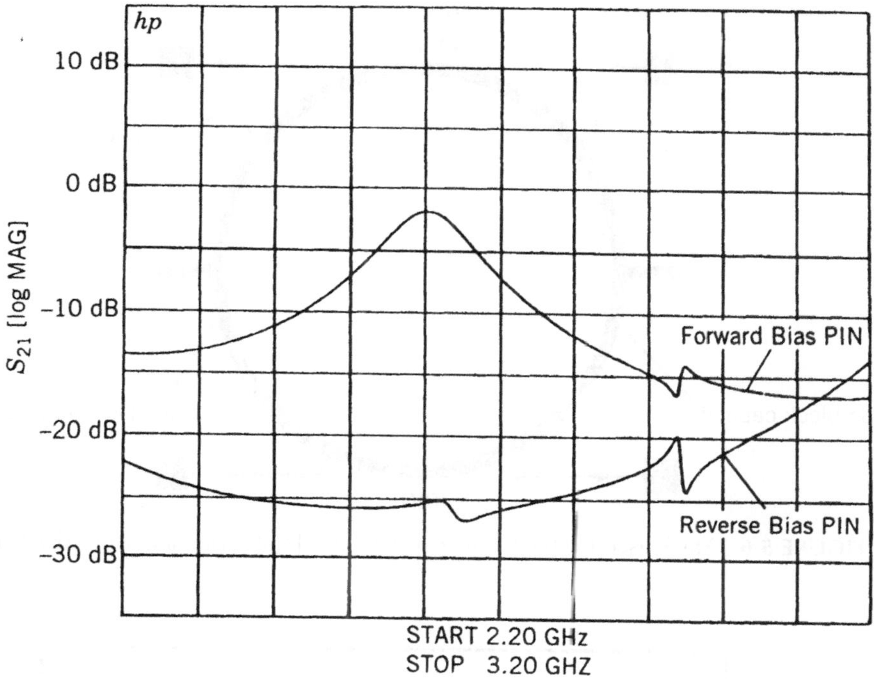

FIGURE 5.5 Theoretical results for the PIN switch/filter ring with a covered coupling region [3]. (Permission from IEEE.)

is similar to the varactor ring; the odd modes disappear and the half-modes appear.

The novel properties of the varactor-tuned ring and the PIN switch/filter can be combined in one circuit. This circuit would not only have an electronically tunable resonant frequency but a resonant frequency that can be switched on and off. Using the equivalent circuit for the ring resonator, varactor diode, and PIN diode described in Chapters 2 and 4, and Section 5.1, respectively, the frequency response of such a circuit could be determined using the transmission-line method. The transmission-line method of analysis has already been used to adequately predict the response of both the varactor-tuned ring and the PIN switch/filter circuit. The same procedure could be used for the theoretical analysis of this combination circuit.

As an example, a circuit was built with the actual mask for the tunable switch/filter circuit as given in Figure 5.6. The varactor and PIN are to be mounted across the gaps cut at 90° radially from the feed lines. Across the remaining two cuts, large dc block capacitors are to be mounted. These capacitors are necessary because the bias on the PIN and varactor will be different.

A theoretical analysis was developed based on the same methods and equivalent circuits from the previous chapters. The theoretical frequency

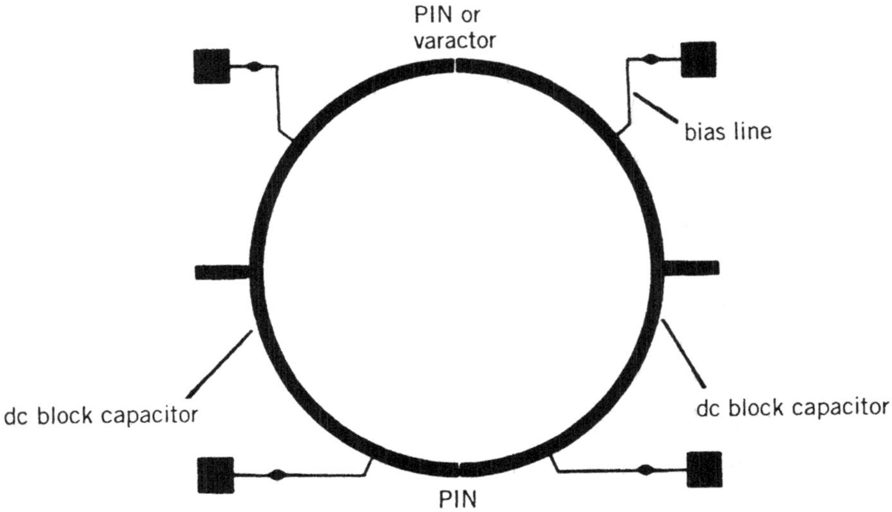

FIGURE 5.6 Mask used for the tunable switch/filter [3]. (Permission from IEEE.)

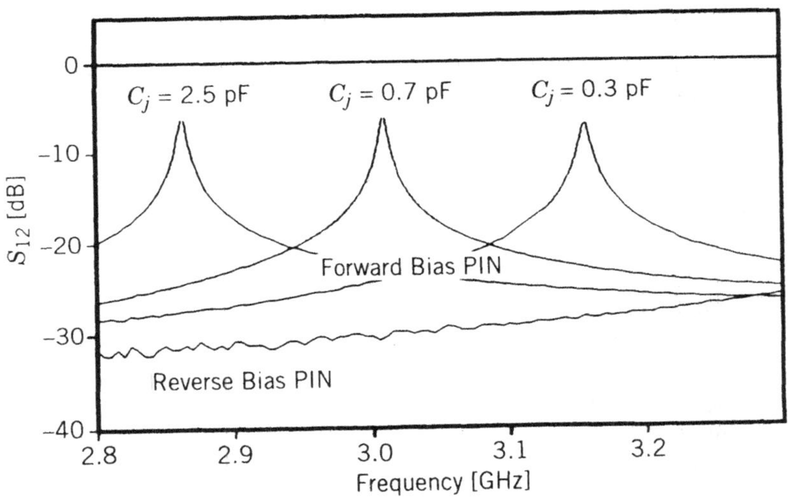

FIGURE 5.7 Theoretical results for the tunable switch/filter.

response fro the circuit shown in Figure 5.6 is given in Figure 5.7. The particular mode of interest is the mode $n = 3.5$. In Figure 5.7 both the forward- and reverse-biased PIN conditions are considered. When the PIN is forward biased, the half-mode is present. The voltage across the varactor can then be varied to adjust the resonant frequency. As the capacitance of the varactor is decreased, the resonant frequency increases. When the PIN is reverse biased,

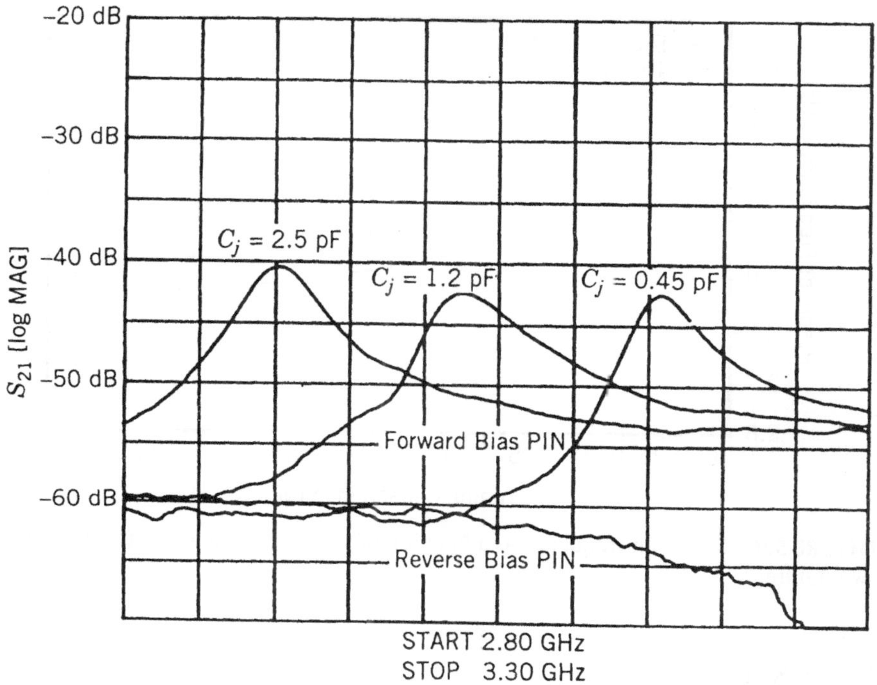

FIGURE 5.8 Experimental results for the tunable switch/filter.

the half-mode is turned off. The predicted isolation for the circuit is approximately 20 dB.

Experimental results were gathered to verify the theoretical analysis. A circuit was manufactured using the mask given in Figure 5.6. A MA 46600 varactor and a MA 47047 PIN were mounted in the circuit. Testing procedures were again carried out on a Hewlett-Packard 8510 network analyzer. The experimental frequency response is given in Figure 5.8. As can be seen, the theoretical and experimental results are quite similar. When the PIN is forward biased, the varactor presents a tuning range from 2.90 to 3.16 GHz. This is approximately a 9% tuning bandwidth. When the PIN is reverse biased, the mode is turned off and gives approximately a 20-dB isolation, which is what was predicted.

The theoretical and experimental tuning ranges are given in Figure 5.9. The results are not in particularly good agreement (error of approximately 3%), but they do have the same trend with the theoretical results shifted slightly lower. This error could have come from choosing too large a value for the approximation of L_s in the theoretical analysis.

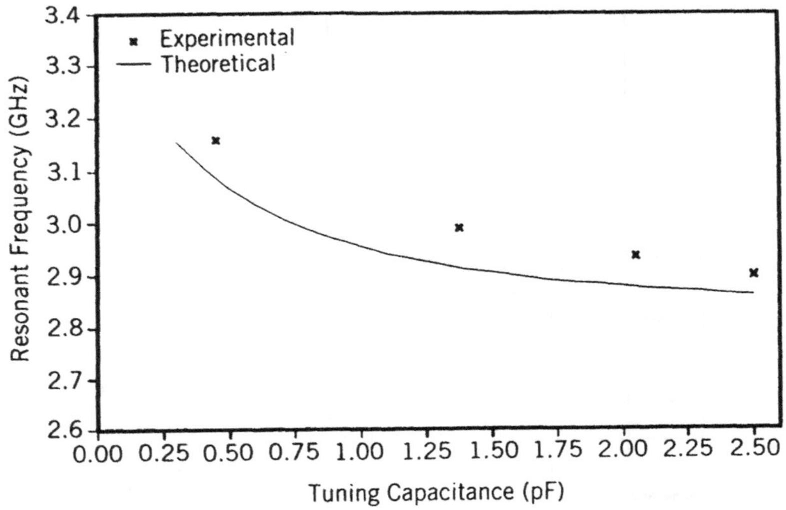

FIGURE 5.9 Resonant frequency as a function of varactor capacitance for the tunable switch/filter.

REFERENCES

[1] T. S. Martin, "A study of microstrip ring resonator and its applications," M.S. thesis, Texas A&M University, College Station, December 1987.

[2] K. Chang, *Microwave Solid-State Circuits and Applications*, Wiley, New York, 1994, Chap. 5.

[3] T. S. Martin, F. Wang, and K. Chang, "Theoretical and experimental investigation of novel varactor-tuned switchable microstrip ring resonator circuits," *IEEE Trans. Microwave Theory Tech.*, Vol. MTT-36, pp. 1733–1739, December 1988.

[4] M/A COM Semiconductor Products Master Catalog, Burlington, Mass.

[5] J. F. White, *Microwave Semiconductor Engineering*, Van Nostrand Reinhold, New York, 1982.

[6] K. Chang and J. Klein, "Dielectric shielded microstrip (DSM) lines," *Electron. Lett.*, Vol. 23, No. 10, pp. 535–537, May 7, 1987.

CHAPTER SIX

Measurement Applications Using Ring Resonators

6.1 INTRODUCTION

The microstrip ring resonator was first proposed by Troughton [1] for the measurements of phase velocity and dispersion of microstrip lines. Compared to the microstrip linear resonator, the microstrip ring resonator does not suffer from open-ended effects and can be used to give more accurate measurements. Since its introduction in 1969, the microstrip ring resonator has found applications in determining optimum substrate thickness [2], discontinuity parameters [3], effective dielectric constant and dispersion [4–8], and loss and Q-measurements [9–11].

This chapter discusses the measurement applications of using ring resonators [12]. Although regular modes are generally used for the measurements, forced modes and split modes can also be used.

6.2 DISPERSION, DIELECTRIC CONSTANT, AND Q-FACTOR MEASUREMENTS

The ring circuit is an ideal tool for dispersion, dielectric constant, and Q-factor measurements [12]. When Troughton first introduced the idea of a microstrip ring resonator, he described techniques used to measure the phase velocity and dispersive characteristics of a microstrip line by observing the resonant frequency of the ring resonator. The ring resonator, shown in Figure 6.1 is

Microwave Ring Circuits and Related Structures, Second Edition,
by Kai Chang and Lung-Hwa Hsieh
ISBN 0-471-44474-X

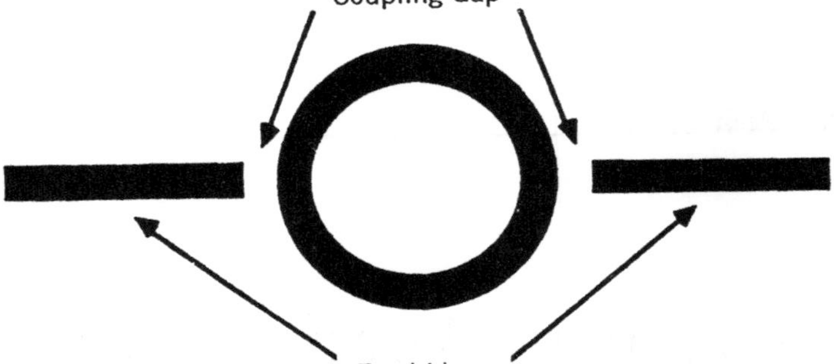

FIGURE 6.1 The microstrip ring resonator.

merely a transmission line formed in a closed loop. The basic circuit consists of the feed lines, coupling gap, and the resonator. The feed lines couple power into and out of the resonator. The feed lines are separated from the resonator by a distance called the coupling gap. The size of the gap should be large enough such that the fields in the resonator are not appreciably perturbed, yet small enough to allow adequate coupling of power. This type of coupling is described in the literature as "loose coupling."

When Troughton used the resonator for his microstrip measurements, he assumed that the structure would only support waves that have an integral multiple of the guided wavelength equal to the mean circumference. This may be expressed as

$$n\lambda_g = 2\pi r \quad \text{for} \quad n = 1, 2, 3, \ldots \tag{6.1}$$

where n is the mode number or number of wavelengths on the ring, λ_g is the guided wavelength, and r is the mean radius.

There exists in a nondispersive medium a linear relationship between the frequency and the phase constant or wavenumber, β, where

$$\beta = 2\pi/\lambda_g \tag{6.2}$$

If the frequency doubles, then likewise the wavenumber doubles. In a dispersive medium this is not true. The microstrip line is a dispersive medium.

The dispersion in a microstrip line can be explained by examining the effective permittivity, ε_{eff}. In microstrip the effective permittivity is a measure of the fields confined in the region beneath the strip. In the case of very narrow lines or a very low frequency the field is almost equally shared by the air ($\varepsilon_r = 1$) and the substrate so that, at this extreme,

$$\varepsilon_{\text{eff}} \approx \frac{1}{2}(\varepsilon_r + 1) \text{ as } f \to 0$$

where ε_r is the relative dielectric of the substrate. For very wide lines or a very high frequency nearly all of the field is confined to the substrate dielectric, and therefore at this extreme,

$$\varepsilon_{\text{eff}} \approx \varepsilon_r \text{ as } f \to \infty$$

It is therefore obvious that the effective permittivity is frequency dependent, increasing as the frequency increases.

The effective permittivity is defined as the square of the ratio of the velocity in free space, c, to the phase velocity, v_p, in microstrip, or

$$\varepsilon_{\text{eff}}(f) = \left(\frac{c}{v_p}\right)^2 \tag{6.3}$$

For any propagating wave, the velocity is given by the appropriate frequency–wavelength product. In the microstrip line, the velocity is $v_p = f\lambda_g$. Substituting for v_p in Equation (6.3) results in the equation

$$\varepsilon_{\text{eff}}(f) = \left(\frac{c}{f\lambda_g}\right)^2 \tag{6.4}$$

If we assume that, as in Equation (6.1), any microstrip resonator will only support wavelengths that are an integral multiple of the total length, then

$$l_t = n\lambda_g \tag{6.5}$$

where l_t is the total length of the resonator. Substituting for λ_g in Equation (6.4) yields the equation

$$\varepsilon_{\text{eff}}(f) = \left(\frac{nc}{fl_t}\right)^2 \tag{6.6}$$

If the total length of a resonator, the resonance order, n, and the resonant frequency are known, then ε_{eff} can be calculated from Equation (6.6).

The accuracy of the dispersion calculation depends on the accuracy of the measurement of the frequency and the total length of the resonator. Until 1969, frequency measurements were made using linear resonators [13]. The linear resonator, shown in Figure 6.2, uses open- or short-circuit terminations to force the bandpass frequency response. Perfect short circuits are hard to achieve in microstrip circuits, and thus most linear resonators utilize open circuits. The open circuit causes radio frequency (RF) power to be radiated. This radiated power is either lost to the outside in open structures, or may lead to unwanted cross-coupling between various circuit elements in a closed housing.

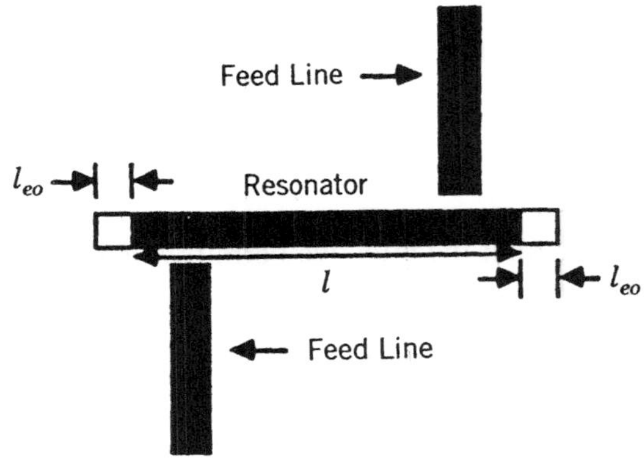

FIGURE 6.2 The microstrip linear resonator.

The effect of the fringing fields at the open circuit is best accounted for by considering the line to be somewhat longer electrically. The total length of the linear resonator can be expressed as

$$\begin{aligned}\frac{l_t}{2} &= l + 2l_{eo} \\ &= n\frac{\lambda g}{2}\end{aligned} \tag{6.7}$$

where l is the physical length and l_{eo} is the additional length representing the open circuit. The length of the fringing field can be calculated from

$$l_{eo} = 0.412h\left(\frac{\varepsilon_{\text{eff}} + 0.3}{\varepsilon_{\text{eff}} - 0.258}\right)\left(\frac{w/h + 0.262}{w/h + 0.813}\right) \tag{6.8}$$

where w and h are the width of the line and the height of the substrate, respectively [4]. If Equation (6.7) is substituted into Equation (6.4), the result is

$$\varepsilon_{\text{eff}}(f) = \left(\frac{nc}{2f(1 + 2l_{eo})}\right)^2 \tag{6.9}$$

If Equation (6.8) is substituted into Equation (6.9), then ε_{eff} will appear on both sides of the equation and it is necessary to iterate for the solution. This in itself makes ring resonators more desirable than linear resonators for dispersion measurements.

A figure of merit for resonators is the circuit Q-factor as defined by expression.

$$Q = \frac{\omega_0 U}{W} \tag{6.10}$$

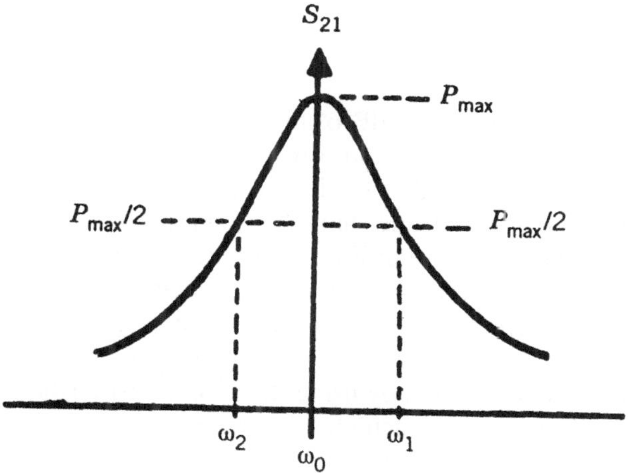

FIGURE 6.3 Resonator frequency response.

where ω_0 is the angular resonant frequency, U is the stored energy per cycle, and W is the average power lost per cycle. The three main losses associated with microstrip circuits are conductor losses, dielectric losses, and radiation losses. The total Q-factor, Q_0, can be expressed as

$$\frac{1}{Q_0} = \frac{1}{Q_c} + \frac{1}{Q_d} + \frac{1}{Q_r} \tag{6.11}$$

where Q_c, Q_d, and Q_r are the individual Q-values associated with the conductor, dielectric, and radiation losses, respectively [14].

For ring and linear resonators of the same length, the dielectric and conductor losses are equal and therefore Q_c and Q_d are equal. The power radiated, W_r, is higher for the linear resonator. This results in a lower Q_r for the linear resonator relative to the ring. We can conclude that because Q_c and Q_d are equal for the two resonators, and that Q_r is higher for the ring, that the ring resonator has a higher Q_0.

The unloaded Q, Q_0, can also be determined by measuring the loaded Q-factor, Q_L, and the insertion loss of the ring at resonance. Figure 6.3 shows a typical resonator frequency response. The loaded Q of the resonator is

$$Q_L = \frac{\omega_0}{\omega_1 - \omega_2} \tag{6.12}$$

where ω_0 is the angular resonant frequency and $\omega_1 - \omega_2$ is the 3-dB bandwidth. Normally a high Q_L is desired for microstrip measurements. A high Q_L requires a narrow 3-dB bandwidth, and thus a sharper peak in the frequency response. This makes the resonant frequency more easily determined.

The unloaded Q-factor can be calculated from

$$Q_0 = \frac{Q_L}{(1 - 10^{-L/20})} \tag{6.13}$$

where L is the insertion loss in dB of the ring at resonance [2]. Because the ring resonator has a higher Q_0 and lower insertion loss than the linear resonator, it will also have a higher loaded Q, Q_L. Therefore the ring resonator has a smaller 3-dB bandwidth and sharper resonance than the linear resonator. This also makes the ring more desirable for microstrip measurements.

Troughton recognized the disadvantages associated with using the linear resonators for measurements and introduced the ring resonator in 1969 [1]. He proposed that the unknown effects of either open- or short-circuit cavity terminations could be avoided by using the ring in dispersion measurements. The equation to be used to calculate dispersion can be found by combining Equations (6.1) and (6.4) to yield

$$\varepsilon_{\text{eff}}(f) = \left(\frac{nc}{2\pi fr}\right)^2 \tag{6.14}$$

Any ill effect introduced by the ring that might falsify the measured value of wavelength or dispersion can be reduced by correctly designing the circuit. There are five sources of error that must be considered:

a. Because the transmission line has a curvature, the dispersion on the ring may not be equal to the straight-line dispersion.
b. Field interactions across the ring could cause mutual inductance.
c. The assumption that the total effective length of the ring can be calculated from the mean radius.
d. The coupling gap may cause field perturbations on the ring.
e. Nonuniformities of the ring width could cause resonance splitting.

To minimize problems (a) through (d) only rings with large diameters should be used. Troughton used rings that were five wavelengths long at the frequency of interest. A larger ring will result in a larger radius of curvature and thus approach the straight-line approximation and diminish the effect of (a). The large ring will reduce (b) and the effect of (d) will be minimized because the coupling gap occupies a smaller percentage of the total ring. The effect of the mean radius, (c), can be reduced by using large rings and narrow line widths.

An increased ring diameter will also increase the chance of variations in the line width, and the possibility of resonance splitting is increased. The only way to avoid resonance splitting is to use precision circuit processing techniques.

Troughton used another method to diminish the effect of the coupling gap. An initial gap of 1 mil was designed. Using swept frequency techniques, Q-factor measurements were made. The gap was etched back until it was obvious that the coupling gap was not affecting the frequency.

The steps Troughton used to measure dispersion can be summarized as follows:

1. Design the ring at least five wavelengths long at the lower frequency of interest.
2. Minimize the effect of the coupling gap by observing the Q-factor and etching back the gap when necessary.
3. Measure the resonant frequency of each mode.
4. Apply Equation (6.4) to calculate ε_{eff}.
5. Plot ε_{eff} versus frequency.

This technique was very important when it was introduced because of the very early stage that the microstrip transmission line was in. Because it was in its early stage, there had been little research that resulted in closed-form expressions for designing microstrip circuits. This technique allowed the frequency dependency of ε_{eff} to be quickly measured and the use of microstrip could be extended to higher frequencies more accurately.

6.3 DISCONTINUITY MEASUREMENTS

One of the most interesting applications of the ring is its use to characterize equivalent circuit parameters of microstrip discontinuities [3, 12]. Because discontinuity parameters are usually very small, extreme accuracy is needed and can be obtained with the ring resonator.

The main difficulty in measuring the circuit parameters of microstrip discontinuities resides in the elimination of systematic errors introduced by the coaxial-to-microstrip transitions. This problem can be avoided by testing discontinuities in a resonant microstrip ring that may be loosely coupled to test equipment. The resonant frequency for narrow rings can be approximated fairly accurately by assuming that the structure resonates if its electrical length is an integral multiple of the guided wavelength. When a discontinuity is introduced into the ring, the electric length may not be equal to the physical length. This difference in the electric and physical length will cause a shift in the resonant frequency. By relating the Z-parameters of the introduced discontinuity to the shift in the resonance frequency the equivalent circuit parameters of the discontinuity can be evaluated.

It has also been explained that the TM_{n10} modes of the microstrip ring are degenerate modes. When a discontinuity is introduced into the ring, the degenerate modes will split into two distinct modes. This splitting can be expressed in terms of an even and an odd incidence on the discontinuity. The even case corresponds to the incidence of two waves of equal magnitude and phase. In the odd case, waves of equal magnitude but opposite phase are incident from both sides. Either mode, odd or even, can be excited or suppressed by an appropriate choice of the point of excitation around the ring.

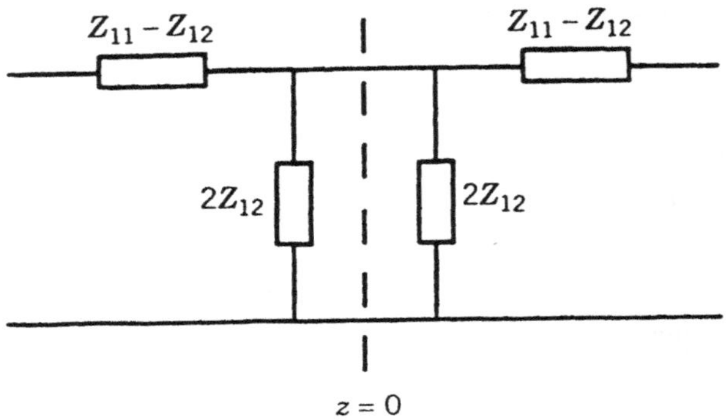

FIGURE 6.4 T equivalent circuit of a discontinuity expressed in terms of its Z-parameters.

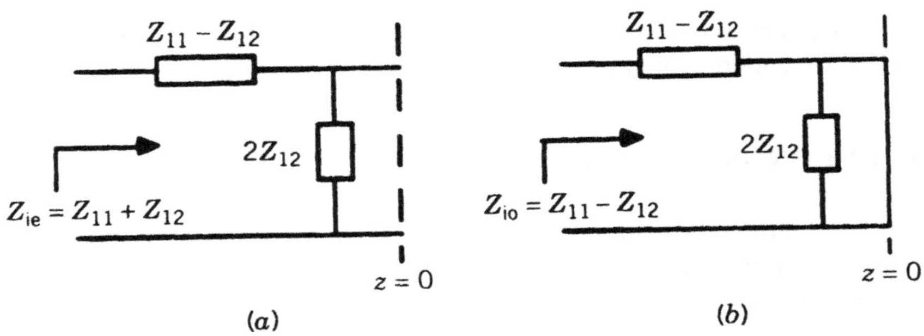

FIGURE 6.5 (*a*) Impedance of a discontinuity with an even-mode incidence, and (*b*) the impedance of a discontinuity with an odd-mode incidence.

A symmetrical discontinuity can be represented by its T equivalent circuit expressed in terms of its Z-parameters. The T equivalent circuit is presented in Figure 6.4. For convenience the circuit is divided into two identical half-sections of zero electrical length. If this circuit is excited in the even mode, it is as if there is an open circuit at the plane of reference $z = 0$. The normalized even input impedance at either port is thus $Z_{ie} = Z_{11} + Z_{12}$ (see Figure 6.5*a*). If this circuit is excited in the odd mode, it is as if there is a short circuit at the plane $z = 0$. The normalized odd input impedance is thus $Z_{io} = Z_{11} - Z_{12}$ (see Figure 6.5*b*). If the discontinuity is lossless, only the resonance frequencies of the perturbed ring are affected since the even and odd impedances are purely reactive. The artificial increase or decrease of the electrical length of the ring, resulting in the decrease of its resonance frequencies, is related to the even and odd impedances by the following expressions:

$$Z_{ie} = Z_{11} + Z_{12} = -j\cot kl_e \tag{6.15}$$

$$Z_{io} = Z_{11} - Z_{12} = j\tan kl_o \tag{6.16}$$

where $k = 2\pi/\lambda_g$ is the propagation constant, and l_e and l_o are the artificial electrical lengths introduced by the even and odd discontinuity impedances. Since at resonance the total electrical length of the resonator is $n\lambda_g$, the resonance conditions are, in the even case,

$$l_{\text{ring}} + 2l_e = n\lambda_{ge} \tag{6.17}$$

and in the odd case,

$$l_{\text{ring}} + 2l_o = n\lambda_{go} \tag{6.18}$$

where l_{ring} is the physical length of the ring, and λ_{ge} and λ_{go} are the guided wavelengths to the even and odd resonance frequency, respectively. Since l_{ring} is known and λ_g can be obtained from measurements, l_e and l_o can be determined from Equations (6.17) and (6.18). The parameters Z_{11} and Z_{12} can be determined by substituting Equations (6.17) and (6.18) into Equations (6.15) and (6.16) to yield [3]

$$Z_{11} + Z_{12} = j\cot\frac{\pi l_{\text{ring}}\sqrt{\varepsilon_{\text{eff}}(f_{re})}f_{re}}{c} \tag{6.19}$$

$$Z_{11} - Z_{12} = -j\tan\frac{\pi l_{\text{ring}}\sqrt{\varepsilon_{\text{eff}}(f_{ro})}f_{ro}}{c} \tag{6.20}$$

where λ_g was replaced by

$$\lambda_g = \frac{c}{f\sqrt{\varepsilon_{\text{eff}}(f)}}$$

and f_{re} and f_{ro} are the measured odd and even resonant frequencies of the perturbed ring.

The procedure described can be altered slightly and used to evaluate lossy discontinuities. Instead of the even an odd modes having open or short circuits at the plane of reference, $z = 0$, there is introduced a termination resistance. The termination resistance can be determined by measuring the circuit Q-factor.

6.4 MEASUREMENTS USING FORCED MODES OR SPLIT MODES

As shown earlier, the guided wavelength of the *regular mode* can be easily obtained from physical dimensions. Because of this advantage, the regular mode has been widely used to measure the characteristics of microstrip line. The *forced modes* and *split modes*, however, can also be applied for such measurements [15].

6.4.1 Measurements Using Forced Modes

The forced mode phenomenon was studied previously in Chapter 3. The shorted forced mode, as illustrated in Figure 6.6 with shorted boundary condition at 90°, is now used to measure the effective dielectric constant of microstrip line. The standing-wave patterns of this circuit is shown in Figure 6.7. According to the design rule mentioned in Chapter 3, the shorted forced modes contains *full-wavelength resonant modes* with odd integer mode numbers and excited *half-wavelength modes* with mode number $\nu = (2m \pm 1)/2$, where $m = 1, 3, 5, \ldots$. The guided wavelength of each resonant mode can be calculated by applying Equation (6.1). The resonant frequencies of each res-

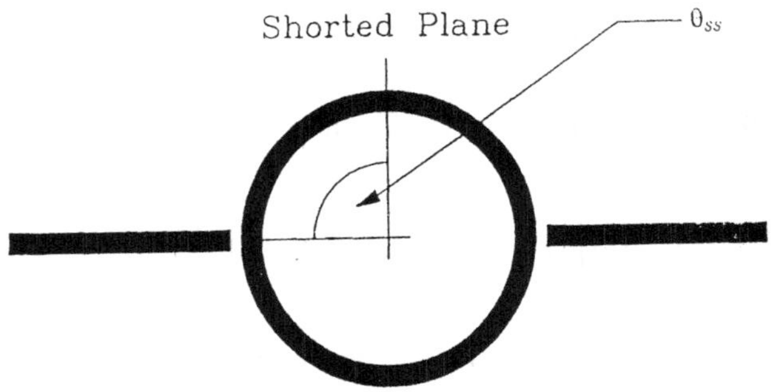

FIGURE 6.6 Coupled annular circuit with short plane at $\theta_{ss} = 90°$.

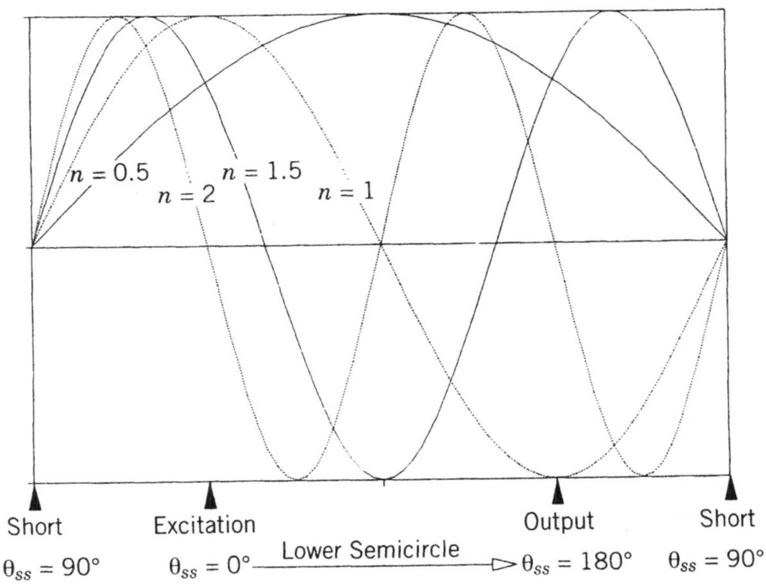

FIGURE 6.7 Standing wave patterns of the shorted forced mode.

onant mode can be measured with an HP8510 network analyzer. The effective dielectric constants for the different resonant frequencies are determined by the following equation:

$$\varepsilon_{\text{eff}} = (\lambda_0/\lambda_g)^2 \tag{6.21}$$

where λ_0 is the wavelength in free space and λ_g is the guided wavelength. Figure 6.8 displays the effective dielectric constants versus frequency that were calculated by the forced mode and regular mode. A comparison of these two results shows that the excited half-wavelength resonant modes have higher dielectric constants than the full-wavelength modes. This phenomenon reveals that the excited half-wavelength modes travel more slowly than the full-wavelength modes inside the annular element.

6.4.2 Measurements Using Split Modes

The idea of using the split mode for dispersion measurement was introduced by Wolff [16]. He used notch perturbation for the measurement and found that the frequency splitting depended on the depth of the notch. The experimental maximum splitting frequency was 53 MHz. Instead of using the notch

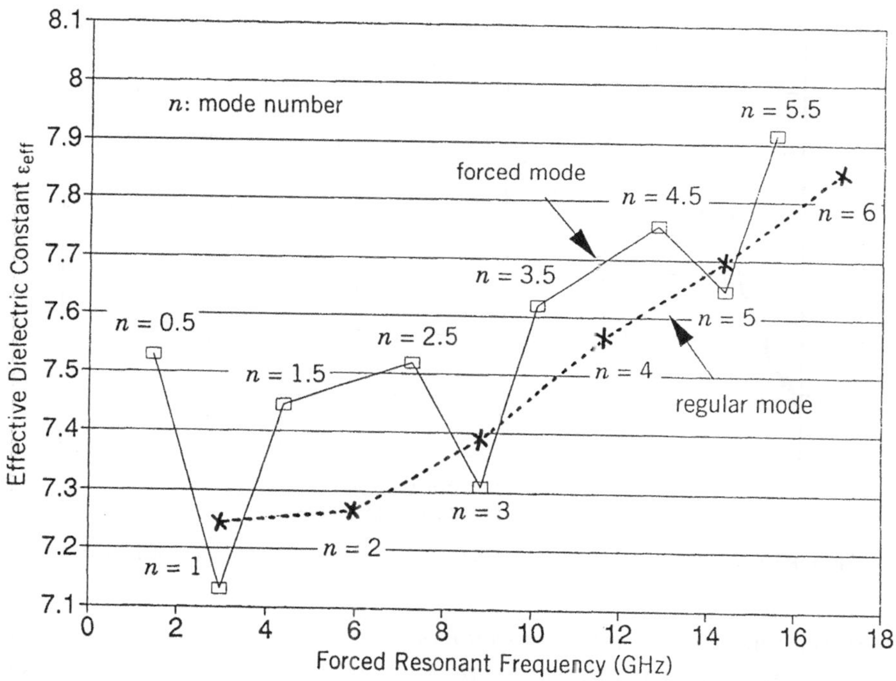

FIGURE 6.8 Effective dielectric constants vs. resonant frequency for the forced mode and regular mode.

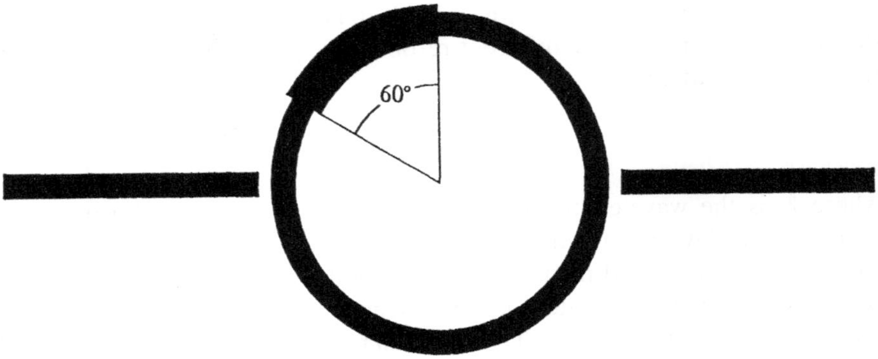

FIGURE 6.9 Layout of annular circuit with 60° LRS resonant sector.

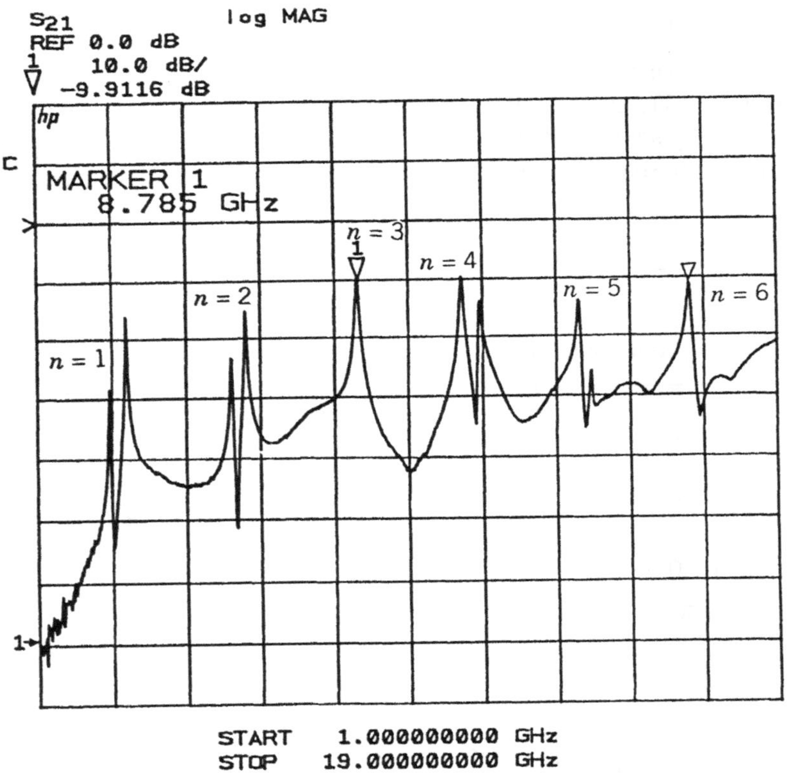

FIGURE 6.10 $|S_{21}|$ vs. frequency for the first six resonant modes of Figure 6.9.

perturbation, the *local resonant split mode* is developed to do the dispersion measurement. As illustrated in Figure 6.9, a 60° *local resonant sector* (LRS) was designed on the symmetric coupled annular ring circuit. The test circuit was built on a RT/Duroid 6010.5 substrate with the following dimensions:

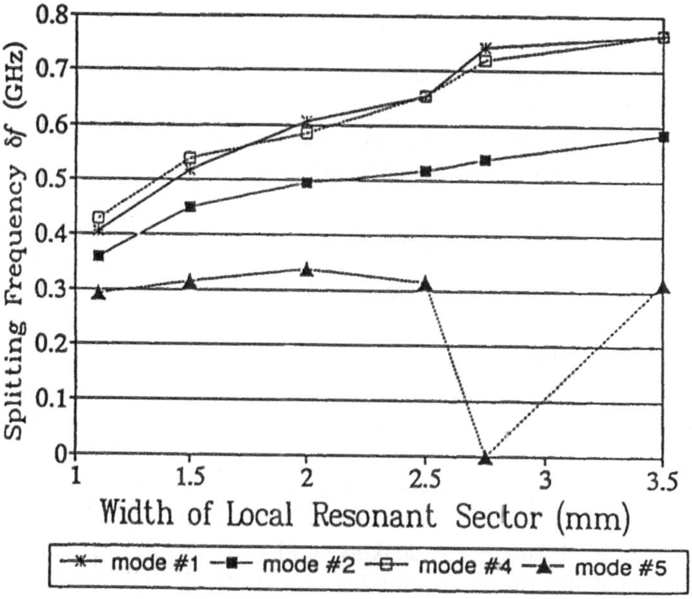

FIGURE 6.11 Splitting frequency vs. width of the 60° LRS.

Substrate thickness = 0.635 mm

Line width = 0.6 mm

LRS line width = 1.1 mm

Coupling gap = 0.1 mm

Ring radius = 6 mm

According to the analysis in Chapter 3, the resonant modes with mode number $n = 3$m, where $m = 1, 2, 3, \ldots$, will not split. Figure 6.10 illustrates the nondisturbed third and sixth resonant modes and the other four split resonant modes that agree with the prediction of standing-wave pattern analysis.

By increasing the perturbation width the frequency-splitting effect will become larger. Figure 6.11 displays the experimental results of the dependence of splitting frequency on the width of the LRS. The largest splitting frequency shown in Figure 6.11 is 765 MHz for the LRS with 3.5 mm width. The use of the local resonant split mode is more flexible than the notch perturbation. The local resonant split mode can also be applied to the measurements of step discontinuities of microstrip lines [17].

REFERENCES

[1] P. Troughton, "Measurement technique in microstrip," *Electron. Lett.*, Vol. 5, No. 2, pp. 25–26, January 23, 1969.

[2] K. Chang, F. Hsu, J. Berenz, and K. Nakano, "Find optimum substrate thickness for millimeter-wave GaAs MMICs," *Microwaves & RF*, Vol. 27, pp. 123–128, September 1984.

[3] W. Hoefer and A. Chattopadhyay, "Evaluation of the equivalent circuit parameters of microstrip discontinuities through perturbation of a resonant ring," *IEEE Trans. Microwave Theory Tech.*, Vol. MTT-23, pp. 1067–1071, December 1975.

[4] T. C. Edwards, *Foundations for Microstrip Circuit Design*, Wiley, Chichester, England, 1981; 2d ed., 1992.

[5] J. Deutsch and J. J. Jung, "Microstrip ring resonator and dispersion measurement on microstrip lines from 2 to 12 GHz," *Nachrichtentech. Z.*, Vol. 20, pp. 620–624, 1970.

[6] I. Wolff and N. Knoppik, "Microstrip ring resonator and dispersion measurements on microstrip lines," *Electron. Lett.*, Vol. 7, No. 26, pp. 779–781, December 30, 1971.

[7] H. J. Finlay, R. H. Jansen, J. A. Jenkins, and I. G. Eddison, "Accurate characterization and modeling of transmission lines for GaAs MMICs," in 1986 *IEEE MTT-S Int. Microwave Symp. Dig.*, New York, pp. 267–270, June 1986.

[8] P. A. Bernard and J. M. Gautray, "Measurement of relative dielectric constant using a microstrip ring resonator," *IEEE Trans. Microwave Theory Tech.*, Vol. MTT-39, pp. 592–595, March 1991.

[9] P. A. Polakos, C. E. Rice, M. V. Schneider, and R. Trambarulo, "Electrical characteristics of thin-film $Ba_2YCu_3O_7$ superconducting ring resonators" *IEEE Microwave Guided Wave Lett.*, Vol. 1, No. 3, pp. 54–56, March 1991.

[10] M. E. Goldfarb and A. Platzker, "Losses in GaAs Microstrip," *IEEE Trans. Microwave Theory Tech.*, Vol. MTT-38, No. 12, pp. 1957–1963, December 1990.

[11] S. Kanamaluru, M. Li, J. M. Carroll, J. M. Phillips, D. G. Naugle, and K. Chang, "Slotline ring resonator test method for high-Tc superconducting films," *IEEE Trans. App. Supercond.*, Vol. ASC-4, No. 3, pp. 183–187, September 1994.

[12] T. S. Martin, "A study of the microstrip ring resonator and its applications," M.S. thesis, Texas A&M University, College Station, December 1987.

[13] P. Troughton, "High Q-factor resonator in microstrip," *Electron. Lett.*, Vol. 4, No. 24, pp. 520–522, November 20, 1968.

[14] E. Belohoubek and E. Denlinger, "Loss considerations for microstrip resonators," *IEEE Trans. Microwave Theory Tech.*, Vol. MTT-23, pp. 522–526, June 1975.

[15] C. Ho and K. Chang, "Mode phenomenons of the perturbed annular ring elements," Texas A&M University Report, College Station, September 1991.

[16] I. Wolff, "Microstrip bandpass filter using degenerate modes of a microstrip ring resonator," *Electron. Lett.*, Vol. 8, No. 12, pp. 302–303, June 15, 1972.

[17] K. C. Gupta, R. Garg, and I. J. Bahl, *Microstrip Lines and Slotlines*, Artech House, Dedham, Mass., pp. 189–192, 1979.

CHAPTER SEVEN

Filter Applications

7.1 INTRODUCTION

As shown in the previous chapters, the ring resonator has bandpass characteristics. If a ring resonator is coupled to input and output transmission lines, the signal will pass through with certain losses at the resonant frequencies of the ring and will be rejected at frequencies outside the resonant frequencies. By cascading several ring resonators in series, various bandpass filtering characteristics can be designed. As discussed in Chapters 2 and 3, the ring resonator can support two degenerate modes if both modes are excited. This forms the base for a compact dual-mode filter. The ring resonators could be designed in microstrip line, slotline, or coplanar waveguide. The ring cavities can be built in waveguides.

7.2 DUAL-MODE RING BANDPASS FILTERS

As described in Chapters 2 and 3, the dual-mode effects are introduced either by skewing one of the feed lines with respect to the other or by introduction of a discontinuity (notch, slit, patch, etc.). The dual-mode bandpass filter was first proposed by Wolff using asymmetric coupling feed lines [1]. Later on, many new configurations using orthogonal feed lines with patch perturbation on a ring resonator were introduced [2–5]. The new configuration with orthogonal feed lines and patch perturbation provides a quasi-elliptic function that has two transmission zeros close to the passband. This property can be used to reject adjacent channel interferences.

Figure 7.1 shows a dual-mode filter. The square ring resonator is fed by a

Microwave Ring Circuits and Related Structures, Second Edition,
by Kai Chang and Lung-Hwa Hsieh
ISBN 0-471-44474-X

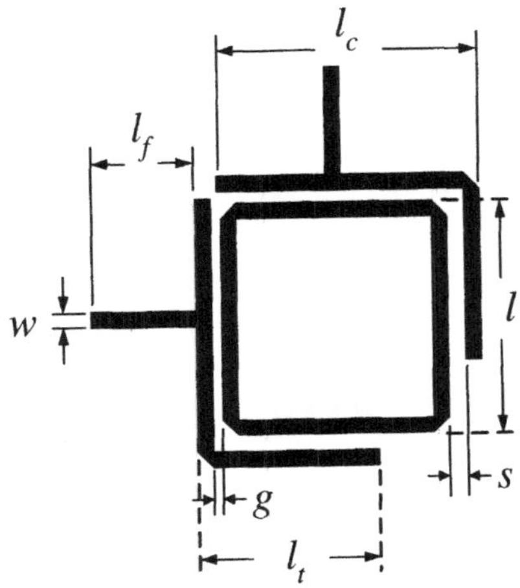

(a)

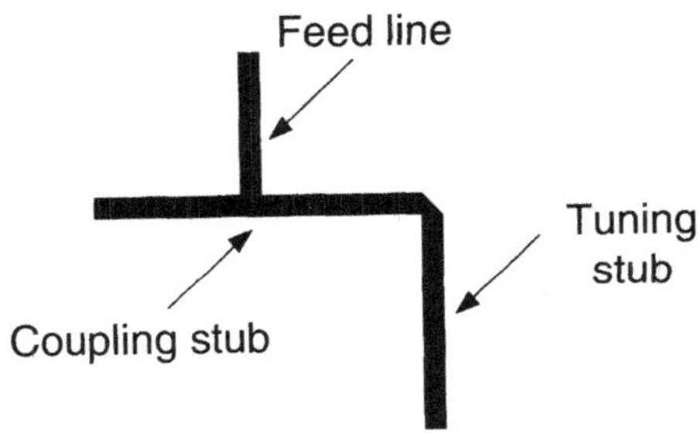

(b)

FIGURE 7.1 Dual-mode bandpass filter with enhanced coupling (a) layout and (b) L-shape coupling arm [6]. (Permission from IEEE.)

pair of orthogonal feed lines, and each feed line is connected to an L-shape coupling arm [6]. Figure 7.1*b* displays the scheme of the coupling arm that consists of a coupling stub and a tuning stub. The tuning stub attached to the end of the coupling stub extends the coupling stub to increase the coupling periphery. In addition, the asymmetrical structure perturbs the field of the ring resonator and excites two degenerate modes [1]. Without the tuning stubs, there is no perturbation on the ring resonator and only a single mode is excited [7]. Comparing the filter in Figure 7.1 with conventional dual-mode filters [1], the conventional filters only provide a dual-mode characteristic without the benefits of enhanced coupling strength and performance optimization.

The filter was designed at the center frequency of 1.75 GHz and fabricated on a 50-mil thickness RT/Duroid 6010.2 substrate with a relative dielectric constant of 10.2. The length of the tuning stubs is l_t, and the gap size between the tuning stubs and the ring resonator is s. The length of the feed lines is l_f = 8 mm; the width of the microstrip line is w = 1.191 mm for a 50-ohm line; the length of the coupling stubs is l_c = 18.839 + s mm; the gap size between the ring resonator and coupling stubs is g = 0.25 mm; the length of one side of the square ring resonator is l = 17.648 mm. The coupling gap g was selected in consideration of strong coupling and etching tolerance. The simulation was completed using an IE3D electromagnetic simulator [8].

By adjusting the length l_t and gap size s of the tuning stubs adequately, the coupling strength and the frequency response can be optimized. Single-mode excitation (Figure 7.2) or dual-mode excitation (Figure 7.3) can be resulted by varying s and l_t. Figures 7.2 and 7.3 show the measured results for five cases from changing the length l_t of tuning stubs with a fixed gap size (s = 0.8 mm) and varying the gap size s with a fixed length (l_t = 13.5 mm). Observing the measured results in Figure 7.2, two cases for l_t = 4.5 and 9 mm with a fixed gap size only excite a single mode.

The coupling between the L arms and the ring can be expressed by external Q (Q_e) as follows [9]:

$$Q_L = \frac{1}{\dfrac{2}{Q_e} + \dfrac{1}{Q_o}} = \frac{f_o}{(\Delta f)_{3dB}} \tag{7.1a}$$

$$Q_o = \frac{Q_L}{(1 - 10^{-L/20})} \tag{7.1b}$$

where Q_L is the loaded Q, Q_o is the unloaded Q of the ring resonator, f_o is the resonant frequency, $(\Delta f)_{3dB}$ is the 3-dB bandwidth, and L is the insertion loss in decibel. The loaded Q is obtained from measurement of f_o and $(\Delta f)_{3dB}$ and unloaded Q (Q_o = 137) is calculated from the Equation (7.1b). From Equation (7.1a), Q_e is given by

$$Q_e = \frac{2Q_oQ_L}{Q_o - Q_L} \tag{7.2}$$

The performance for these two single-mode ring resonators is shown in Table 7.1.

The coupling coefficient between two degenerate modes is given by [10]

$$K = \frac{f_{p2}^2 - f_{p1}^2}{f_{p2}^2 + f_{p1}^2} \tag{7.3}$$

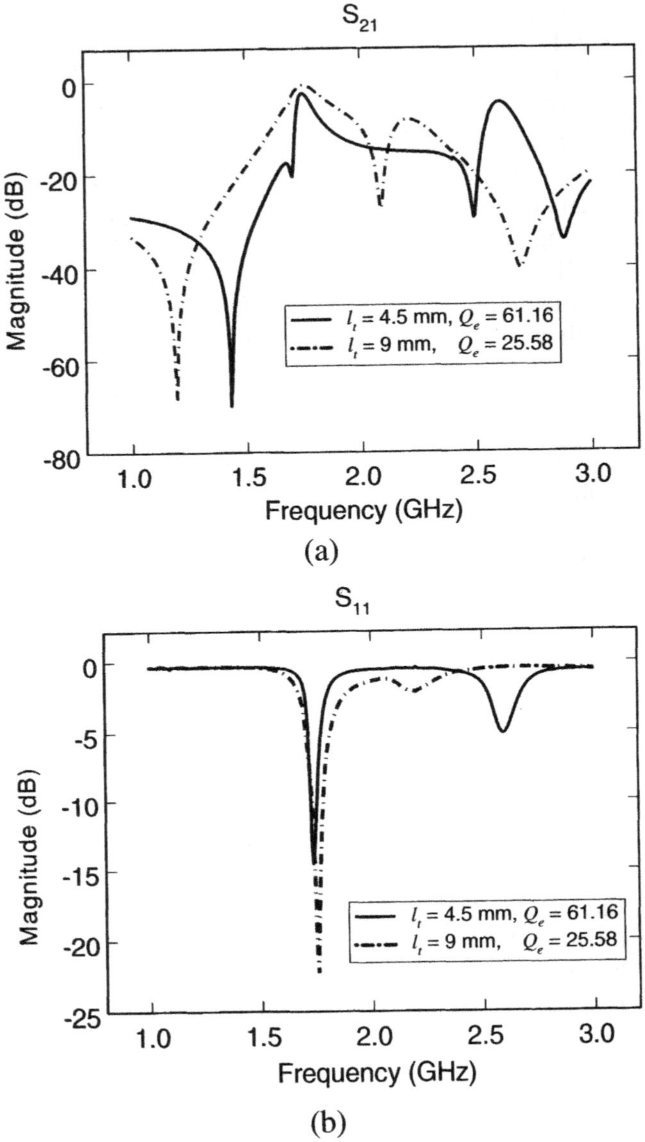

FIGURE 7.2 Measured (a) S_{21} and (b) S_{11} by adjusting the length of the tuning stub l_t with a fixed gap size ($s = 0.8$ mm) [6]. (Permission from IEEE.)

where f_{p1} and f_{p2} are the resonant frequencies. In addition, the midband insertion loss L corresponding to Q_o, Q_e, and K can be expressed as [9]

$$L = 20\log\left[\frac{(1+Q_e/Q_o)^2}{2KQ_e} + \frac{KQ_e}{2}\right]\text{dB} \tag{7.4}$$

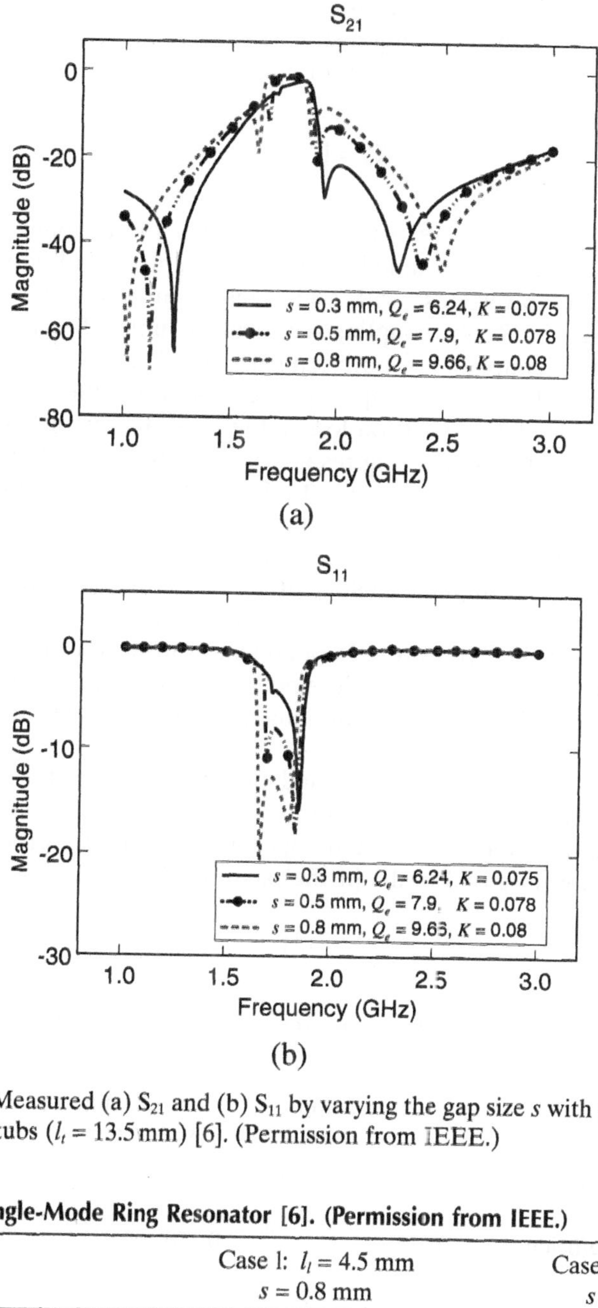

FIGURE 7.3 Measured (a) S_{21} and (b) S_{11} by varying the gap size s with a fixed length of the tuning stubs ($l_t = 13.5\,\text{mm}$) [6]. (Permission from IEEE.)

TABLE 7.1 Single-Mode Ring Resonator [6]. (Permission from IEEE.)

	Case 1: $l_t = 4.5$ mm $s = 0.8$ mm	Case 2: $l_t = 9$ mm $s = 0.8$ mm
Resonant Frequency f_o	1.75 GHz	1.755 GHz
Insertion Loss IL	2.69 dB	0.97 dB
3-dB Bandwidth	70 MHz	150 MHz
Loaded Q	25	11.7
External Q	61.16	25.58

The external Q can be obtained from Equation (7.4) through measured L, K, and Q_o. Moreover, the coupling coefficient between two degenerate modes shows three different coupling conditions.

$$\text{Let } K_o = 1/Q_e + 1/Q_o. \tag{7.5}$$

If the coupling coefficient satisfies $K > K_o$, then the coupling between two degenerate modes is overcoupled. In this overcoupled condition, the ring resonator has a hump response with a high insertion loss in the middle of the passband [5]. If $K = K_o$, the coupling is critically coupled. Finally, if $K < K_o$, the coupling is undercoupled. For both critically coupled and undercoupled coupling conditions, there is no hump response. Also, when the coupling becomes more undercoupled, the insertion loss in the passband increases [9]. The performance for the dual-mode ring resonators is displayed in Table 7.2.

Observing the single-mode ring in Table 7.1, it shows that a higher external Q produces higher insertion loss and narrower bandwidth. In addition, for the dual-mode ring resonator in Table 7.2, its insertion loss and bandwidth depend on the external Q, coupling coefficient K, and coupling conditions. For an undercoupled condition, the more undercoupled, the more the insertion loss and the narrower the bandwidth. To obtain a low insertion-loss and wide-band pass band characteristic, the single-mode ring resonator should have a low external Q, which implies more coupling periphery between the feeders and the ring resonator.

Figure 7.4 shows the simulated and measured results for the optimized quasi-elliptic bandpass filter. Two transmission zeros locate on either side of the passband to suppress unwanted adjacent channel interferences. The filter has an insertion loss of 1.04 dB in the passpband with a 3-dB bandwidth of 192.5 MHz.

TABLE 7.2 Dual-Mode Ring Resonator [6]. (Permission from IEEE.)

	Case 1: $l_t = 13.5$ mm $s = 0.3$ mm	Case 2: $l_t = 13.5$ mm $s = 0.5$ mm	Case 3: $l_t = 13.5$ mm $s = 0.8$ mm
Resonant Frequencies (f_{p1}, f_{p2})	(1.72, 1.855) GHz	(1.7, 1.84) GHz	(1.67, 1.81) GHz
Coupling Coefficient K	0.075	0.078	0.08
External Q	6.24	7.9	9.66
Midband Insertion Loss IL	2.9 dB	1.63 dB	1.04 dB
3-dB Bandwidth	160 MHz	175 MHz	192.5 MHz
Coupling Condition	undercoupled	undercoupled	undercoupled

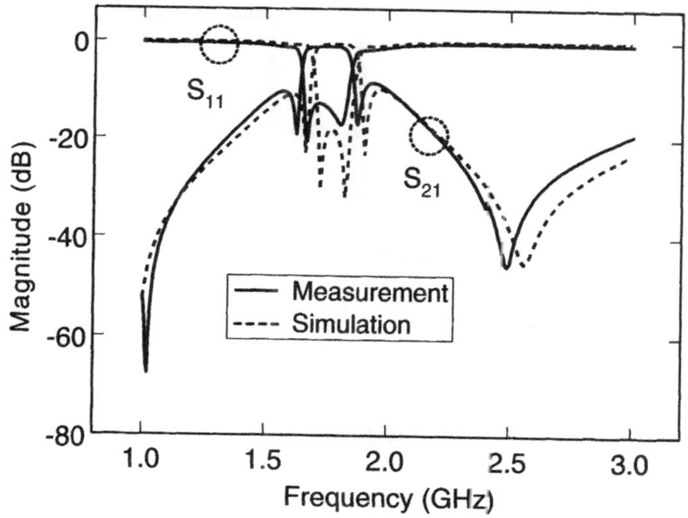

FIGURE 7.4 Simulate and measured results for the case of l_t = 13.5 mm and s = 0.8 mm [6]. (Permission from IEEE.)

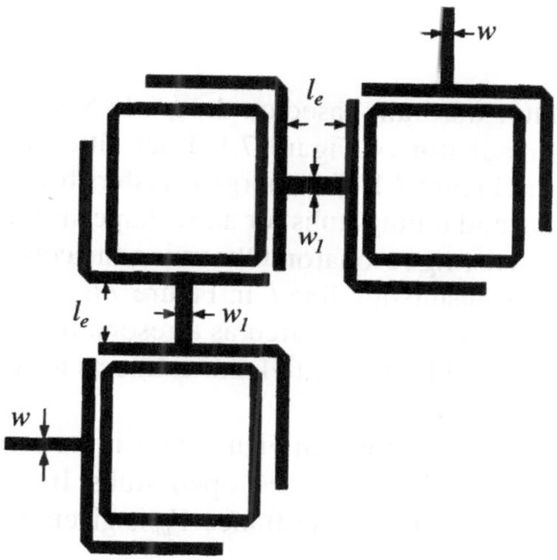

FIGURE 7.5 Layout of the filter using three resonators with L-shape coupling arms [6]. (Permission from IEEE.)

Cascaded multiple ring resonators have advantages in acquiring a much narrower and shaper rejection. Figure 7.5 illustrates the filter using three cascaded ring resonators. Any two of three resonators are linked by an L-shape arm with a short transmission line l_e of 6.2 mm with a width w_1 = 1.691 mm.

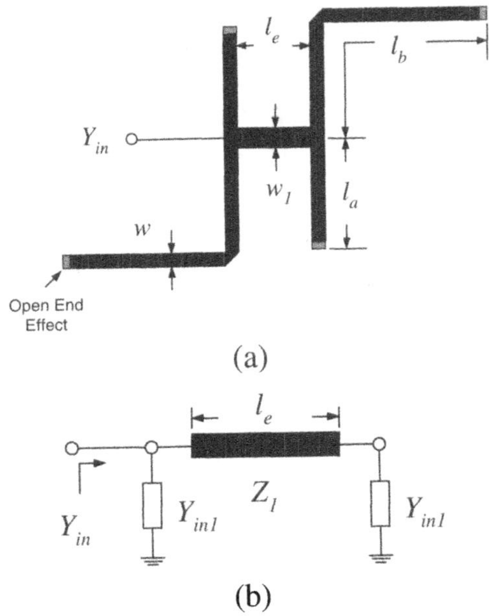

FIGURE 7.6 Back-to-back L-shape resonator (a) layout and (b) equivalent circuit. The lengths l_a and l_b include the open end effects.

This bandpass filter was built based on the $l_t = 13.5$ mm and $s = 0.8$ mm case of the single ring resonator of Figure 7.1. Each filter section has identical dimensions as that in Figure 7.1. The energy transfers from one ring resonator through the coupling and tuning stubs (or an L-shape arm) and the short transmission line to another ring resonator. Observing the configuration for the L-shape and the short transmission line l_e in Figure 7.6, it not only perturbs the ring resonator, but also it can be treated as a resonator. A short transmission line l_c of 6.2 mm with a width $w_1 = 1.691$ mm connects to the coupling stubs to link the two ring resonators.

Considering this type resonator in Figure 7.6*a*, it is consisted of a transmission line l_e and two parallel-connected open stubs. Its equivalent circuit is shown in Figure 7.6*b*. The input admittance Y_{in} is given by

$$Y_{in} = Y_{in1} + Y_1 \left[\frac{Y_{in1} + jY_1 \tan(\beta l_e)}{Y_1 + jY_{in1} \tan(\beta l_e)} \right]$$

$$Y_{in1} = jY_o[\tan(\beta l_a) + \tan(\beta l_a)], \quad \beta\text{: phase constant}$$
$$Y_l = 1/Z_l,\ Y_o = 1/Z_o. \tag{7.6}$$

Y_l is the characteristic admittance of the transmission line l_e, and Y_o is the characteristic admittances of the transmission lines l_a, and l_b. Letting $Y_{in} = 0$, the resonant frequencies of the resonator can be predicted. The resonant fre-

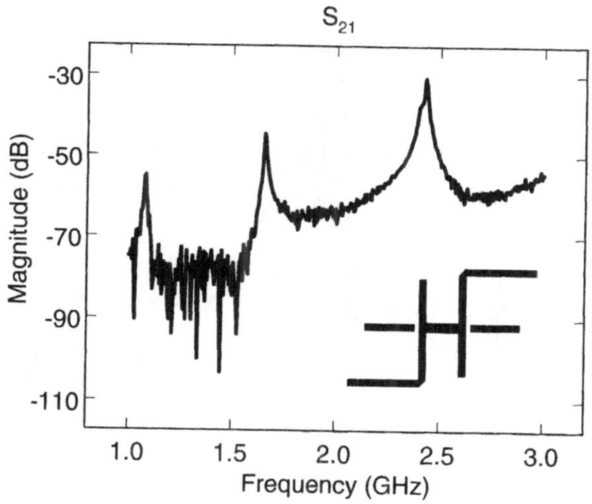

FIGURE 7.7 Measured S_{21} for the back-to-back *L*-shape resonator [6]. (Permission from IEEE.)

quencies of the resonator are calculated as f_{o1} = 1.067, f_{o2} = 1.654, and f_{o3} = 2.424 GHz within 1–3 GHz. To verify the resonant frequencies, an end-to-side coupling circuit is built as shown in Figure 7.7.

Also, the measured resonant frequencies can be found as f_{mo1} = 1.08, f_{mo2} = 1.655, and f_{mo3} = 2.43 GHz, which show a good agreement with calculated results. Inspecting the frequency responses in Figures 7.6 and 7.7, the spike at f_{mo3} = 2.43 GHz is suppressed by the ring resonators and only one spike appears at low frequency (f_{mo1} = 1.08 GHz) with a high insertion loss, which dose not influence the filter performance. Furthermore, the resonant frequency (f_{mo2} = 1.655 GHz) of the resonator in Figure 7.6 couples with the ring resonators. By changing the length l_e, the resonant frequencies will move to different locations. For a shorter length l_e, the resonant frequencies move to higher frequency and for a longer length l_e, the resonant frequencies shift to lower frequency. Considering the filter performance, a proper length l_e should be carefully chosen. The simulated and measured results of the three cascaded ring filter are shown in Figure 7.8. The filter has a measured insertion loss of 2.39 dB in the passpband with a 3-dB bandwidth of 145 MHz.

7.3 RING BANDSTOP FILTERS

The bandstop characteristic of the ring circuit can be realized by using two orthogonal feed lines with coupling gaps between the feed lines and the ring resonator [11]. For odd-mode excitation, the output feed line is coupled to a position of the zero electric field along the ring resonator and shows a short

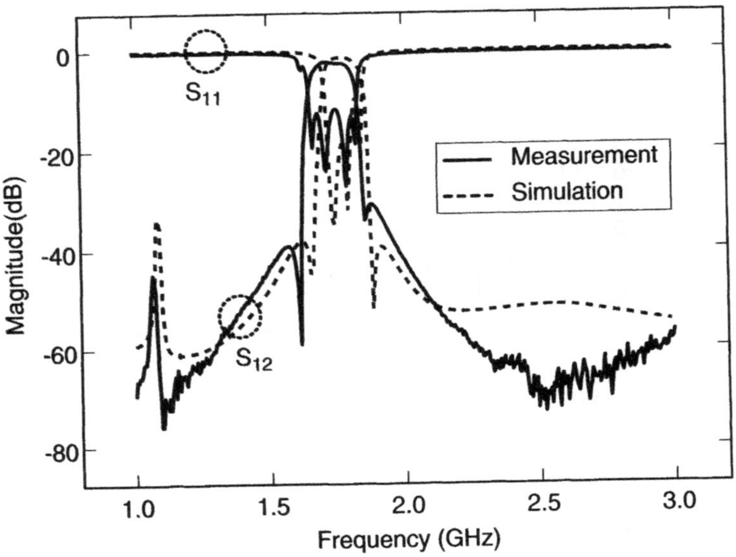

FIGURE 7.8 Simulated and measured results for the filter using three resonators with L-shape coupling arms [6]. (Permission from IEEE.)

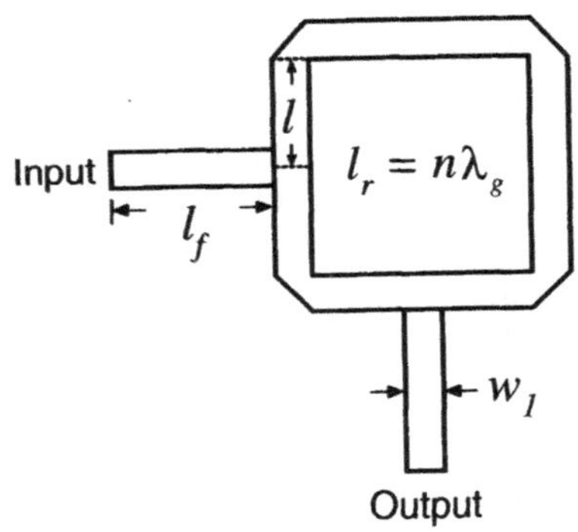

FIGURE 7.9 A ring resonator using direct-connected orthogonal feeders [12]. (Permission from IEEE.)

circuit [12]. Therefore, no energy is extracted from the ring resonator, and the ring circuit provides a stopband. A ring resonator directly connected to a pair of orthogonal feed lines is shown in Figure 7.9 [13]. In this case, no coupling gaps are used between the resonator and the feed lines for low insertion loss. The circumference l_r of the ring resonator is expressed as

$$l_r = n\lambda_g \tag{7.7}$$

where n is the mode number and λ_g is the guided wavelength. In order to investigate the behavior of this ring circuit, an EM simulator [8] and a transmission line model are used.

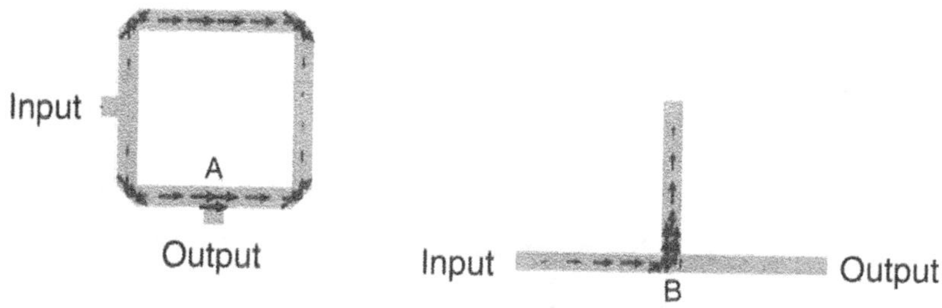

FIGURE 7.10 Simulated electric current at the fundamental resonant frequency for the ring and open-stub bandstop circuits [12]. (Permission from IEEE.)

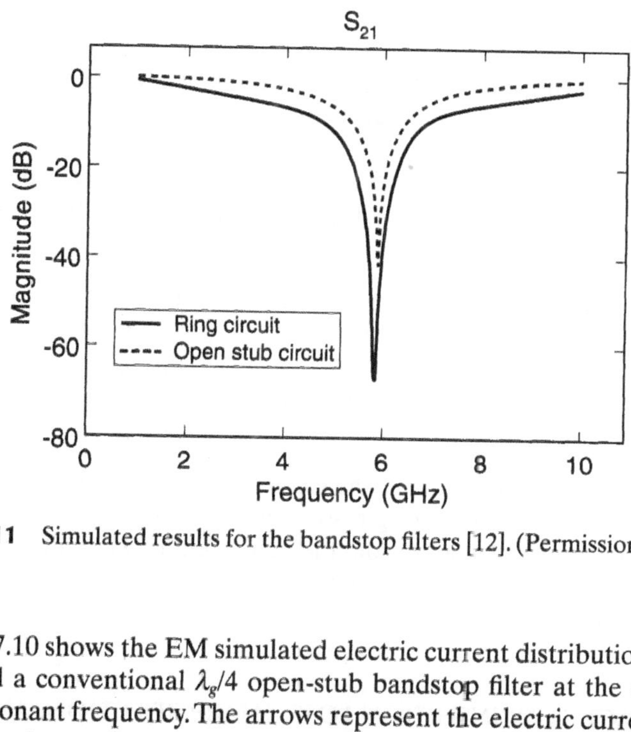

FIGURE 7.11 Simulated results for the bandstop filters [12]. (Permission from IEEE.)

Figure 7.10 shows the EM simulated electric current distribution of the ring circuit and a conventional $\lambda_g/4$ open-stub bandstop filter at the same fundamental resonant frequency. The arrows represent the electric current. The simulated electric current shows minimum electric fields at positions A and B, which correspond to the maximum magnetic fields. Thus, both circuits provide bandstop characteristics by presenting zero voltages to the outputs at the fundamental resonant frequency that can be observed by their simulated frequency response of S_{21} as shown in Figure 7.11.

The ring resonator and the conventional $\lambda_g/4$ open-stub bandstop filter are designed at a fundamental resonant frequency of $f_o = 5.6$ GHz and fabricated on a RT/Duriod 6010.2 substrate with a thickness $h = 25$ mil and a relative dielectric constant $\varepsilon_r = 10.2$. The dimensions of the ring are $l_f = 5$ mm, $l_r = 20.34$ mm, and $w_1 = 0.6$ mm.

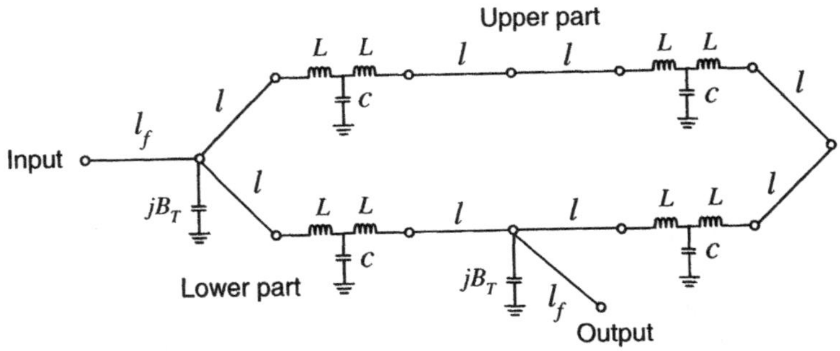

FIGURE 7.12 Equivalent circuit of the ring using direct-connected orthogonal feed lines [12]. (Permission from IEEE.)

The equivalent ring circuit shown in Figure 7.12 is divided by the input and output ports to form a shunt circuit denoted by the upper and lower parts, respectively.

The equivalent circuits of the 45-degree-mitered bend are represented by two inductors L and a capacitor C [14] those are expressed by

$$C = 0.001h\left[(3.39\varepsilon_r + 0.62)\left(\frac{w_1}{h}\right)^2 + 7.6\varepsilon_r + 3.8\left(\frac{w_1}{h}\right)\right] \quad \text{pF} \tag{7.8a}$$

$$L = 0.22h\left\{1 - 1.35\exp\left[-0.18\left(\frac{w_1}{h}\right)^{1.39}\right]\right\} \quad \text{nH} \tag{7.8b}$$

where h and w_1 are in millimeters. The capacitance jB_T is the T-junction effect between the feed line and the ring resonator [15]. The frequency response of the ring circuit can be calculated from the equivalent ring circuit using *ABCD*, *Y*, and *S* parameters. Figure 7.13 shows the calculated and measured results with good agreement.

7.4 COMPACT, LOW INSERTION LOSS, SHARP REJECTION, AND WIDEBAND BANDPASS FILTERS

Figure 7.14 shows a compact, low insertion loss, sharp rejection, wideband microstrip bandpass filter. This bandpass filter is developed from the bandstop filter introduced in Section 7.3 [13]. Two tuning stubs are added to the bandstop filter to create a wide passband. Without coupling gaps between feed lines and rings, there are no mismatch and radiation losses between them [16]. Thus, the filter can obtain a low insertion loss, and the major losses of the filter are contributed by conductor and dielectric losses. In Figure 7.14, the ring resonator is loaded with two tuning stubs of length $l_t = \lambda_g/4$ at $\Phi = 90°$ and

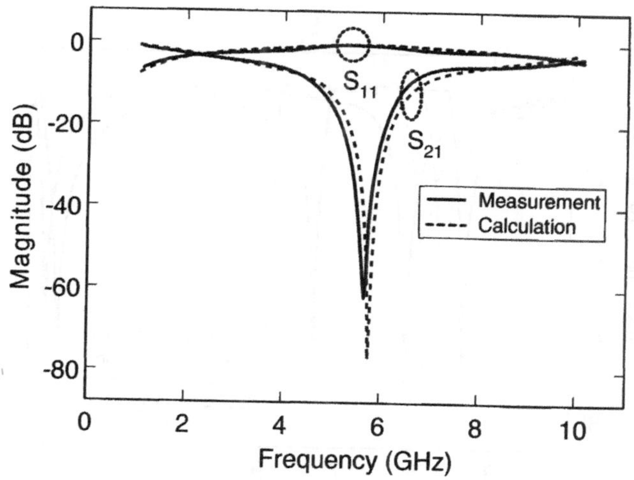

FIGURE 7.13 Calculated and measured results of the ring using direct-connected orthogonal feed lines [12]. (Permission from IEEE.)

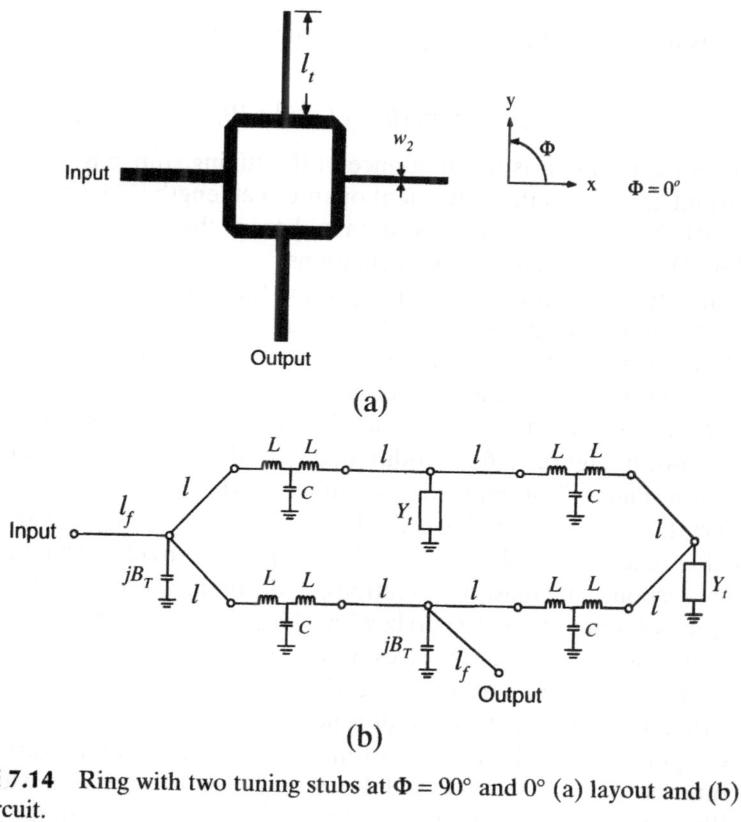

FIGURE 7.14 Ring with two tuning stubs at $\Phi = 90°$ and $0°$ (a) layout and (b) equivalent circuit.

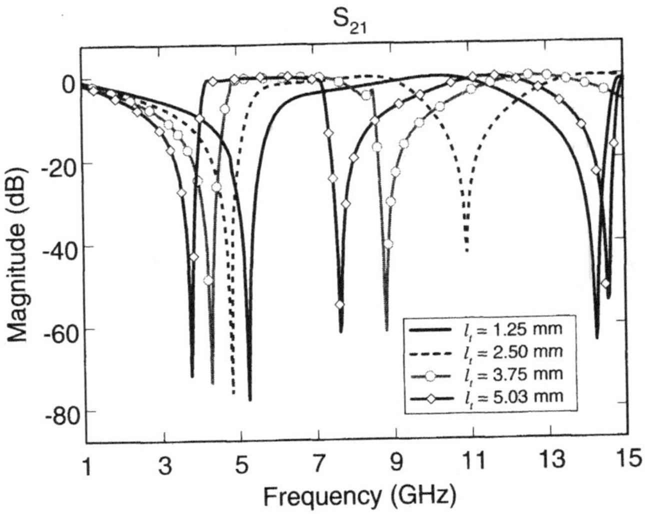

FIGURE 7.15 Calculated results of the ring with various lengths of the tuning stub at $\Phi = 90°$ and $0°$ [12]. (Permission from IEEE.)

$\Phi = 0°$. Y_t is the admittance looking into the tuning stub and can be expressed by

$$Y_t = Y_o \tanh[\beta(l_t + l_{open})] + jB_{T1} \tag{7.9}$$

where Y_o is the characteristic admittance of the tuning stub, β is the propagation constant, and l_{open} is the equivalent open-effect length [17]. The frequency response of the ring circuit can be obtained from the equivalent circuit by using *ABCD*, *Y*, and *S* parameter calculations.

By changing the lengths of two tuning stubs, the frequency response of the ring circuit will be varied. Observing the calculated results in Figure 7.15, two attenuation poles starting from the center frequencies of the fundamental and the third modes move to the lower frequencies and form a wide passband. The measured and calculated results of the filter with the tuning stubs of length $\lambda_g/4$ are shown in Figure 7.16. In addition, due to the symmetric structure, the ring circuit in Figure 7.14 only excites a single mode.

Observing the results in Figure 7.16, the effects of adding two tuning stubs with a length of $l_t = \lambda_g/4$ at $\Phi = 90°$ and $\Phi = 0°$ provide a sharper cutoff frequency response, increase attenuations, and obtain a wide passband. Two attenuation poles are at $f_1 = 3.81$ GHz with –46-dB rejection and $f_2 = 7.75$ GHz with –51-dB rejection. The differences between the measurement and the calculation on f_1 and f_2 are due to fabrication tolerances that cause a slightly asymmetric layout and excite small degenerate modes.

The key point behind this new filter topology is that two tuning stubs loaded on the ring resonator at $\Phi = 90°$ and $\Phi = 0°$ are used to achieve a wide passband with a sharp cutoff characteristic. In some cases, an undesired passband

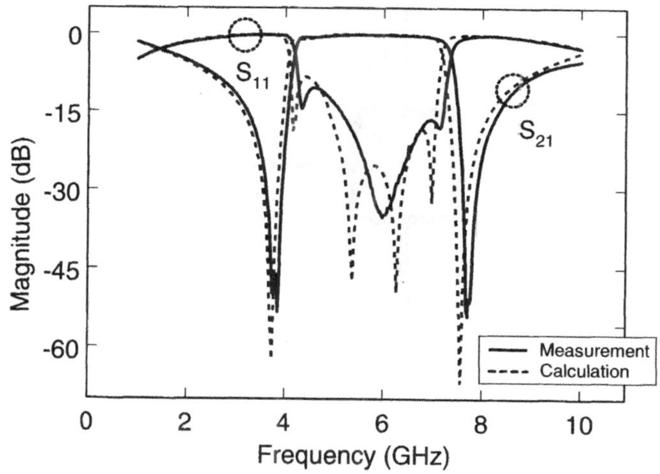

FIGURE 7.16 Calculated and measured results of the ring with two tuning stubs of $l_t = \lambda_g/4 = 5.026$ mm at $\Phi = 90°$ and $0°$ [12]. (Permission from IEEE.)

below the main passband may require a high passband section to be used in conjunction with this approach.

In Figure 7.16, the two stopbands of the filter show a narrow bandwidth. To increase the narrow stopbands, a dual-mode design can be used [4]. A square perturbation stub at $\Phi = 45°$ is incorporated on the ring resonator in Figure 7.17*a*. The square stub perturbs the fields of the ring resonator so that the resonator can excite a dual mode around the stopbands in order to improve the narrow stopbands. By increasing (decreasing) the size of the square stub, the distance (stopband bandwidth) between two modes is increased (decreased). The equivalent circuits of the square stub and the filter are displayed in Figure 7.17*b* and *c*, respectively. As seen in Figure 7.17*b*, the geometry at the corner of $\Phi = 45°$ is approximately equal to the square section of width $w_1 + w_p$, subtracting an isometric triangle of height w_1. Also, the equivalent L-C circuit of this approximation is shown in Figure 7.17*c*, where $C_{pf} = C_r - C$ and $L_p = LL_r/(L - L_r)$. The equivalent capacitance and inductance of the right angle bend, C_r and L_r, are given by [14]

$$C_r = 0.001h\left[(10.35\varepsilon_r + 2.5)\left(\frac{w_1 + w_p}{h}\right)^2 + 2.6\varepsilon_r + 5.64\left(\frac{w_1 + w_p}{h}\right)\right] \text{ pF} \quad (7.10a)$$

$$L = 0.22h\left\{1 - 1.35\exp\left[-0.18\left(\frac{w_1 + w_p}{h}\right)^{1.39}\right]\right\} \text{ nH} \quad (7.10b)$$

The asymmetric step capacitance C_s is [18]

$$C_s = w_p(0.012 + 0.0039\varepsilon_r)\text{pF} \quad (7.11)$$

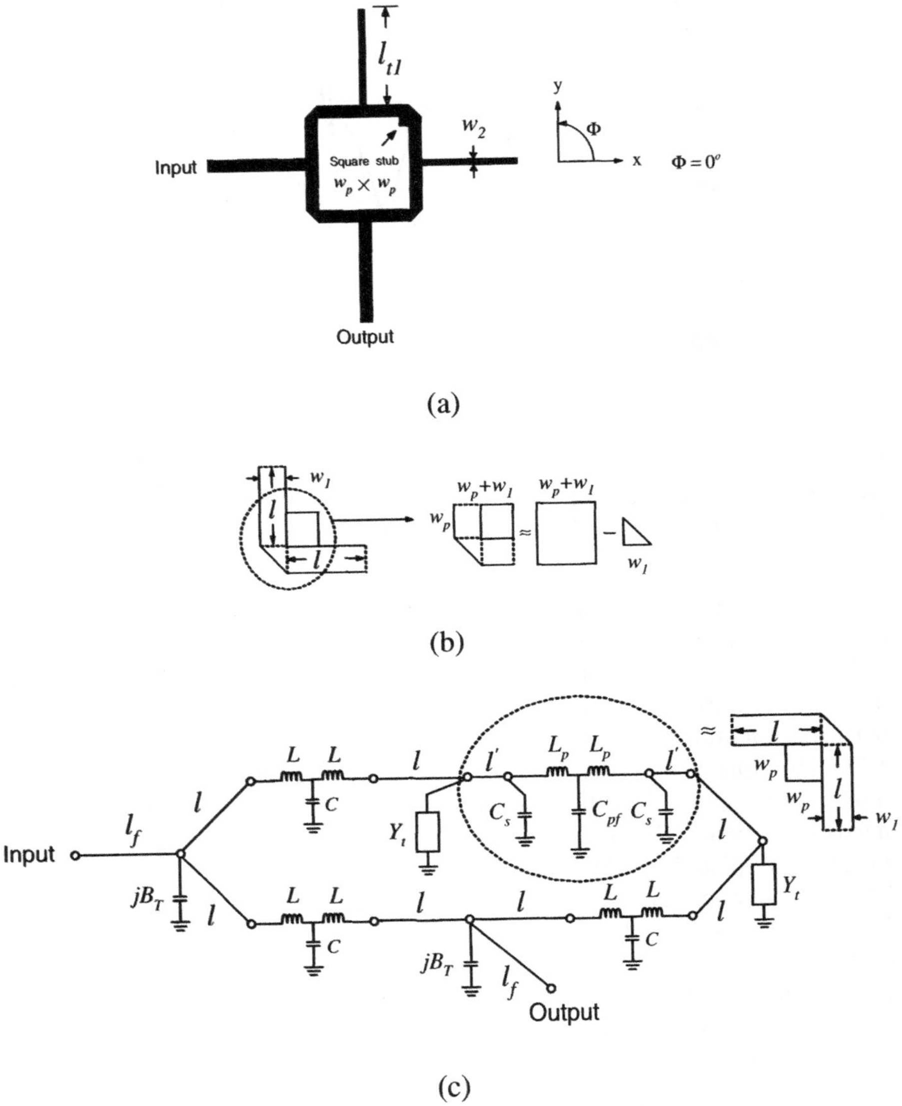

FIGURE 7.17 The dual-mode filter (a) layout, (b) equivalence of the perturbed stub, and (c) overall equivalent circuit [12]. (Permission from IEEE.)

In the above equations, all lengths are in millimeters. The length of the tuning stubs and the size of the square stub are $l_{t1} = 4.83\,\text{mm}$ and $w_p \times w_p = 0.5 \times 0.5\,\text{mm}^2$.

The calculated and measured results of the filter are shown in Figure 7.18. The square stub generates two transmission zeros (which are marked as x in Figure 7.18) or dual modes located on either side of the passband at 3.66, 7.62

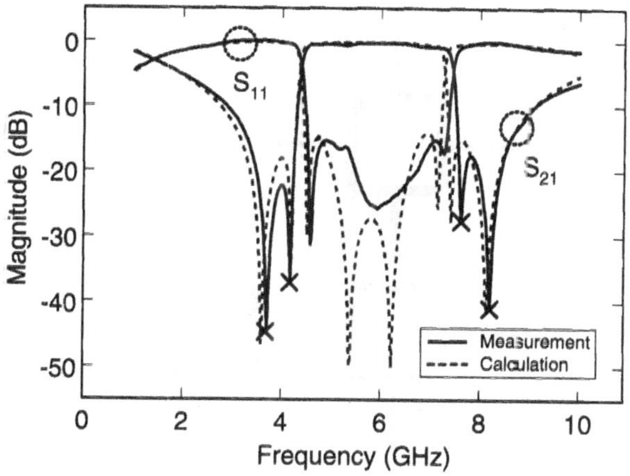

FIGURE 7.18 Calculated and measured results of the dual-mode ring filter. The crosses (x) show the two transmission zero locations [12]. (Permission from IEEE.)

and 7.62, 8.07 GHz, respectively. Comparing S_{21} with that in Figure 7.16, the dual-mode effects or transmission zeros increase the stopband bandwidth and improve the return loss in the edges of the passband. The filter has 3-dB fractional bandwidth of 51.6%, an insertion loss of better than 0.7 dB, two rejections of greater than 18 dB within 3.43–4.3 GHz and 7.57–8.47 GHz, and an attenuation rate for the sharp cutoff frequency responses of 137.58 dB/GHz (calculated from 4.173 GHz with −36.9 dB to 4.42 GHz with −2.85 dB) and 131.8 dB/GHz (calculated from 7.44 GHz with 3.77 dB to 7.62 GHz with −27.5 dB). In addition, comparing this filter with some compact and low insertion loss filters [19, 20], those filters only show gradual rejections. To obtain a sharp cutoff frequency response, the filters need to increase the number of resonators. However, increasing the number of resonators will increase the insertion loss and the size of the filter and result in narrow passband bandwidth [21, 22].

To obtain even higher rejection, a filter using three cascaded ring resonators is shown in Figure 7.19. In this configuration, the three ring resonators are connected by a short transmission line of length $l_c = \lambda_g/4 = 4.89$ mm. The different length $l_{t2} = 4.85$ mm, $l_{t3} = 4.88$ mm, and $l_{t4} = 4.83$ mm for the tuning stubs are optimized for a good return loss.

Figure 7.20 shows the calculated and measured results. The calculation also uses the transmission-line model with *ABCD*, *Y*, and *S* parameter operations. The 3-dB fractional bandwidth of the filter is 49.3%. The filter has an insertion loss better than 1.6 dB and return loss greater than 13.3 dB in the passband from 4.58 to 7.3 GHz. Two stopbands are located at 2.75–4.02 GHz and 7.73–9.08 GHz with rejection greater than 40 dB. The attenuation rate of the filter for the sharp cutoff frequency responses is 99.75 dB/GHz (calculated from 4.17 GHz with −34.9 dB to 4.49 GHz with −2.98 dB) and 101.56 dB/GHz

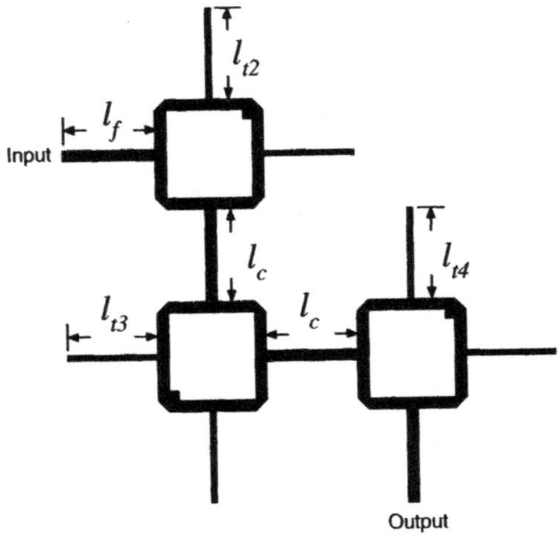

FIGURE 7.19 Configuration of the cascaded dual-mode ring resonator [12]. (Permission from IEEE.)

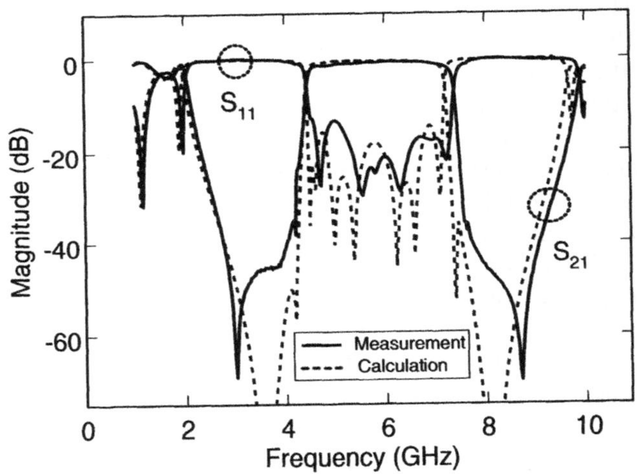

FIGURE 7.20 Calculated and measured results of the cascaded dual-mode ring resonator filter [12]. (Permission from IEEE.)

(calculated from 7.32 GHz with –3.4 dB to 7.64 GHz with –35.9 dB). The group delay of this wideband bandpass filter can be calculated by

$$\tau = -\frac{\partial \angle S_{21}}{\partial \omega} \tag{7.12}$$

where $\angle S_{21}$ is the insertion loss phase and ω is the frequency in radians per second. Figure 7.21 shows the group delay of the filter. Within the passband, the group delay is below 2 nS.

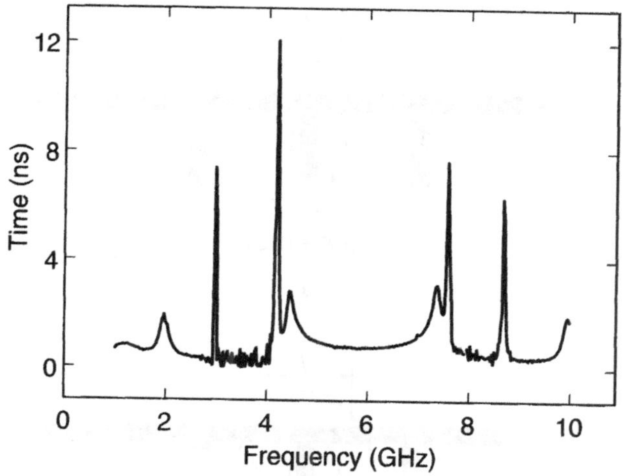

FIGURE 7.21 Group delay of the cascaded dual-mode ring resonator filter [12]. (Permission from IEEE.)

7.5 RING SLOW-WAVE BANDPASS FILTERS

The conventional slow-wave bandpass filter using a microstrip line periodically loaded by capacitive or inductive loads. As shown in Figure 7.22*a* [16], a transmission line is periodically loaded with identical open stub elements. Each unit element includes a length of d transmission line with a length of l open stub, where Z_{in1} is the input impedance looking into the open stub.

The conventional slow-wave periodic structure usually works as a lowpass or stopband filter [23, 24]. Also, using higher order modes, the conventional slow-wave periodic structure can act as a wide band bandpass filter, by constructing two consecutive stopbands close to the passband [25]. Considering the slow-wave periodic structure in Figure 7.22*b*, a loading impedance Z_L is connected at the end of the open stub. The input impedance Z_{in2} is given by

$$Z_{in2} = Z_o \frac{Z_L + jZ_o \tan(\beta l)}{Z_o + jZ_L \tan(\beta l)} \quad \text{for lossless line} \tag{7.13}$$

where Z_o and β are the characteristic impedance and phase constant of the open stub, respectively. If $Z_L = \infty$ or 0 with a very small value of $\tan(\beta l)$, the input impedance $Z_{in2} \to \infty$ or 0, respectively. Under these cases, the slow-wave periodic structure loaded by Z_{in2} in Figure 7.22*b* provides passband ($Z_{in2} \to \infty$) and stopband ($Z_{in2} \to 0$) characteristics. For example, the conventional capacitance-load Kuroda-identity periodic structure is the case of $Z_L = \infty$ with $l = \lambda_g/8$ [26].

Figure 7.23 shows lossless parallel and series resonant circuits. At resonance, the input impedance Z_{LC} of the parallel and series resonant circuits is

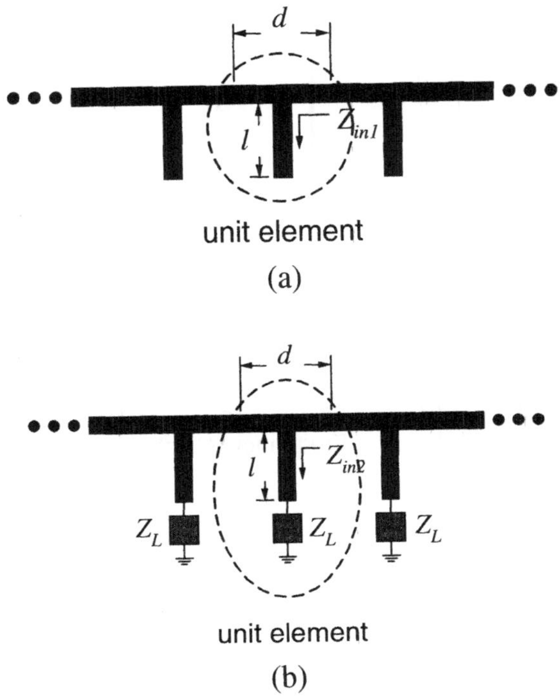

FIGURE 7.22 Slow-wave periodic structure (a) conventional type and (b) with loading Z_L at open end [16]. (Permission from IEEE.)

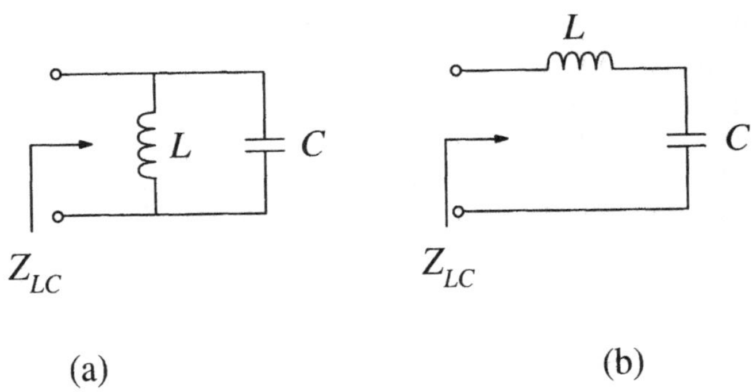

FIGURE 7.23 Lossless (a) parallel and (b) series resonant circuits.

∞ and 0, respectively. The input impedance Z_{LC} of the resonant circuits can act as the loading impedance Z_L in Figure 22*b* for the passband and stopband characteristics of a slow-wave periodic structure. In practice, for the high Q ring and hairpin resonators, the input impedance of the resonators shows very large and small values at parallel and series resonant frequencies, respectively. Thus, a slow-wave periodic structure loaded by ring or hairpin resonators with two series resonant frequencies close to a parallel resonant frequency [27] can be designed for a bandpass filter at fundamental mode.

The key point behind this new slow-wave filter topology is that both the series and the parallel resonances of the loading circuit are used to achieve bandpass characteristics. The approach can, in fact, be interpreted as using the stop bands of two series resonances in conjunction with the pass band of a parallel resonance to achieve a bandpass frequency response. It is noted, however, that in some cases, undesired pass bands below and above the main pass band may require a high pass or bandpass section to be used in conjunction with this approach.

Figure 7.24 shows a transmission line loaded by a square ring resonator with a line-to-ring coupling structure and its simple equivalent circuit, where Z_{in3} is

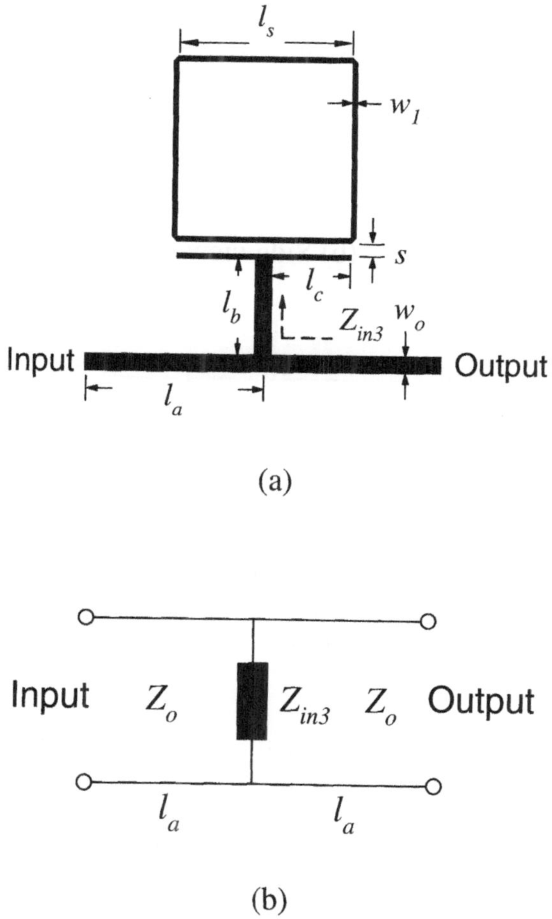

FIGURE 7.24 Slow-wave bandpass filter using one ring resonator with one coupling gap (a) layout and (b) simplified equivalent circuit [16]. (Permission from IEEE.)

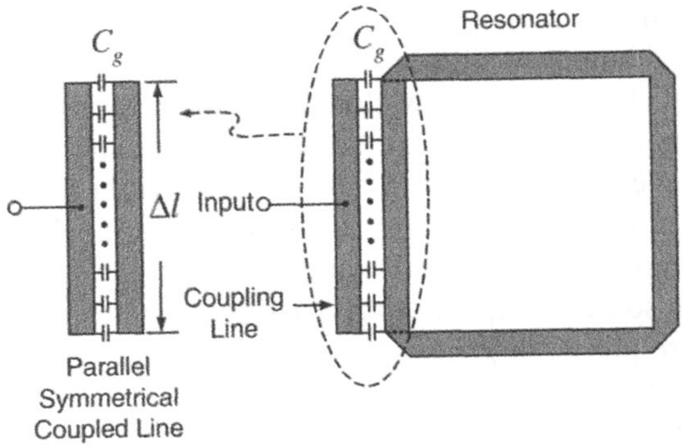

(a) Top view

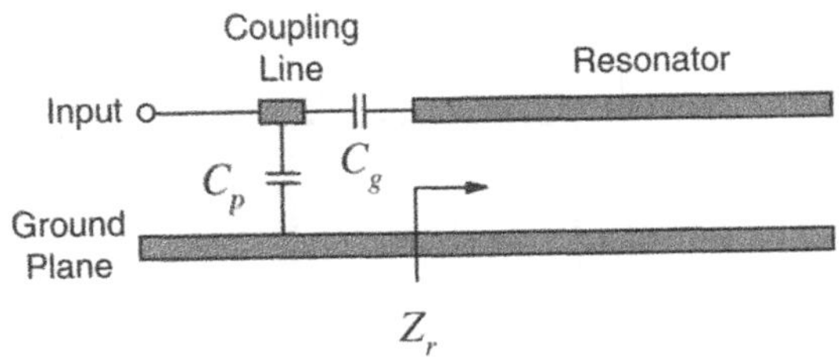

(b) Side view

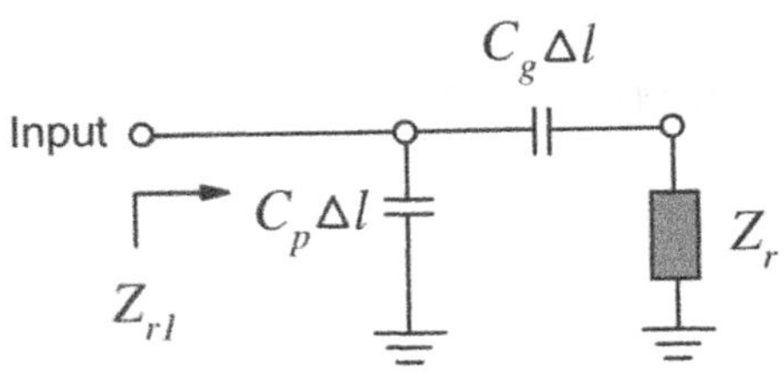

(c) Equivalent circuit

FIGURE 7.25 Line-to-ring coupling structure (a) top view, (b) side view, and (c) equivalent circuit [16]. (Permission from IEEE.)

the input impedance looking into the transmission line l_b toward the ring resonator with the line-to-ring coupling.

As seen in Figure 7.25*a*, the coupling structure includes the coupling line, one side of the square ring resonator and a coupling gap. This coupling structure can be treated as symmetrical coupled lines [28]. The coupling gap between the symmetrical coupled lines is modeled as a capacitive L-network as shown in Figure 7.25*b* [29]. C_g is the gap capacitance per unit length, and C_p is the capacitance per unit length between the strip and ground plane.

These capacitances, C_g and C_p, can be found from the even- and odd-mode capacitances of symmetrical coupled lines [30]. Figure 7.25*c* shows the equivalent circuit of the capacitive network, where the input impedance of the ring resonator Z_r can be obtained from [29]. The input impedance Z_{r1} looks into the line-to-ring coupling structure toward the ring resonator. The input impedance Z_{in3} is

$$Z_{in3} = \frac{Z_{r1} + jZ_o \tan(\beta l_b)}{Z_o + jZ_{r1} \tan(\beta l_b)} \tag{7.14}$$

where $Z_{r1} = (Z_r + Z_g) \parallel Z_p$, $Z_g = \dfrac{1}{j\omega C_g \Delta l}$, $Z_p = \dfrac{1}{j\omega C_p \Delta l}$, and ω is the angular frequency. The parallel (f_p) and series (f_s) resonances of the ring resonator can be obtained by setting

$$|Y_{in3}| = |1/Z_{in3}| \cong 0 \text{ and } |Z_{in3}| \cong 0. \tag{7.15}$$

The frequency response of the ring circuit can be calculated using the equivalent circuit in Figure 24*b*. The *ABCD* matrix of the ring circuit is

$$\begin{aligned}\begin{bmatrix} A & B \\ C & D \end{bmatrix} &= \begin{bmatrix} \cos(\beta l_a) & jZ_o \sin(\beta l_a) \\ jY_o \sin(\beta l_a) & \cos(\beta l_a) \end{bmatrix} \begin{bmatrix} 1 & 0 \\ Y_{in3} & 1 \end{bmatrix} \begin{bmatrix} \cos(\beta l_a) & jZ_o \sin(\beta l_a) \\ jY_o \sin(\beta l_a) & \cos(\beta l_a) \end{bmatrix} \\ &= \begin{bmatrix} 1 - 2\sin^2(\beta l_a) + jZ_o Y_{in3} \sin(\beta l_a)\cos(\beta l_a) & -Z_o^2 Y_{in3} \sin^2(\beta l_a) + j2Z_o \sin(\beta l_a)\cos(\beta l_a) \\ Y_{in3} \cos^2(\beta l_a) + j2Y_o \sin(\beta l_a)\cos(\beta l_a) & 1 - 2\sin^2(\beta l_a) + jZ_o Y_{in3} \sin(\beta l_a)\cos(\beta l_a) \end{bmatrix}\end{aligned} \tag{7.16}$$

where $Y_o = 1/Z_o$. Using $Y_{in3}(f_p)$ and $Z_{in3}(f_s)$ at resonances, the passband and stopband of the ring circuit can be obtained by calculating S_{11} and S_{21} from the *ABCD* matrix in Equation (7.16).

The ring circuit was designed at the center frequency of 2.4 GHz and fabricated on a RT/Duroid 6010.5 substrate with a thickness $h = 50$ mil and a relative dielectric constant $\varepsilon_r = 10.5$. The dimensions of the filter are $l_s = 12.07$ mm, $s = 0.2$ mm, $l_a = 12.376$ mm, $l_b = 6.5$ mm, $w_o = 1.158$ mm, and $w_l = 0.3$ mm. These parameter values are synthesized from the design equations using numerical optimization to construct a bandpass filter with attenuation poles centered at ±330 MHz about the parallel resonant frequency.

Figure 7.26*a* shows the calculated input impedance Z_{in3} with parallel and two series resonances of the ring resonator at different lengths of l_b. The parallel (f_p), lower (f_{sL}), and higher (f_{sH}) series resonances corresponding to the passband and stopband of the ring circuit in Figure 7.24 are denoted by Δ, ○, and □, respectively. By adjusting the length of l_b properly, the parallel resonance can be centered between two series resonances. Also, Figure 7.26*b* is an expanded view showing the series resonances. The measured and calculated

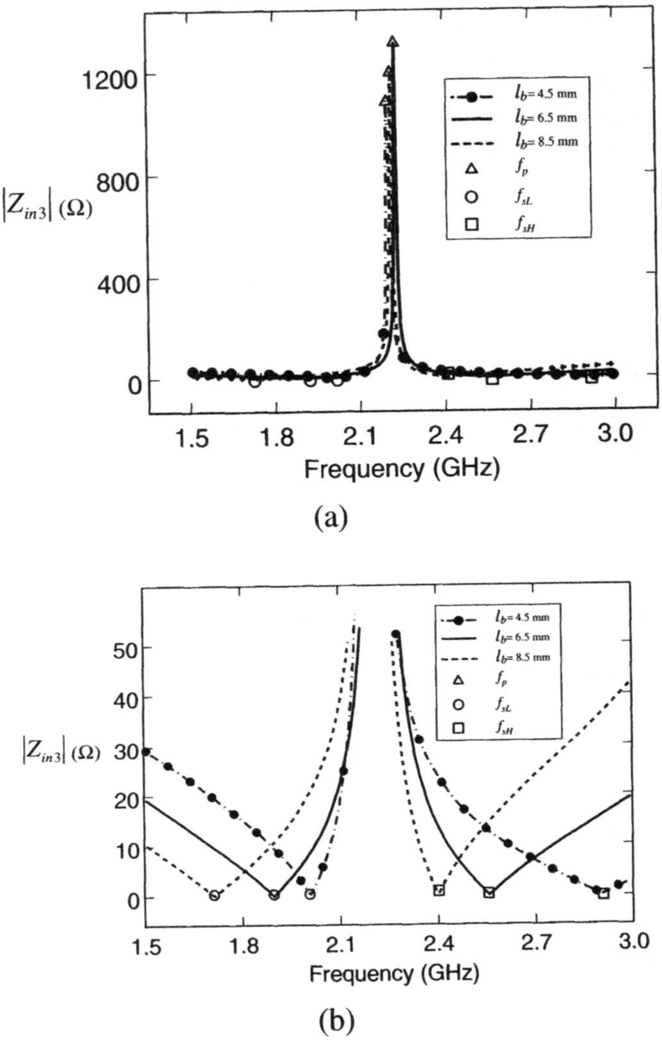

FIGURE 7.26 Variation in input impedance $|Z_{in3}|$ for different lengths of l_b showing (a) parallel and series resonances and (b) an expanded view for the series resonances [16]. (Permission from IEEE.)

frequency response of the ring circuit is illustrated in Figure 7.27. The filter has a fractional 3-dB bandwidth of 15.5%. The insertion and return losses are 0.53 dB and 25.7 dB at 2.3 GHz, respectively. Two attenuation poles are at 1.83 and 2.59 GHz with attenuation level of 35.2 and 31.3 dB, respectively. The measured unloaded Q of the closed-loop ring resonator is 122.

To improve the passband and rejection, a slow-wave bandpass filter using three ring resonators has also been built. As seen in Figure 7.28, the

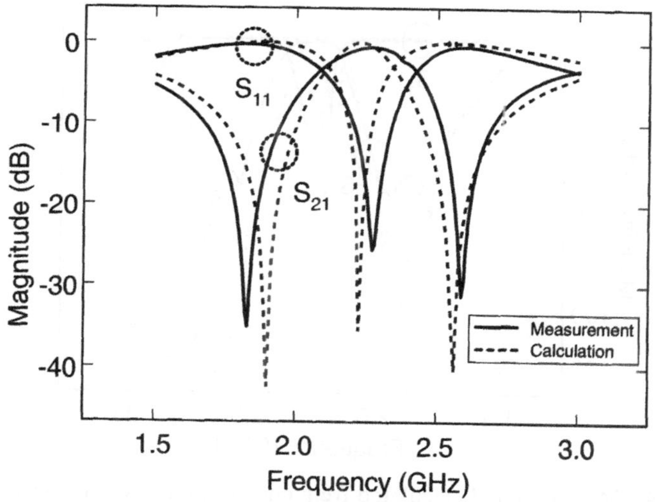

FIGURE 7.27 Measured and calculated frequency response for the slow-wave bandpass filter using one square ring resonator [16]. (Permission from IEEE.)

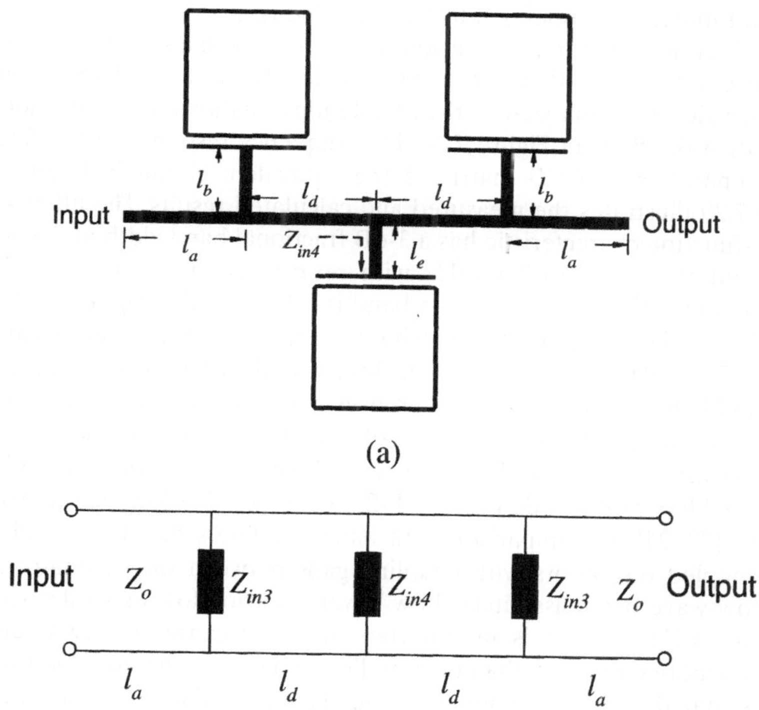

FIGURE 7.28 Slow-wave bandpass filter using three ring resonators (a) layout and (b) simplified equivalent circuit [16]. (Permission from IEEE.)

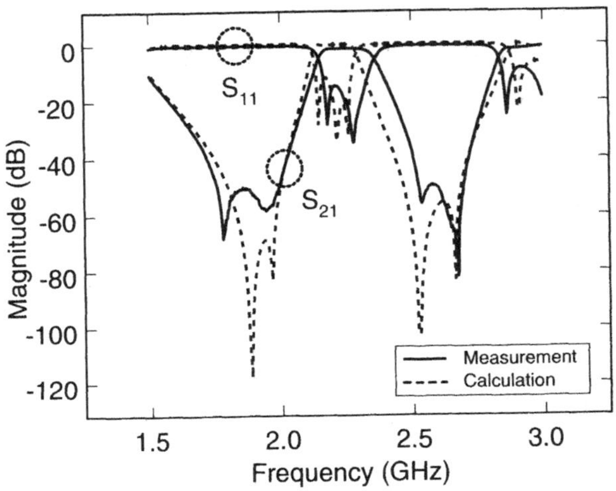

FIGURE 7.29 Measured and calculated frequency response for slow-wave bandpass filter using three square ring resonators [16]. (Permission from IEEE.)

transmission line is loaded periodically by three ring resonators, where Z_{in4} is the input impedance looking into l_e toward the ring.

The filter uses the same dimensions as the filter with a single ring resonator in Figure 7.24, but with the transmission lengths l_d = 15.686 mm and l_e = 5.5 mm, which are optimized by the calculation equations to obtain wider stop bands than the filter in Figure 7.24. The frequency response of the filter can be obtained from *ABCD* matrix of the equivalent circuit in Figure 7.28*b*. Figure 7.29 illustrates the measured and calculated results. The filter with an elliptic-function characteristic has a 3-dB fractional bandwidth of 8.5% and a pass band from 2.16 to 2.34 GHz with return loss better than 10 dB. The maximum insertion loss in the pass band is 1.45 dB with a ripple of ±0.09 dB. In addition, the two stop bands exhibit a rejection level larger than 50 dB within 1.76–2 GHz and 2.52–2.7 GHz. Observing the frequency response of the filters in Figures 27 and 29, the differences between the calculated and measured results are partly due to the use of a lossless calculation model.

The new slow-wave bandpass filters use higher order modes to build up a bandpass filter with a wide passband [25] or to provide lowpass or bandstop features [23, 24]. In comparison with bandpass filters that use parallel- and cross-coupled resonators with coupling gaps between the resonators, these new slow-wave bandpass filters show lower insertion loss at similar resonant frequencies [31–33]. This is an important finding because the new filter structure uses more conductor than the parallel- and cross-coupled structures. This implies that the new filter topology significantly reduces the insertion loss caused in parallel- and cross-coupled bandpass structures by eliminating coupling gaps between resonators.

7.6 RING BANDPASS FILTERS WITH TWO TRANSMISSION ZEROS

To achieve the high selectivity characteristic, Levy introduced filters using cross-coupled structure [34]. The cross coupling between nonadjacent resonators creates transmission zeros that improve the skirt rejection of the microstrip filters [35–37]. Figure 7.30 shows the configuration of the filter using two hairpin resonators with asymmetric feed lines tapping the resonators [38]. The input and output feed lines divide the resonators into two sections of l_1 and l_2. The total length of the resonator is $l = l_1 + l_2 = \lambda_g/2$, where λ_g is the guided-wavelength at fundamental resonance. The coupling between the two open ends of the resonators is simply expressed by the gap capacitance C_{s1} [36, 39].

Inspecting Figure 7.30, the whole circuit represents a shunt circuit, which consists of upper and lower sections. Each section is composed of l_1, l_2, and C_{s1}. The *ABCD* matrixes for the upper and lower sections of the lossless shunt circuit are

$$\begin{bmatrix} A & B \\ C & D \end{bmatrix}_{upper} = M_1 M_2 M_3 \tag{7.17a}$$

$$\begin{bmatrix} A & B \\ C & D \end{bmatrix}_{lower} = M_3 M_2 M_1 \tag{7.17b}$$

with $M_1 = \begin{bmatrix} \cos\beta l_1 & jZ_o \sin\beta l_1 \\ jY_o \sin\beta l_1 & \cos\beta l_1 \end{bmatrix}$, $M_2 = \begin{bmatrix} 1 & z_c \\ 0 & 1 \end{bmatrix}$, and

$M_3 = \begin{bmatrix} \cos\beta l_2 & jZ_o \sin\beta l_2 \\ jY_o \sin\beta l_2 & \cos\beta l_2 \end{bmatrix}$, where β is the propagation constant, $z_c = 1/j\omega C_{s1}$ is the impedance of the gap capacitance C_{s1}, ω is the angular frequency, and Z_o

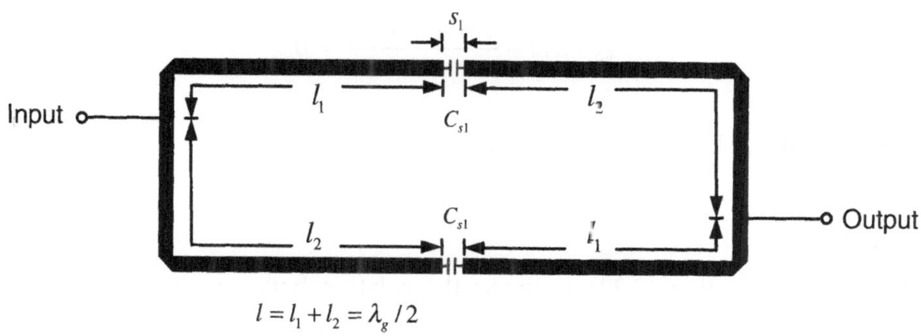

FIGURE 7.30 Configuration of the filter using two hairpin resonators with asymmetric tapping feed lines [38]. (Permission from IEEE.)

$= 1/Y_o$ is the characteristic impedance of the resonator. The Y parameters of the upper and lower sections are obtained from (7.17a) and (7.17b) and given by

$$\begin{bmatrix} Y_{11} & Y_{12} \\ Y_{21} & Y_{22} \end{bmatrix}_j = \begin{bmatrix} D_j/B_j & (B_jC_j - A_jD_j)/B_j \\ -1/B_j & A_j/B_j \end{bmatrix} \tag{7.18}$$

where j = *upper* or *lower* is for upper or lower sections. In addition, the total Y parameter of the whole circuit is expressed as

$$\begin{bmatrix} Y_{11} & Y_{12} \\ Y_{21} & Y_{22} \end{bmatrix} = \begin{bmatrix} Y_{11} & Y_{12} \\ Y_{21} & Y_{22} \end{bmatrix}_{upper} + \begin{bmatrix} Y_{11} & Y_{12} \\ Y_{21} & Y_{22} \end{bmatrix}_{lower} \tag{7.19}$$

The insertion loss S_{21} of the circuit can then be calculated from the total Y-parameters and is expressed as

$$\begin{aligned} S_{21} &= \frac{-2Y_{21}Y_o}{(Y_{11}+Y_o)(Y_{22}+Y_o)-Y_{12}Y_{21}} \\ &= \frac{j4\left(Z_o \sin\beta l - \dfrac{\cos\beta l_1 \cos\beta l_2}{\omega C_{s1}}\right)Y_o}{\left[2\cos\beta l + \dfrac{Y_o \sin\beta l}{\omega C_{s1}} + j\left(Z_o \sin\beta l - \dfrac{\cos\beta l_1 \cos\beta l_2}{\omega C_{s1}}\right)Y_o\right]^2 - 4} \end{aligned} \tag{7.20}$$

Comparing Equations (7.17)–(7.20) with (12), (13), and (16) in [38], (12), (13), and (16) in [38] only present a special case of the two hairpin resonators with two asymmetric feed lines tapped at the center. Equations (7.17) to (7.20) given here are more general for the asymmetric feed lines tapped at arbitrary positions on the resonators. The transmission zeros can be found by letting $S_{21} = 0$, namely

$$z_o \sin\beta l - \frac{\cos\beta l_1 \cos\beta l_2}{\omega C_{s1}} = 0 \tag{7.21}$$

For a small C_{s1}, Equation (7.21) can be approximated as

$$\cos\beta l_1 \cos\beta l_2 \cong 0 \tag{7.22}$$

Inspecting Equation (7.22), it shows the relation between the transmission zeros and the tapping positions. Substituting $\beta = \dfrac{2\pi f\sqrt{\varepsilon_{eff}}}{c}$ into Equation (7.22), the transmission zeros corresponding to the tapping positions are

$$f_1 = \frac{nc}{4l_1\sqrt{\varepsilon_{eff}}} \quad \text{and} \quad f_2 = \frac{nc}{4l_2\sqrt{\varepsilon_{eff}}} \qquad \text{n} = 1, 3, 5 \ldots \tag{7.23}$$

where f is the frequency, ε_{eff} is the effective dielectric constant, n is the mode number, c is the speed of light in free space, and f_1 and f_2 are the frequencies of the two transmission zeros corresponding to the tapping positions of the lengths of l_1 and l_2 on the resonators. At the transmission zeros, $S_{21} = 0$ and there is maximum rejection.

Figure 7.31 shows the measured results for different tapping positions on the hairpin resonators in Figure 7.30. The filter was designed at the fundamental frequency of 2 GHz and fabricated on a RT/Duroid 6010.2 substrate with a thickness $h = 25$ mil and a relative dielectric constant $\varepsilon_r = 10.2$. Table 7.3 shows the measured and the calculated results for the transmission zeros

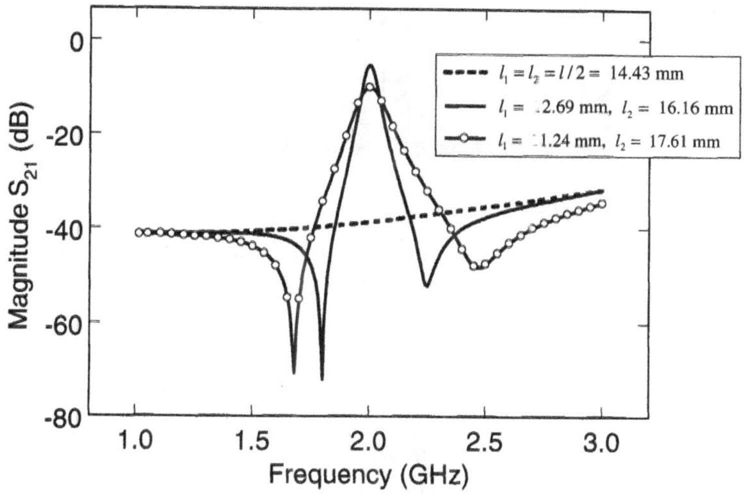

FIGURE 7.31 Measured results for different tapping positions with coupling gap $s_1 = 0.35$ mm [38]. (Permission from IEEE.)

TABLE 7.3 Measured and calculated results of the hairpin resonators for different tapping positions [38]. (Permission from IEEE.)

	Measurements	Calculations
$l_1 = l_2 = l/2 = 14.43$ mm	No passband at 2 GHz	$f_1 = f_2 = 2$ GHz
$l_1 = 12.69$ mm, $l_2 = 16.16$ mm	$f_1 = 1.8$ GHz, $f_2 = 2.25$ GHz	$f_1 = 1.79$ GHz, $f_2 = 2.27$ GHz
$l_1 = 11.24$ mm, $l_2 = 17.61$ mm	$f_1 = 1.68$ GHz, $f_2 = 2.48$ GHz	$f_1 = 1.64$ GHz, $f_2 = 2.57$ GHz

corresponding to the different tapping positions. Inspecting the results, the measurements agree well with the calculations.

Figure 7.32 shows the filter using two open-loop ring resonators [38]. This type resonator with two folded arms is more compact than the filter in Figure 7.30. This filter has the same dimensions as the filter in Figure 7.30, except for the two additional 45-degree chamfered bends and the coupling gap g = 0.5 mm between the two open ends of the ring.

Figure 7.33 shows the measured results for the different tapping positions on the rings. The measured locations of the transmission zeros are listed in Table 7.4. Comparing with Table 7.3, the locations of the transmission zeros of

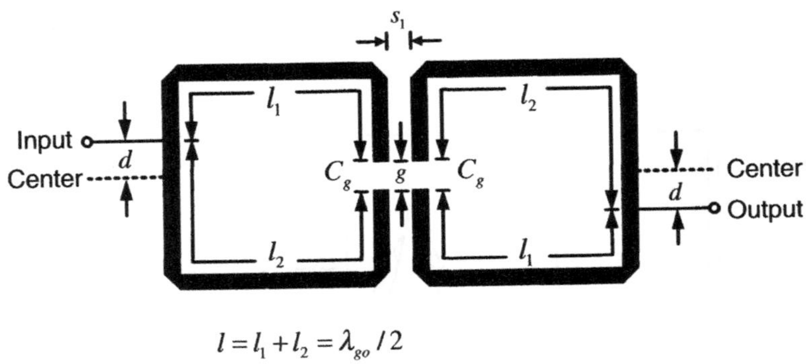

FIGURE 7.32 Layout of the filter using two open-loop ring resonators with asymmetric tapping feed lines [38]. (Permission from IEEE.)

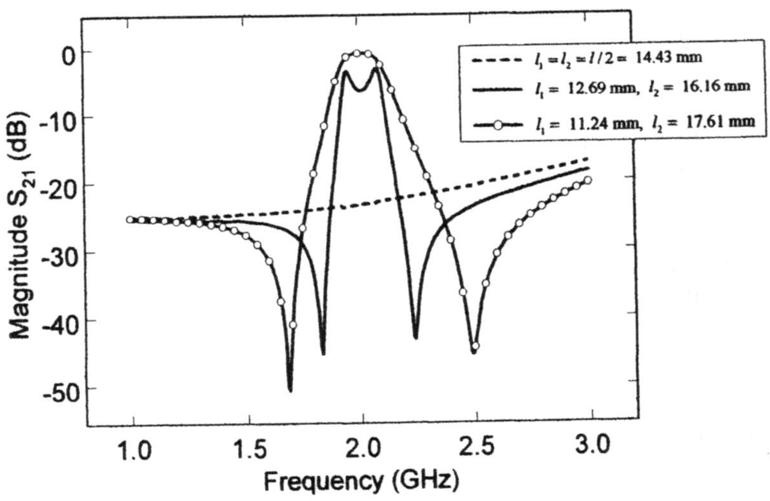

FIGURE 7.33 Measured results for different tapping positions with coupling gap $s_1 = 0.35$ mm [38]. (Permission from IEEE.)

TABLE 7.4 Measured Results of the Open-loop Ring Resonators for Different Tapping Positions [38]. (Permission from IEEE.)

	Measurements
$l_1 = l_2 = l/2 = 14.43$ mm	No passband at 2 GHz
$l_1 = 12.69$ mm, $l_2 = 16.16$ mm	$f_1 = 1.83$ GHz, $f_2 = 2.24$ GHz
$l_1 = 11.24$ mm, $l_2 = 17.61$ mm	$f_1 = 1.69$ GHz, $f_2 = 2.5$ GHz

the filters using open-loop rings are very close to those of the filters using hairpin resonators. This implies that the coupling effects between the two rings and the effects of two additional 45-degree chamfered bends only slightly affect the locations of the two transmission zeros. Thus, Equation (7.23) can also be used to predict the locations of the transmission zeros of the filters using open-loop rings.

Observing the measured results in Figures 7.31 and 7.33, the tapping positions also affect the couplings between two resonators. The case of $l_1 = 12.69$ mm and $l_2 = 16.16$ mm in Figure 7.33 shows an overcoupled condition [6, 9], which has a hump within the passband. The overcoupled condition is given by

$$K > \frac{1}{Q_o} + \frac{1}{Q_e} \tag{7.24}$$

where K is the coupling coefficient, Q_o is the unloaded Q of either of the two resonators, and Q_e is the external Q. The coupling condition of the filter can be found using the measured K, Q_o, and Q_e. The measured K can be calculated from Equation (7.3). The measured external Q is given by [40]

$$Q_e = \frac{f_o}{\Delta f_{\pm 90^\circ}} \tag{7.25}$$

where $\Delta f_{\pm 90^\circ}$ is the bandwidth about the resonant frequency, over which the phase varies from -90° to $+90^\circ$.

Figure 7.32 shows the tapping positions at a distance d from the center of the resonators to the input and output ports. When d becomes shorter or the tapping position moves toward the center, the external Q becomes larger [41]. The larger external Q allows the filter to approach the overcoupled condition in Equation (7.24), causing a hump within the passband. In addition, observing Equations (7.23) and (7.24), for a shorter d, the two transmission zeros appear close to the passband, providing a high selectivity nearby the passband. But this may easily induce an overcoupled condition. Beyond the coupling effects caused by the tapping positions, the coupling gap s_1 also influences the couplings between two resonators [31]. Therefore, to avoid overcoupling, the proper tapping positions and gap size should be carefully chosen.

Figure 7.34 shows the measured results of the filter for the case of l_1 = 11.24mm and l_2 = 17.61mm. This filter with $K = 0.02 < 1/Q_o + 1/Q_e = 1/130 + 1/15.4$ shows an undercoupled condition [6, 9], which does not have a hump in the passband. The filter has an insertion loss of 0.95dB at 2.02GHz, a return loss of greater than 20dB from 1.98 to 2.06GHz, and two transmission zeros at 1.69GHz with –50.7-dB rejection and 2.5GHz with –45.5-dB rejection, respectively. The 3-dB fractional bandwidth of the filter is 10.4%. Comparing with the insertion losses of the cross-coupling filters at similar fundamental resonant frequencies (2.2dB in [31] and 2.8dB in [36]), the filter in Figure 7.34 has a lower insertion loss of 0.95dB.

The filter using cascaded resonators is shown in Figure 7.35. The filter uses

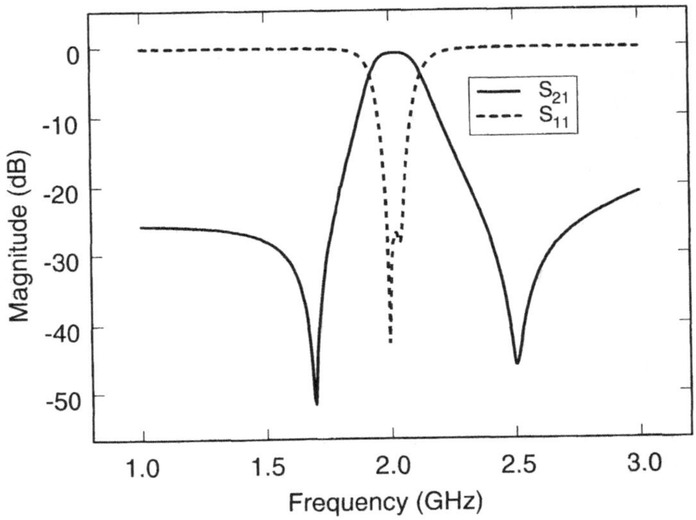

FIGURE 7.34 Measured results of the open-loop ring resonators for the case of tapping positions of l_1 = 11.24mm and l_2 = 17.61mm [38]. (Permission from IEEE.)

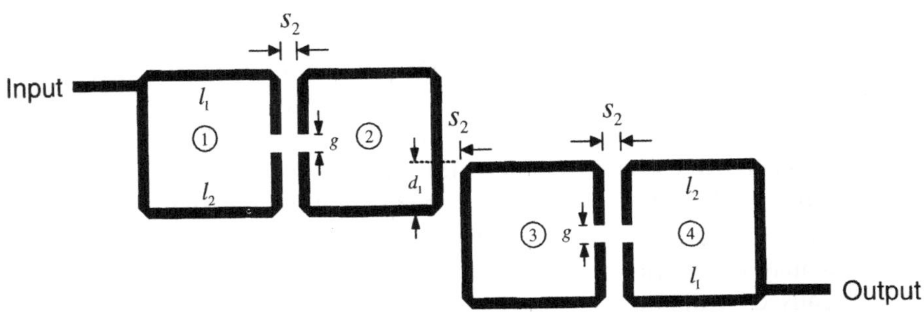

FIGURE 7.35 Configuration of the filter using four cascaded open-loop ring resonators [38]. (Permission from IEEE.)

the same dimensions as the open-loop ring in Figure 7.32 with the tapping positions of $l_1 = 11.24$ mm and $l_2 = 17.61$ mm at the first and last resonators. Also, the offset distance d_1 between the rings 2 and 3 is designed for asymmetric feeding between rings 1, 2 and rings 3, 4 to maintain the sharp cutoff frequency response. Therefore, the positions of the two transmission zeros of the filter can be predicted at around 1.69 and 2.5 GHz, respectively. The coupling gap size between rings is s_2. The coupling gap $s_2 = 0.5$ mm and the offset distance $d_1 = 2.88$ mm are optimized by EM simulation [8] to avoid the over-coupled condition.

The measured external Q and the mutual coupling K can be calculated from Equations (7.3), and they are

$$Q_e = 15.4 \text{ and } K = \begin{bmatrix} K_{11} & K_{12} & K_{13} & K_{14} \\ K_{21} & K_{22} & K_{23} & K_{24} \\ K_{31} & K_{32} & K_{33} & K_{34} \\ K_{41} & K_{42} & K_{43} & K_{44} \end{bmatrix} = \begin{bmatrix} 0 & -0.037 & 0 & 0 \\ -0.037 & 0 & 0.035 & 0 \\ 0 & 0.035 & 0 & -0.037 \\ 0 & 0 & -0.037 & 0 \end{bmatrix}$$

where $K_{i,j} = \left(\dfrac{f_{p2}^2 - f_{p1}^2}{f_{p2}^2 + f_{p1}^2} \right)_{i,j}$ is the mutual coupling between ith ring and jth ring, $(f_{p2})_{i,j}$ and $(f_{p1})_{i,j}$ are the resonant frequencies of ith ring and jth ring, and the negative sign in coupling matrix is for electrical coupling [32]. Figure 7.36 shows the simulated and measured results. The filter has a fractional 3-dB bandwidth of 6.25%. The insertion loss is 2.75 dB at 2 GHz, and the return loss

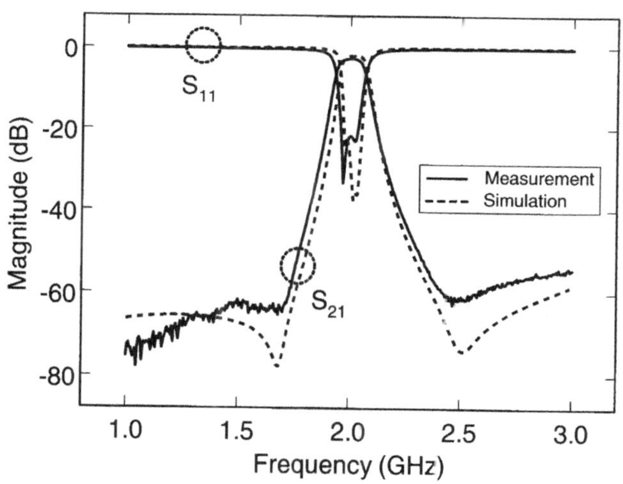

FIGURE 7.36 Measured and simulated results of the filter using four cascaded open-loop ring resonators [38]. (Permission from IEEE.)

is greater than 13.5 dB within 1.95–2.05 GHz. The out-of-band rejection is better than 50 dB extended to 1 and 3 GHz and beyond.

7.7 PIEZOELECTRIC TRANSDUCER–TUNED BANDPASS FILTERS

Electronically tunable filters have many applications in transmitters and receivers. As shown in Figure 7.37, the tunable filter circuit consists of the filter using cascaded resonators, a piezoelectric transducer (PET), and an attached dielectric perturber above the filter [42]. As described in Chapter 4, Section 4.9, the PET moves the perturber and varies the effective dielectric constant of the filter, allowing the passband of the filter to shift toward the higher or lower frequencies. Figure 7.38 shows the measured results for the tuning range of the passband. With the maximum applied voltage of 90 V and a perturber of dielectric constant $\varepsilon_r = 10.8$ and thickness $h = 50$ mil, the tuning range of the

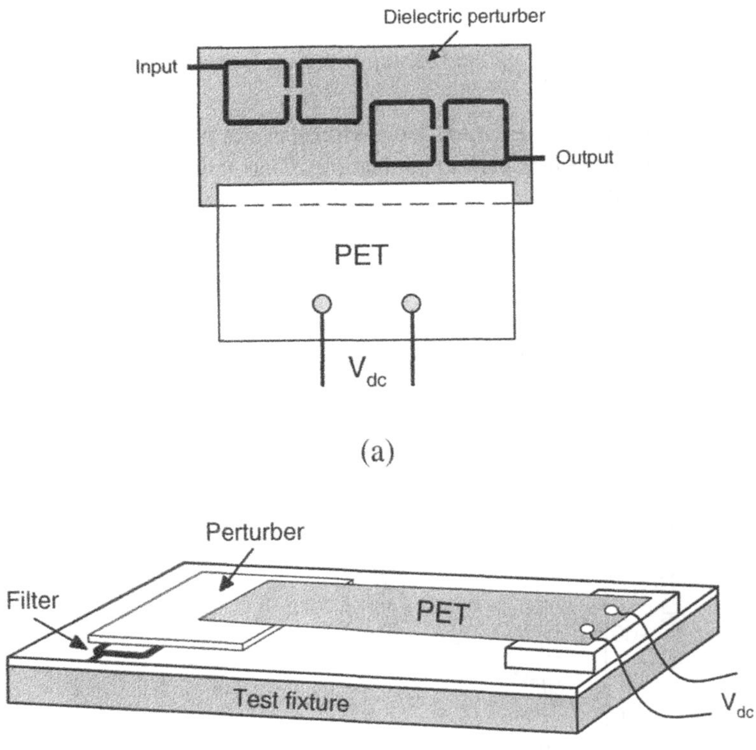

(b)

FIGURE 7.37 Configuration of the tunable bandpass filter (a) top view and (b) 3D view [38]. (Permission from IEEE.)

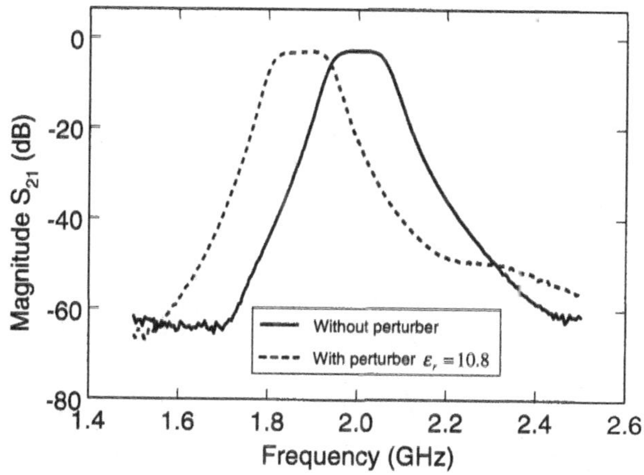

FIGURE 7.38 Measured results of the tunable bandpass filter with a perturber of $\varepsilon_r = 10.8$ and $h = 50$ mil [38]. (Permission from IEEE.)

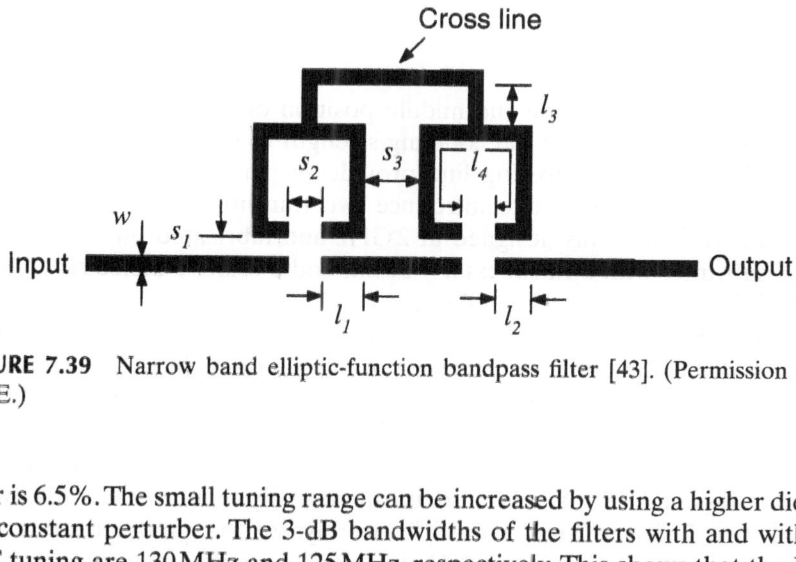

FIGURE 7.39 Narrow band elliptic-function bandpass filter [43]. (Permission from IEEE.)

filter is 6.5%. The small tuning range can be increased by using a higher dielectric constant perturber. The 3-dB bandwidths of the filters with and without PET tuning are 130 MHz and 125 MHz, respectively. This shows that the PET tuning has little effect on bandwidth. The size of the PET is 70 mm × 32 mm × 0.635 mm. The overall size of the filter including the perturber and PET is 90 mm × 50 mm × 3.85 mm.

7.8 NARROW BAND ELLIPTIC-FUNCTION BANDPASS FILTERS

The narrow band elliptic-function bandpass filter is shown in Figure 7.39 [43]. The filter is constructed by two identical open-loop ring resonators, coupled

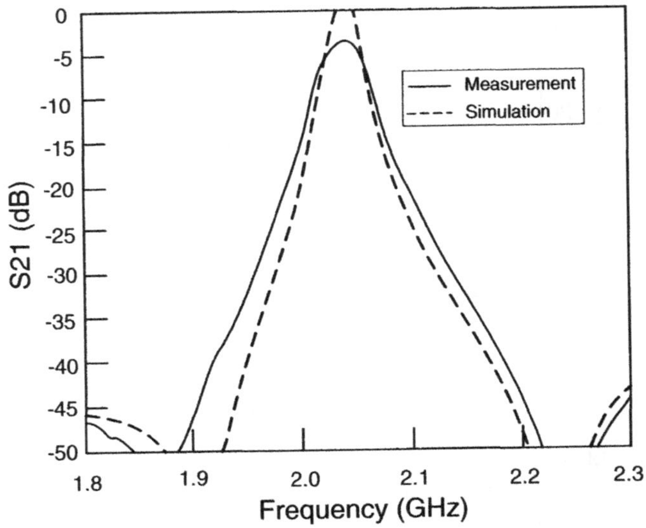

FIGURE 7.40 Simulated and measured results of the filter [43]. (Permission from IEEE.)

lines, and a crossing line at the middle position of the two resonators. The coupled lines can enhance the coupling strength to reduce the insertion loss of the filter. Also, the crossing line provides a perturbation at the current maximum of the resonator to introduce two transmission zeros next to the passband. The filter was designed at 2 GHz and fabricated on a RT/Duriod 6010.5 substrate with a thickness $h = 50$ mil and a relative dielectric constant $\varepsilon_r = 10.5$. The dimensions of the filter are $w = 1.145$ mm, $s_1 = 0.15$ mm, $s_2 = 3.435$ mm, $s_3 = 4.58$ mm, $l_1 = 3.29$ mm, $l_2 = 2.9$ mm, $l_3 = 3.435$ mm, and $l_4 = 27.61$ mm.

The simulated and measured results of the filter are shown in Figure 7.40. Two deep transmission zeros located in the stopband can suppress adjacent channel interferences. The filter has a 3-dB bandwidth of 1.96% at the frequency of 2.039 GHz. The size of the filter is 2.5 cm × 1.5 cm. Although the insertion loss of 3.7 dB is measured, it can be easily reduced to 2.6 dB by just placing two 2-mm × 2-mm dielectric overlays of the same substrate over interstage coupling gaps. Figure 7.41 shows the measured results for with and without dielectric overlays. In Figure 13, the 3-dB bandwidth is increased slightly from 1.96% to 2.21% by overlays. Also, the insertion loss has been improved.

7.9 SLOTLINE RING FILTERS

As mentioned earlier, the resonant modes with odd mode numbers cannot exist in the asymmetrically coupled microstrip ring structure. However, by

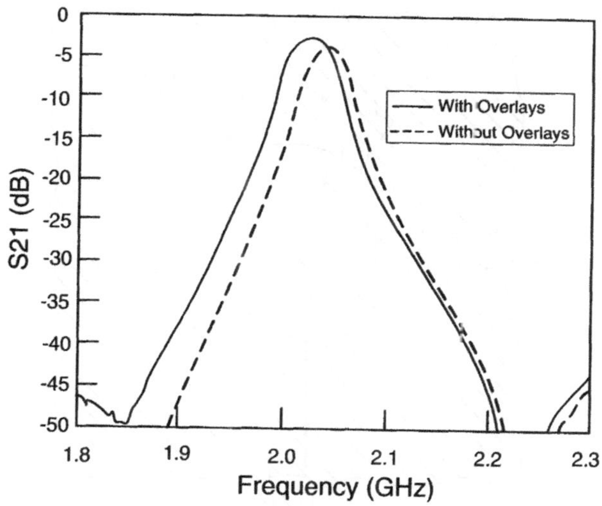

FIGURE 7.41 Measured results for the filter with and without dielectric overlays [43]. (Permission from IEEE.)

applying a perturbation at 45° or 135°, the dual resonant mode can be excited. The same dual-mode characteristic can also be found in the slotline ring structure with the perturbation of backside microstrip tuning stubs [44, 45].

By using microstrip tuning stubs on the backside of the slotline ring at 45° and 135°, the dual resonant mode can be excited. Figure 7.42 shows the physical configuration of the slotline ring dual-mode filter. Figure 7.43 shows the measured frequency responses of insertion loss and return loss for the slotline ring dual-mode filter with mode number $n = 3$. The test circuit was built on a RT/Duroid 6010.5 substrate with the following dimensions: substrate thickness $h = 0.635\,\text{mm}$, characteristic impedance of the input/output microstrip feed lines $Z_{m0} = 50\,\Omega$, input/output microstrip feed lines line width $W_{m0} = 0.57\,\text{mm}$, characteristic impedance of the slotline ring $Z_s = 70.7\,\Omega$, slotline ring line width $W_S = 0.2\,\text{mm}$, and slotline ring mean radius $r = 18.21\,\text{mm}$. The S-parameters were measured using standard SMA connectors with an HP-8510 network analyzer.

The slotline ring dual-mode filter was obtained with a bandwidth of 7.4%, a stopband attenuation of more than 40 dB, a mode purity of 1.86 GHz around the center frequency, 3.657 GHz, and a sharp gain slope transition, Compared with the microstrip ring dual-mode filter, which was published in [11], the slotline ring dual-mode filter has better in-band and out-band performance. Also, the slotline ring dual-mode filter has the advantages of flexible tuning and ease of adding series and shunt components.

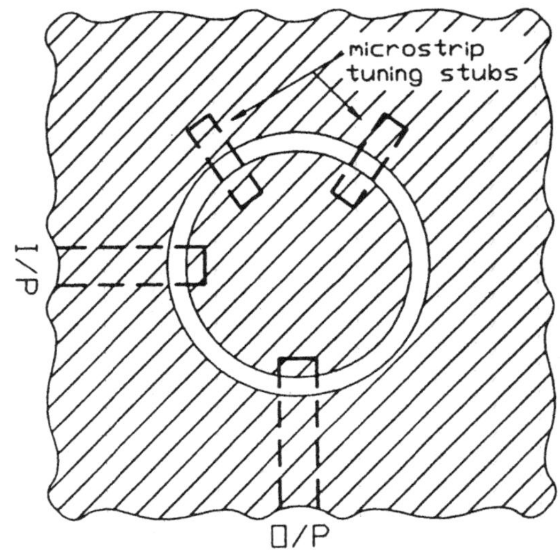

FIGURE 7.42 Physical configuration of the slotline ring dual-mode bandpass filter. [45]. (Permission from IEEE.)

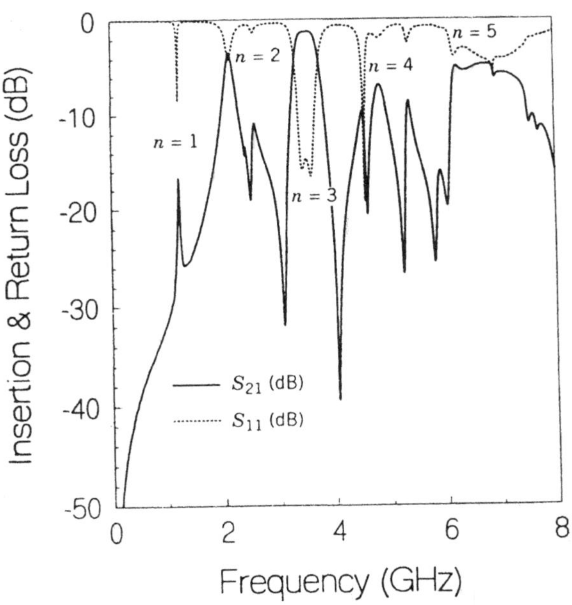

FIGURE 7.43 Measured frequency responses of insertion loss and return loss for the slotline ring dual-mode filter with backside microstrip tuning stubs at 45° and 135° [45]. (Permission from IEEE.)

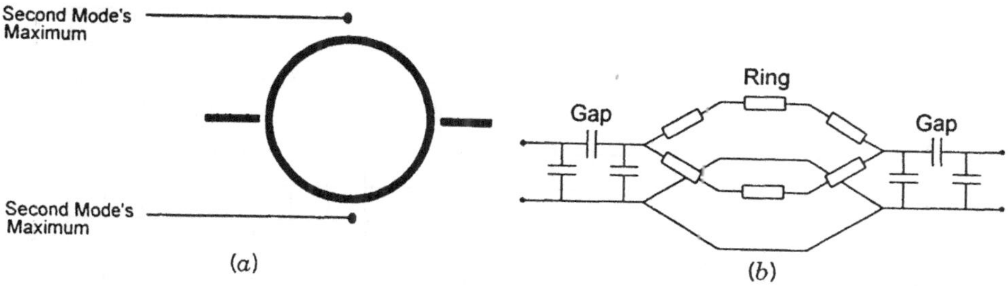

FIGURE 7.44 Normal microstrip ring resonator topology: (*a*) circuit layout, and (*b*) transmission-line model [46]. (Permission from *Electronics Letters.*)

7.10 MODE SUPPRESSION

The utility of ring resonators as filters or tunable resonators can be limited by their rejection bandwidth, which is determined by the occurrence of multiple modes. Suppression of the neighboring modes could improve the rejection bandwidth [46–48]. One method for mode suppression is the incorporation of a stepped impedance low-pass filter directly into the ring resonator [46].

Figure 7.44 shows a normal ring resonator and its transmission-line equivalent circuit. Certain frequencies of the traveling waves can be attenuated with the use of filters placed before or after the ring resonator. However, the filters can be easily incorporated into the transmission lines of the ring resonator to attenuate certain frequencies traveling through the ring. The filters must be carefully placed at an unwanted mode's maximums so as to affect it. Other modes are undisturbed if the filters are at their minima points. An example of incorporating a filter into a ring resonator is shown below. The desirable mode was the ring resonator's fundamental, while the second mode was designed to be suppressed.

A 50-Ω microstrip ring resonator was designed to have a fundamental resonance at 1.25 GHz on 0.635-mm Duroid substrate ($\varepsilon_r = 10.6$). Figure 7.45 shows the computer-aided design (CAD) package's simulation of the lightly coupled ring using the transmission-line model shown in Figure 7.44. We wish to suppress, without increasing the circuit size, the undesirable second mode that appears at about 2.5 GHz. For this purpose, a three-pole stepped-impedance low-pass filter (LPF) with a cutoff frequency of 2 GHz was designed using microstrip transmission lines. The filter cutoff was placed far enough above the first resonance so as not to affect its traveling waves while still attenuating the second mode by 7 dB. Figure 7.45 shows the three-pole filter's theoretical response across the modes of the ring resonator. A three-pole filter was used because it needed to be compact enough not to disturb the fundamental mode's maximums that occur at the ring's gaps. Stepped impedance microstrip filters mimic capacitive and inductive filter elements with wide, low-impedance and thin, high-impedance microstrip lines. The

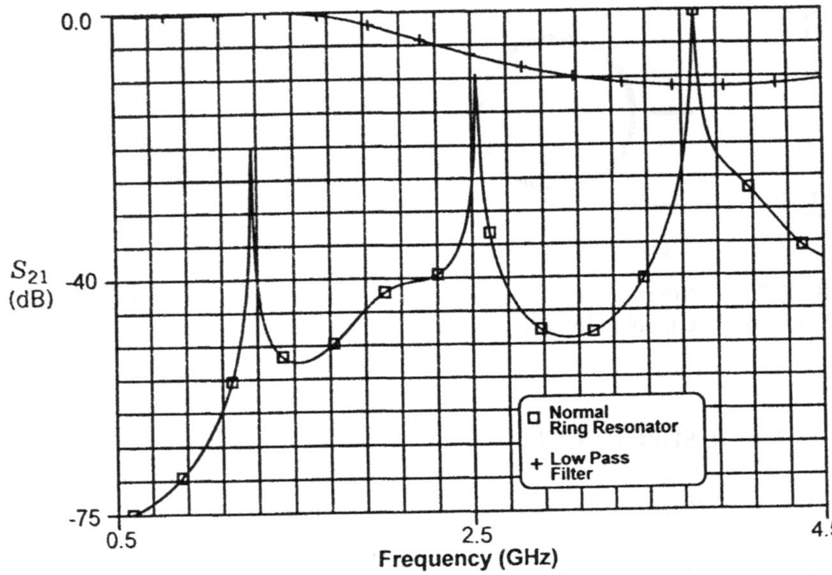

FIGURE 7.45 Normal ring resonator and stepped-impedance filter responses [46]. (Permission from *Electronics Letters.*)

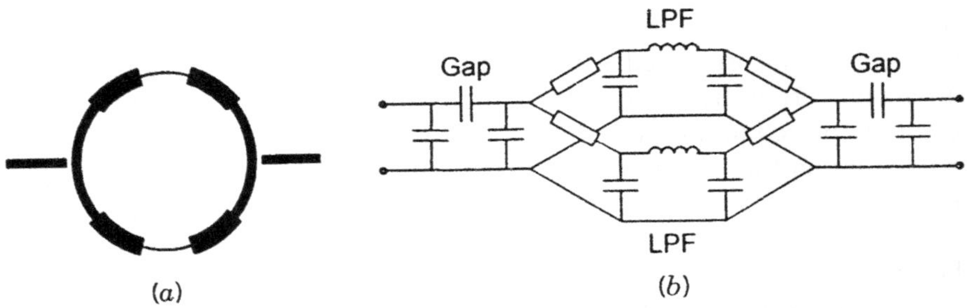

FIGURE 7.46 Microstrip mode-suppression ring resonator topology: (*a*) circuit layout, and (*b*) transmission-line model [46]. (Permission from *Electronics Letters.*)

impedances for the microstrip lines were 30 Ω and 100 Ω for the capacitive and inductive elements, respectively. Two out of the three filter elements were selected to be wide, capacitive lines because they have less conductive loss than thin, inductive microstrip lines.

The LPF filter schematic and microstrip implementations can be seen in Figure 7.46. The LPF was placed at both maxima indicated in Figure 7.44a to assure proper suppression of the second mode. Figure 7.47 shows the CAD simulations and measured results of the ring resonator with the incorporation of the LPF. It can be seen that the CAD simulations of the topology in Figure 7.46b predict the measured results very well. The second mode was completely

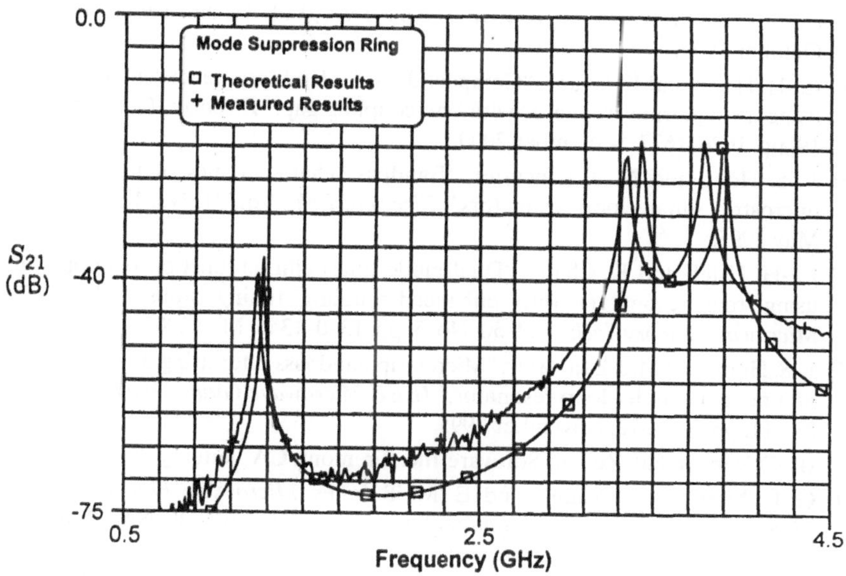

FIGURE 7.47 Theoretical and measured mode-suppression ring resonator responses [46]. (Permission from *Electronics Letters.*)

suppressed by the LPF, with additional losses in the fundamental frequency. Notice that the second mode was not just attenuated by the LPF but completely suppressed. This occurred because of the placement of the LPF at the affected mode's maxima, which disrupted the resonance. The fundamental losses are thought to be due to mismatching and conduction losses associated with the inductive LPF element. The third mode was affected in two ways, both of which were modeled accurately by the transmission-line model. First, the third mode was split due to the LPF discontinuities. A similar split was observed for a notch discontinuity [49]. Secondly, the LPF attenuated the split third mode by more than 12 dB. The third mode was not completely suppressed because the LPF was not placed at the third mode's maxima. However, the third-mode resonance was significantly attenuated by the LPF.

REFERENCES

[1] I. Wolff, "Microstrip bandpass filter using degenerate modes of a microstrip ring resonator," *Electron. Lett.*, Vol. 8, No. 12, pp. 302–303, June 1972.

[2] U. Karacaoglu, I. D. Robertson, and M. Guglielmi, "A dual-mode microstrip ring resonator filter with active devices for loss compensation," in *1993 IEEE Int. Microwave Symp. Dig.*, pp. 189–192, 1993.

[3] L. Zhu and K. Wu, "A joint field/circuit model of line-to-ring coupling structures and its application to the design of microstrip dual-mode filters and ring resonator

circuits," *IEEE Trans. Microwave Theory Tech.*, Vol. 47, No. 10, pp. 1938–1948, October 1999.

[4] L. -H. Hsieh and K. Chang, "Compact dual-mode elliptic-function bandpass filter using a single ring resonator with one coupling gap," *Electron. Lett.*, Vol. 36, No. 19, pp. 1626–1627, September 2000.

[5] J. -S. Hong and M. J. Lancaster, "Bandpass characteristics of new dual-mode microstrip square loop resonators," *Electron. Lett.*, Vol. 31, No. 11, pp. 891–892, May 1995.

[6] L. -H. Hsieh and K. Chang, "Dual-mode quasi-elliptic-function bandpass filters using ring resonators with enhanced-coupling tuning stubs," *IEEE Trans. Microwave Theory Tech.*, Vol. 50, No. 5, pp. 1340–1345, May 2002.

[7] J. -S. Hong and M. J. Lancaster, "Microstrip bandpass filter using degenerate mode of a novel meander loop resonator," *IEEE Microwave Guided Wave Lett.*, Vol. 5, No. 11, pp. 371–372, November 1995.

[8] IE3D Version 8.0, Zeland Software Inc., Fremont, CA, January 2001.

[9] G. L. Matthaei, L. Young, and E. M. T. Jones, *Microwave Filters, Impedance-Matching Networks, and Coupling Structures*, McGraw-Hill, New York, 1980, Chaps. 10 and 11.

[10] F. Rouchaud, V. Madrangeas, M. Aubourg, P. Guillon, B. Theron, and M. Maignan, "New classes of microstrip resonators for HTS microwave filters applications," *IEEE MTT-S Int. Microwave Symp. Dig.*, pp. 1023–1026, 1998.

[11] M. Gulielmi and G. Gatti, "Experimental investigation of dual-mode microstrip ring resonator," *Proc. 20th Eur. Microwave Conf.*, pp. 901–906, September 1990.

[12] L. -H. Hsieh and K. Chang, "Compact, low insertion loss, sharp rejection and wide-band microstrip bandpass filters," *IEEE Trans. Microwave Theory Tech.*, Vol. 51, No. 4, pp. 1241–1246, April 2003.

[13] D. K. Paul, P. Gardner, and K. P. Tan, "Suppression of even modes in microstrip ring resonators," *Electron. Lett.*, Vol. 30, No. 30, pp. 1772–1774, October 1994.

[14] M. Kirschning, R. H. Jansen, and N. H. L. Koster, "Measurement and computer-aided modeling of microstrip discontinuities by an improved resonator method," *IEEE MTT-S Int. Microwave Symp. Dig.*, pp. 495–497, 1983.

[15] E. Hammerstad, "Computer-aided design for microstrip couplers with accurate discontinuity models," *IEEE MTT-S Int. Microwave Symp. Dig.*, pp. 54–56, 1981.

[16] L. -H. Hsieh and K. Chang, "Slow-wave bandpass filters using ring or stepped-impedance hairpin resonators," *IEEE Trans. Microwave Theory Tech.*, Vol. 50, No. 7, pp. 1795–1800, July 2002.

[17] K. C. Gupta, R. Garg, I. Bahl, and P. Bhartia, *Microstrip Lines and Slotlines*, 2nd ed., Artech House, Boston, MA, 1996, p. 181.

[18] B. C. Wadell, *Transmission Line Design Handbook*, Artech House, Boston, MA, 1991, p. 321.

[19] L. Zhu, H. Bu, and K. Wu, "Broadband and compact multi-pole microstrip bandpass filters using ground plane aperture technique," *IEE Proceedings, Microwaves, Antennas and Propagation*, Vol. 149, No. 1, pp. 71–77, February 2002.

[20] J. -S. Park, J. -S. Yun, and D. Ahn, "A design of the novel coupled-line bandpass filter using defected ground structure with wide stopband performance," *IEEE Trans. Microwave Theory Tech.*, Vol. 50, No. 9, pp. 2037–2043, September 2002.

[21] J. J. Yu, S. T. Chew, M. S. Leong, and B. L. Ooi, "New class of microstrip miniaturized filter using triangular stub," *Electron. Lett.*, Vol. 37, No. 37, pp. 1169–1170, September 2001.

[22] J. -R. Lee, J. -H. Cho, and S. -W. Yun, "New compact bandpass filter using microstrip $\lambda/4$ resonators with open stub inverter," *IEEE Wireless Compon. Lett.*, Vol. 10, No. 12, pp. 526–527, December 2000.

[23] F. R. Yang, K. P. Ma, Y. Qian, and T. Itoh, "A uniplanar compact photonic-bandgap (UC-PBG) structure and its applications for microwave circuits," *IEEE Trans. Microwave Theory Tech.*, Vol. 47, No. 8, pp. 1509–1514, August 1999.

[24] H. C. Bell, "Narrow bandstop filters," *IEEE Trans. Microwave Theory Tech.*, Vol. 39, No. 12, pp. 2188–2191, December 1991.

[25] J. P. Hsu, T. Anada, H. Tsubaki, Y. Migita, and K. Nagashima, "Synthesis of planar microwave band-pass filter based on foster-type network and normal mode expansion method," *IEEE MTT-S Int. Microwave Symp. Dig.*, pp. 1199–1202, 1992.

[26] D. M. Pozar, *Microwave Engineering*, 2nd ed., Wiley, New York, 1998, Chap. 8.

[27] M. Makimoto and S. Yamashita, *Microwave Resonators and Filters for Wireless Communication Theory, Design and Application*, Springer, Berlin, 2001, Chap. 4.

[28] G. Kumar and K. C. Gupta, "Broad-band microstrip antennas using additional resonators gap-coupled to the radiating edges," *IEEE Trans. Antenna Propagat.*, Vol. 32, No. 12, pp. 1375–1379, December 1984.

[29] C. -C. Yu and K. Chang, "Transmission-line analysis of a capacitively coupled microstrip-ring resonator," *IEEE Trans. Microwave Theory Tech.*, Vol. 45, No. 11, pp. 2018–2024, November 1997.

[30] R. Mongia, I. Bahl, and P. Bhartia, *RF and Microwave Coupled-Line Circuits*, Artech House, Boston, MA, 1999, Chap. 3.

[31] J. -S. Hong and M. J. Lancaster, "Couplings of microstrip square open-loop resonators for cross-coupled planar microwave filters," *IEEE Trans. Microwave Theory Tech.*, Vol. 44, No. 11, pp. 2099–2019, November 1996.

[32] M. Sagawa, K. Takahashi, and M. Makimoto, "Miniaturized hairpin resonator filters and their application to receiver front-end MIC's," *IEEE Trans. Microwave Theory Tech.*, Vol. 37, No. 12, pp. 1991–1997, December 1989.

[33] J. -T. Kuo, M. -J. Maa, and P. -H. Lu, "A microstrip elliptic function filter with compact miniaturized hairpin resonators," *IEEE Microwave Guided Wave Lett.*, Vol. 10, No. 3, pp. 94–95, March 2000.

[34] R. Levy, "Filters with single transmission zeros at real or imaginary frequencies," *IEEE Trans. Microwave Theory Tech.*, Vol. 24, No. 4, pp. 172–181, April 1976.

[35] K. T. Jokela, "Narrow-band stripline or microstrip filters with transmission zeros at real and imaginary frequencies," *IEEE Trans. Microwave Theory Tech.*, Vol. 28, No. 6, pp. 542–547, June 1980.

[36] S. -Y. Lee and C. -M. Tsai, "New cross-coupled filter design using improved hairpin resonators," *IEEE Trans. Microwave Theory Tech.*, Vol. 48, No. 12, pp. 2482–2490, December 2000.

[37] C. -M. Tsai, S. -Y. Lee, and C. -C. Tsai, "Hairpin filters with tunable transmission zeros," *IEEE MTT-S Int. Microwave Symp. Dig.*, pp. 2175–2178, 2001.

[38] L. H. Hsieh and K. Chang, "Tunable microstrip bandpass filters with two transmission zeros," *IEEE Trans. Microwave Theory Tech.*, Vol. 51, No. 2, pp. 520–525, Feburary 2003.

[39] K. C. Gupta, R. Garg, I. Bahl, and P. Bhartia, *Microstrip Lines and Slotlines*, 2nd ed., Artech House, Boston, MA, 1996, Chap. 3.

[40] R. S. Kwok and J. F. Liang, "Characterization of high-Q resonators for microwave-filter applications," *IEEE Trans. Microwave Theory Tech.*, Vol. 47, No. 1, pp. 111–114, January 1999.

[41] J. S. Wong, "Microsrip tapped-line filter design," *IEEE Trans. Microwave Theory Tech.*, Vol. 27, No. 1, pp. 44–50, January 1979.

[42] T. -Y. Yun and K. Chang, "A low loss time-delay phase shift controlled by piezoelectric transducer to perturb microstrip line," *IEEE Microwave and Guided Wave Lett.*, Vol. 10, No. 3, pp. 96–98, March 2000.

[43] C. -C. Yu and K. Chang, "Novel compact elliptic-function narrow-band bandpass filters using microstrip open-loop resonators with coupled and crossing lines," *IEEE Trans. Microwave Theory Tech.*, Vol. 46, No. 7, pp. 952–958, July 1998.

[44] C. Ho, "Slotline, CPW ring circuits and waveguide ring cavities for coupler and filter applications," Ph.D. dissertation, Texas A&M University, College Station, May 1994.

[45] C. Ho, L. Fan, and K. Chang, "Slotline annular ring resonator and its applications to resonator, filter, and coupler design," *IEEE Trans. Microwave Theory Tech.*, Vol. 41, pp. 1648–1650, September 1993.

[46] J. M. Carroll and K. Chang, "Microstrip mode suppression ring resonator," *Electron. Lett.*, Vol. 30, No. 22, pp. 1861–1862, October 1994.

[47] U. Karacaoglu, D. Sanchez-Hernandez, I. D. Robertson, and M. Guglielmi, "Harmonic suppression in microstrip dual-mode ring-resonator bandpass filters," *Microwave Symp. Dig., 1996., IEEE MTT-S Int.*, Vol. 3, pp. 1635–1638, June 1996.

[48] J. Marti and A. Griol, "Harmonic suppressed microstrip multistage coupled ring bandpass filters," *Electron. Lett.*, Vol. 34, No. 22, pp. 2140–2142, October 1998.

[49] G. K. Gopalakrishnan and K. Chang, "Bandpass characteristics of split-modes in asymmetric ring resonators," *Electron. Lett.*, Vol. 26, No. 12, pp. 774–775, June 1990.

CHAPTER EIGHT

Ring Couplers

8.1 INTRODUCTION

Hybrid couplers are indispensable components in various microwave integrated circuit (MIC) applications such as balanced mixers, balanced amplifiers, frequency discriminators, phase shifters, and feeding networks in antenna arrays. Some of the more commonly used are 180° hybrid-ring and 90° branch-line couplers. Rat-race hybrid rings [1] and reverse-phase hybrid rings [2–4] are well-known examples of 180° hybrid-ring couplers. Some other hybrid-ring couplers with improved bandwidth have also been reported [5, 6]. 90° branch-line couplers have been analyzed in references [7–12]. A computer-aided design technique that is suitable for the optimum design of multisection branch-line couplers was described in [13]. Some other optimized methods that included compensation for the junction discontinuities were also reported [14–17]. Another class of MIC 90° branch-line coupler, that is, de Ronde's coupler, using a combination of microstrip lines and slotlines was proposed in reference [18] and optimized in references [19–24]. This chapter presents the ring circuits for coupler applications [25].

8.2 180° RAT-RACE HYBRID-RING COUPLERS

8.2.1 Microstrip Hybrid-Ring Couplers

The microstrip rat-race hybrid-ring coupler [25] has been widely used in microwave power dividers and combiners. Figure 8.1 shows the physical configuration of the microstrip rat-race hybrid-ring coupler. To analyze the hybrid-

Microwave Ring Circuits and Related Structures, Second Edition,
by Kai Chang and Lung-Hwa Hsieh
ISBN 0-471-44474-X

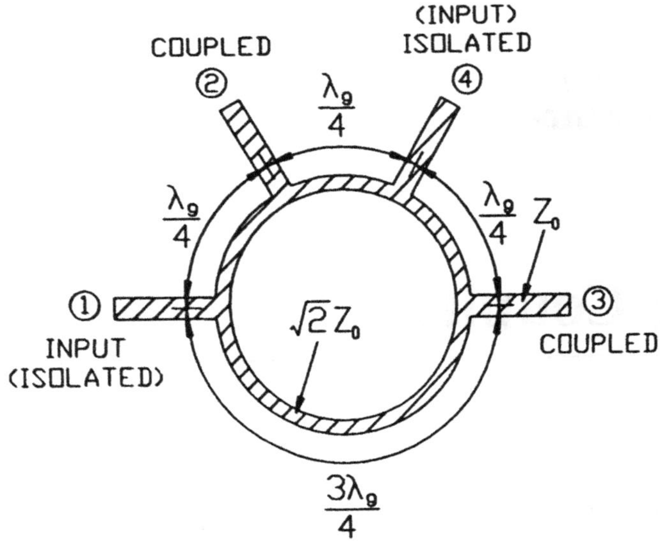

FIGURE 8.1 Physical layout of the microstrip rat-race hybrid-ring coupler.

ring coupler, an even–odd-mode method is used. When a unit amplitude wave is incident at port 4 of the hybrid-ring coupler, this wave is divided into two components at the ring junction. The two component waves arrive in phase at ports 2 and 3, and 180° out of phase at port 1. By using the even–odd-mode analysis technique, this case can be decomposed into a superposition of two simpler circuits, as shown in Figures 8.2 and 8.3. The amplitudes of the scattered waves from the hybrid-ring are given by [26]

$$B_1 = \frac{1}{2}T_e - \frac{1}{2}T_o \tag{8.1a}$$

$$B_2 = \frac{1}{2}\Gamma_e - \frac{1}{2}\Gamma_o \tag{8.1b}$$

$$B_3 = \frac{1}{2}T_e + \frac{1}{2}T_o \tag{8.1c}$$

$$B_4 = \frac{1}{2}\Gamma_e + \frac{1}{2}\Gamma_o \tag{8.1d}$$

where $\Gamma_{e,o}$ and $T_{e,o}$ are the even- and odd-mode reflection and transmission coefficients, and B_1, B_2, B_3, and B_4 are the amplitudes of the scattered waves at ports 1, 2, 3, and 4, respectively. Using the *ABCD* matrix for the even- and odd-mode two-port circuits shown in Figures 8.2 and 8.3, the required reflection and transmission coefficients in Equation (8.1) are [26]

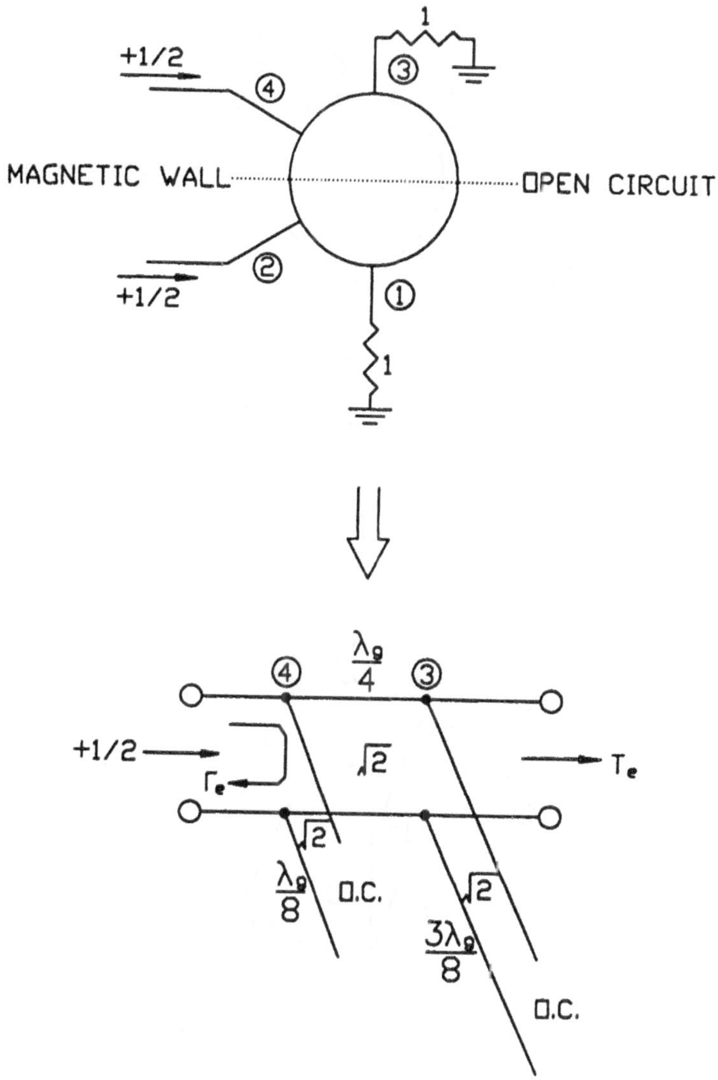

FIGURE 8.2 Even-mode decomposition of the rat-race hybrid-ring coupler when port 4 is excited with a unit amplitude incident wave.

$$\Gamma_e = \frac{-j}{\sqrt{2}} \tag{8.2a}$$

$$T_e = \frac{-j}{\sqrt{2}} \tag{8.2b}$$

$$\Gamma_o = \frac{j}{\sqrt{2}} \tag{8.2c}$$

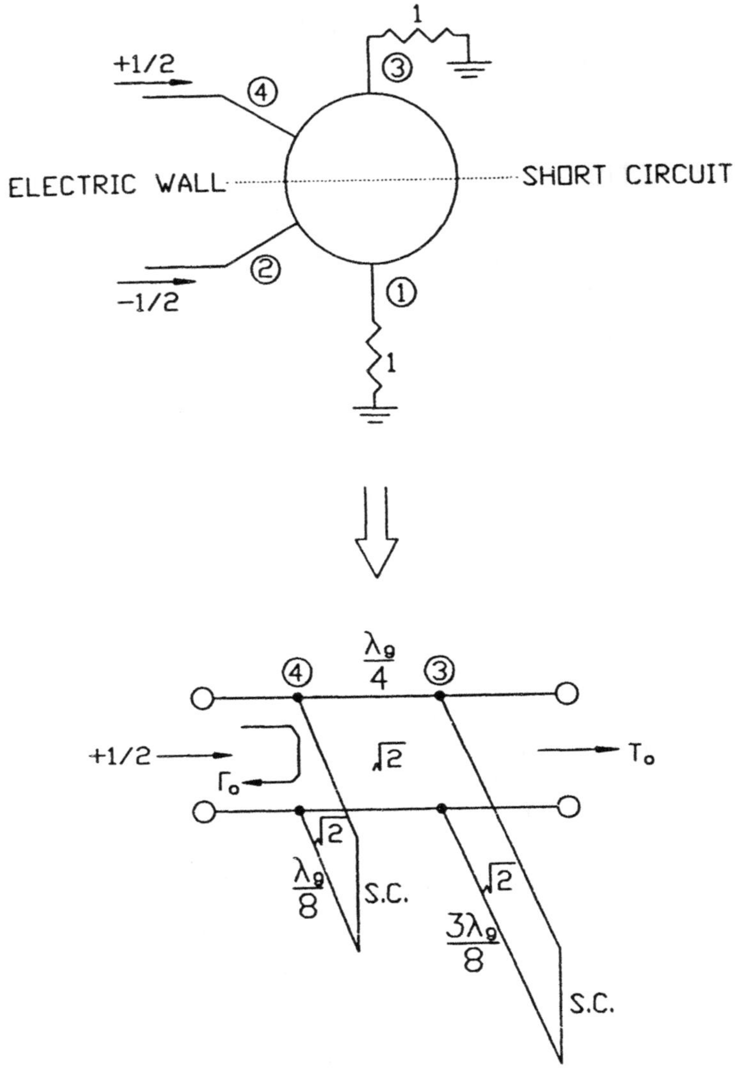

FIGURE 8.3 Odd-mode decomposition of the rat-race hybrid-ring coupler when port 4 is excited with a unit amplitude incident wave.

$$T_o = \frac{-j}{\sqrt{2}} \tag{8.2d}$$

Using these results in Equation (8.1) gives

$$B_1 = 0 \tag{8.3a}$$

$$B_2 = \frac{-j}{\sqrt{2}} \tag{8.3b}$$

$$B_3 = \frac{-j}{\sqrt{2}} \tag{8.3c}$$

$$B_4 = 0 \tag{8.3d}$$

which shows that the input port (port 4) is matched, port 1 is isolated from port 4, and the input power is evenly divided in phase between ports 2 and 3. For impedance matching, the square of the characteristic impedance of the ring is two times the square of the termination impedance.

Consider a unit amplitude wave incident at port 1 of the hybrid-ring coupler in Figure 8.1. The wave divides into two components, both of which arrive at ports 2 and 3 with a net phase difference of 180°. The two component waves are 180° out of phase at port 4. This case can be decomposed into a superposition of two simpler circuits and excitations, as shown in Figures 8.4 and 8.5. The amplitudes of the scattered waves will be [26]

$$B_1 = \frac{1}{2}\Gamma_e + \frac{1}{2}\Gamma_o \tag{8.4a}$$

$$B_2 = \frac{1}{2}T_e + \frac{1}{2}T_o \tag{8.4b}$$

$$B_3 = \frac{1}{2}\Gamma_e - \frac{1}{2}\Gamma_o \tag{8.4c}$$

$$B_4 = \frac{1}{2}T_e - \frac{1}{2}T_o \tag{8.4d}$$

Using the *ABCD* matrix for the even- and odd-mode two-port circuits shown in Figure 8.3, the required reflection and transmission coefficients in Equation (8.4) are [26]

$$\Gamma_e = \frac{j}{\sqrt{2}} \tag{8.5a}$$

$$T_e = \frac{-j}{\sqrt{2}} \tag{8.5b}$$

$$\Gamma_o = \frac{-j}{\sqrt{2}} \tag{8.5c}$$

$$T_o = \frac{-j}{\sqrt{2}} \tag{8.5d}$$

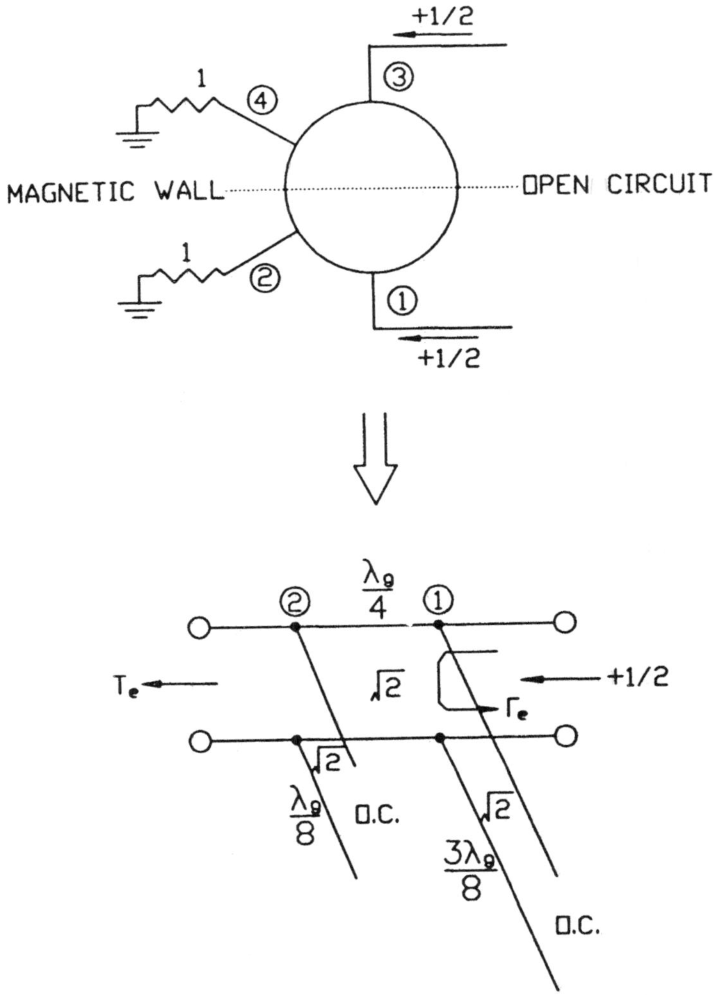

FIGURE 8.4 Even-mode decomposition of the rat-race hybrid-ring coupler when port 1 is excited with a unit amplitude incident wave.

Using these results in Equation (8.4) gives

$$B_1 = 0 \tag{8.6a}$$

$$B_2 = \frac{-j}{\sqrt{2}} \tag{8.6b}$$

$$B_3 = \frac{+j}{\sqrt{2}} \tag{8.6c}$$

$$B_4 = 0 \tag{8.6d}$$

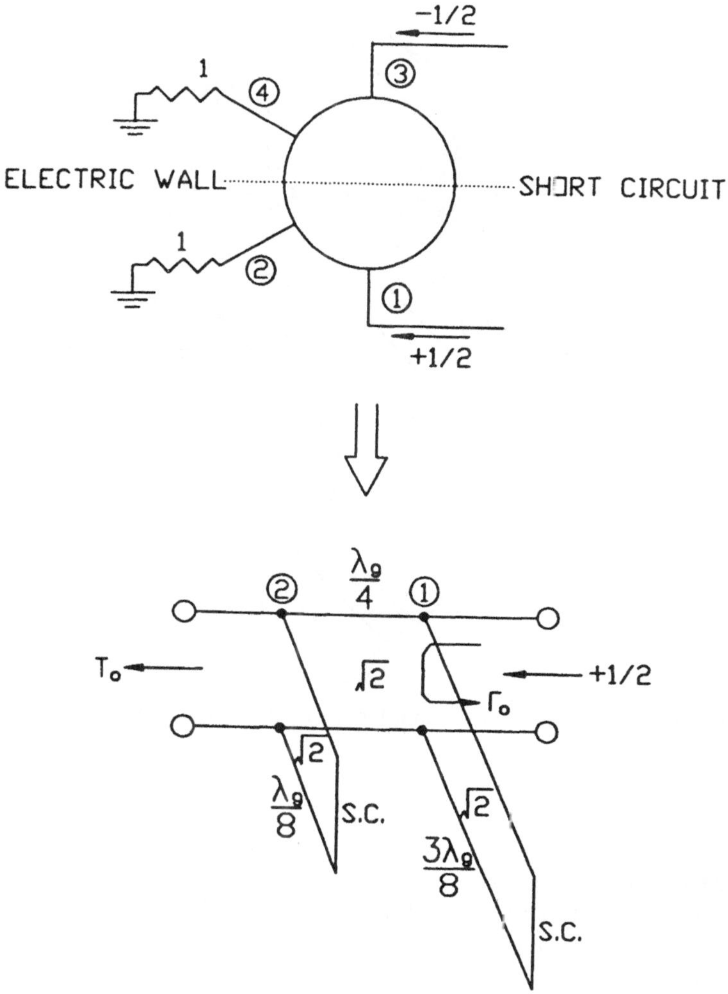

FIGURE 8.5 Odd-mode decomposition of the rat-race hybrid-ring coupler when port 1 is excited with a unit amplitude incident wave.

which shows that the input port (port 1) is matched, port 4 is isolated from port 1, and the input power is evenly divided between ports 2 and 3 with a 180° phase difference.

8.2.2 Coplanar Waveguide-Slotline Hybrid-Ring Couplers

The uniplanar rat-race hybrid-ring coupler was developed based on the same operating principle as the microstrip rat-race hybrid-ring coupler. Figure 8.6 shows the circuit diagram of the uniplanar slotline hybrid-ring coupler with coplanar waveguide (CPW) feeds. The uniplanar slotline ring coupler consists

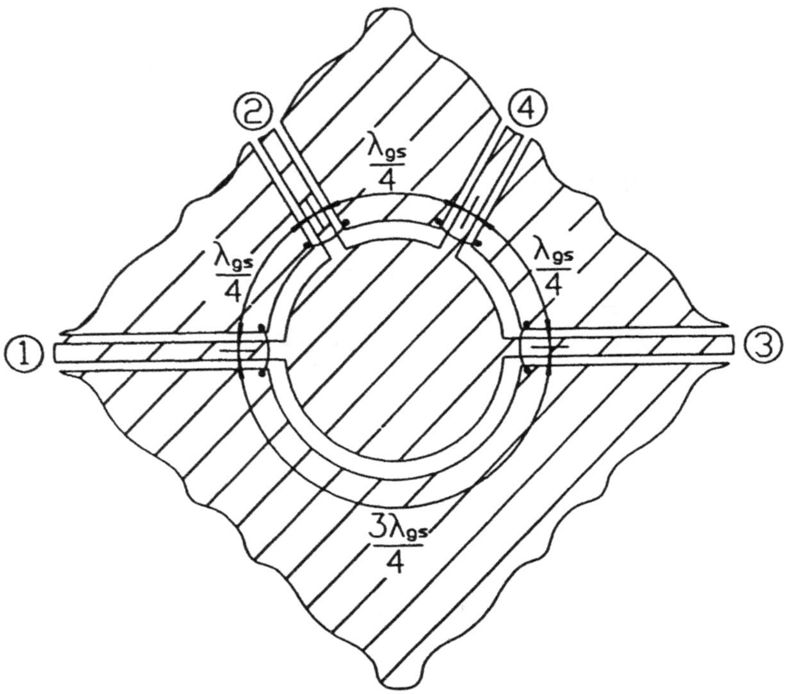

FIGURE 8.6 Circuit layout of the uniplanar rat-race slotline hybrid-ring coupler with CPW feeds.

of three quarter-wavelength slotline sections, one phase-delay section, and four CPW to slotline T-junctions. Figure 8.7 shows the equivalent circuit of the slotline hybrid-ring coupler. The characteristic impedance of the slotline ring Z_S is determined by

$$Z_S = \sqrt{2} Z_{C0} \tag{8.7}$$

where Z_{C0} is the characteristic impedance of the CPW feed lines. The mean radius r of the slotline is also determined by

$$2\pi r = \frac{3}{2}\lambda_{gs} \tag{8.8}$$

where λ_{gs} is the guide wavelength of the slotline. The measured and calculated results of the uniplanar slotline rat-race hybrid-ring coupler are shown in Figures 8.8 and 8.9, respectively. The calculated results were based on the transmission-line model of Figure 8.7. As shown in Figure 8.8, the uniplanar slotline hybrid-ring coupler has a bandwidth of 18.6% with a maximum ampli-

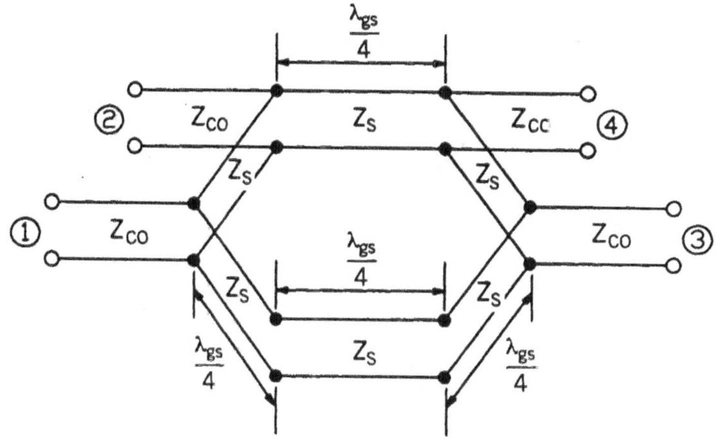

FIGURE 8.7 Equivalent circuit of the uniplanar rat-race slotline hybrid-ring coupler with CPW feeds.

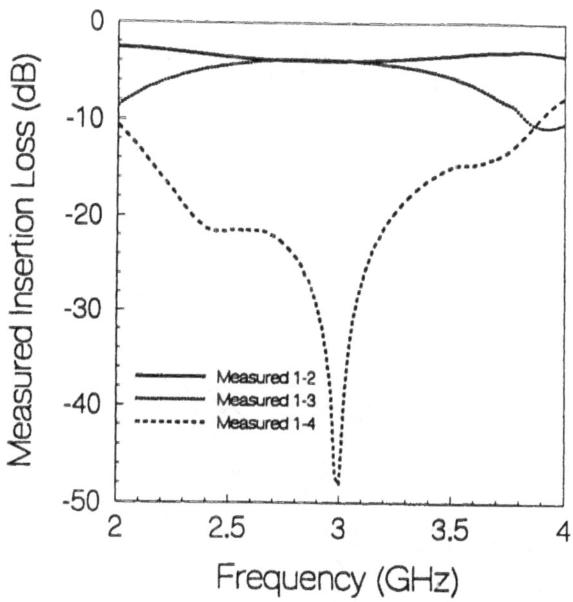

FIGURE 8.8 Measured results of power dividing and isolation for the uniplanar rat-race slotline hybrid-ring coupler with CPW feeds.

tude imbalance of 1 dB and an isolation of over 20 dB. For an ideal 3-dB coupler, the insertion loss should be 3 dB. The 1.2 dB extra loss is mainly due to the CPW-slotline T-junctions.

Figure 8.10 shows the circuit layout of a double-sided slotline rat-race

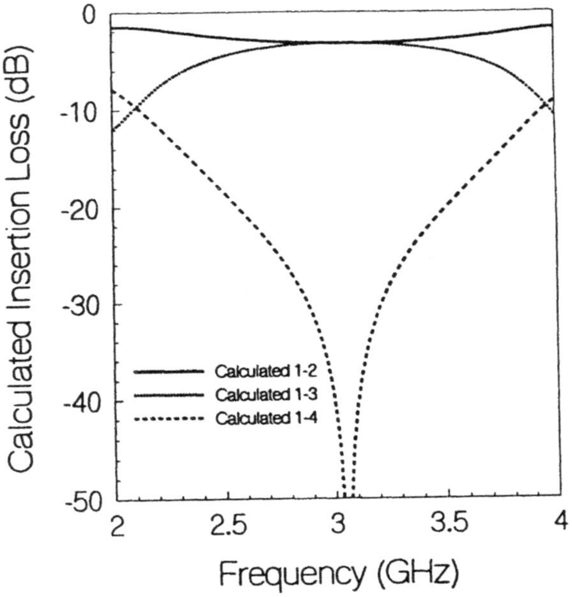

FIGURE 8.9 Calculated results of power dividing and isolation for the uniplanar rat-race slotline hybrid-ring coupler with CPW feeds.

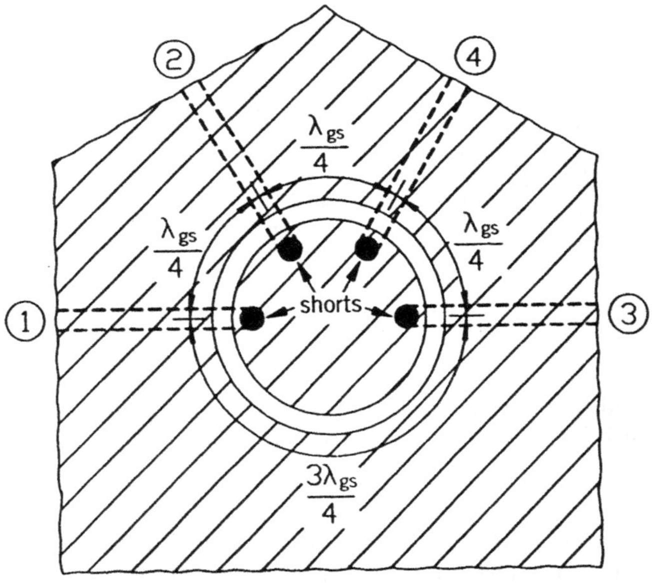

FIGURE 8.10 Physical configuration of the double-sided rat-race slot-line hybrid-ring coupler with microstrip feeds [27]. (Permission from IEEE.)

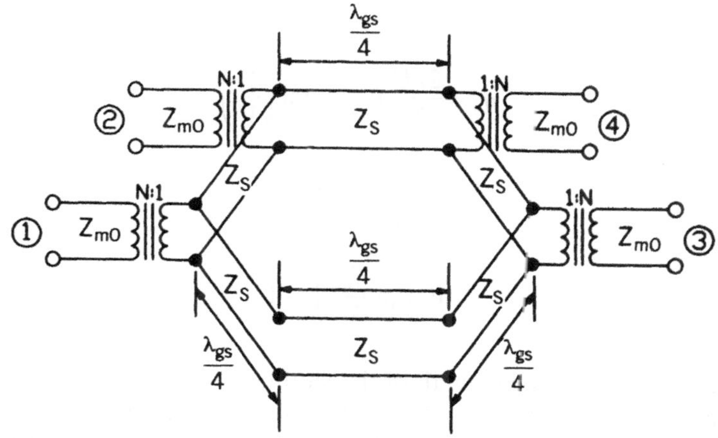

FIGURE 8.11 Equivalent circuit of the double-sided rat-race slotline hybrid-ring coupler with microstrip feeds.

hybrid-ring coupler with microstrip feeds [27]. Figure 8.11 shows the equivalent transmission-line model. The impedance of the slotline ring and microstrip feed lines is obtained from the following equation:

$$\frac{Z_S^2}{Z_{m0}^2} = 2N^2 \tag{8.9}$$

where Z_S and Z_{m0} are the characteristic impedance of the slotline ring and microstrip feed lines, respectively, and N is the turn ratio of the equivalent transformer. The turn ratio was reported by [28] to be

$$N = \frac{V(h)}{V_0} \tag{8.10}$$

where

$$V(h) = -\int_{-b/2}^{b/2} E_y(h)dy \tag{8.11}$$

h is the thickness of the substrate, $b/2$ is the length of the microstrip feed to the slotline, V_0 is the voltage across the slot, and $E_y(h)$ is the electric field of the slotline on the dielectric surface of the opposite side. From Cohn's analysis [29],

$$E_y(h) = -\frac{V_0}{b}\left[\cos\left(\frac{2\pi U}{\lambda_0}h\right) - \cot(q_0)\sin\left(\frac{2\pi U}{\lambda_0}h\right)\right] \tag{8.12}$$

where

$$q_0 = \frac{2\pi U}{\lambda_0} h + \arctan\left(\frac{U}{V}\right) \tag{8.13}$$

$$U = \sqrt{\varepsilon_r - \left(\frac{\lambda_0}{\lambda_{gs}}\right)^2} \tag{8.14}$$

$$V = \sqrt{\left(\frac{\lambda_0}{\lambda_{gs}}\right)^2 - 1} \tag{8.15}$$

where λ_{gs} is the guide wavelength of the slotline.

The mean radius of the double-sided slotline rat-race hybrid-ring is determined by Equation (8.8). Figures 8.12 and 8.13 show the measured and calculated results of the double-sided slotline rat-race hybrid-ring coupler. The theoretical results were calculated from the equivalent transmission-line model of Figure 8.11. The test circuit was built on a RT/Duroid 6010.8 substrate with the following dimensions: substrate thickness h = 1.27 mm,

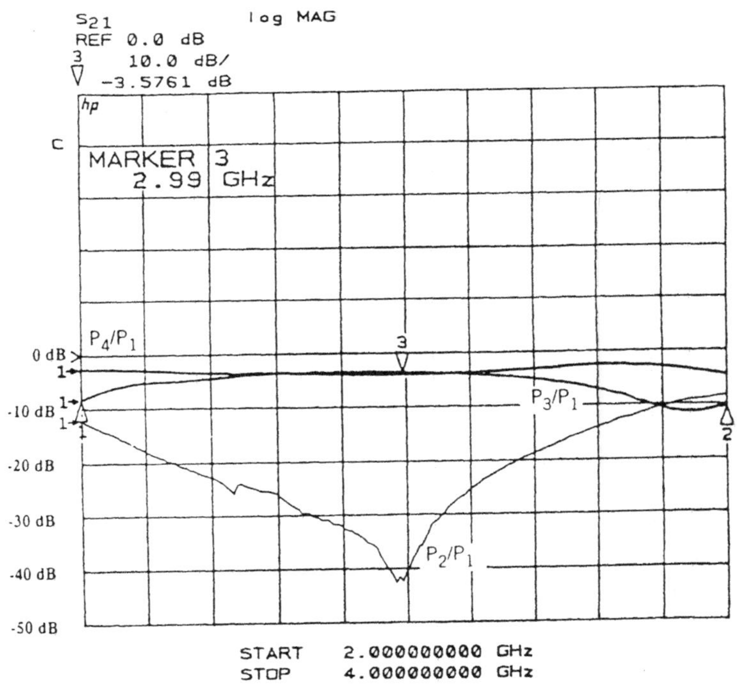

FIGURE 8.12 Measured results of power dividing and isolation for the double-sided rat-race slotline hybrid-ring coupler with microstrip feeds [27]. (Permission from IEEE.)

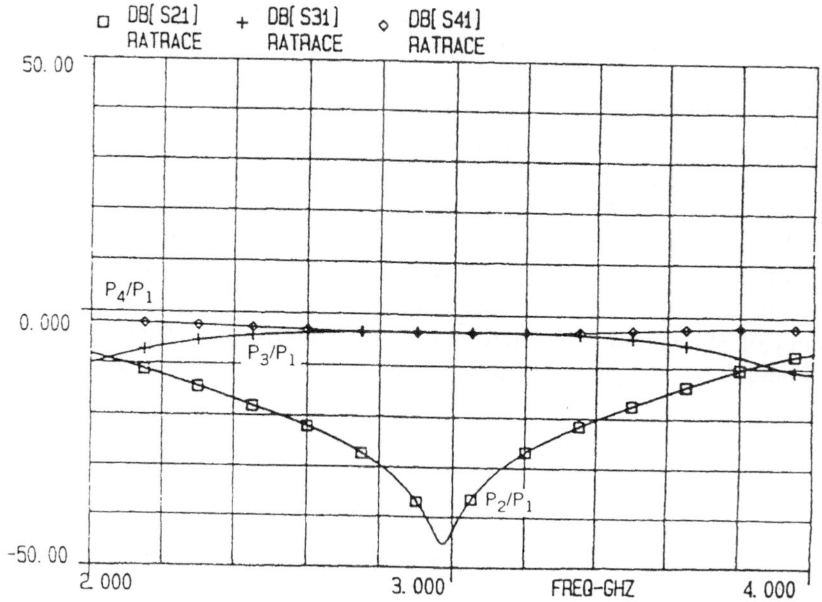

FIGURE 8.13 Calculated results of power dividing and isolation for the double-sided rat-race slotline hybrid-ring coupler with microstrip feeds [27]. (Permission from IEEE.)

microstrip impedance $Z_{m0} = 50\,\Omega$, microstrip line width $W_{m0} = 1.09$ mm, slotline impedance $Z_S = 70.7\,\Omega$, slotline line width $W_s = 0.85$ mm, and slotline ring mean radius $r = 12.78$ mm.

As shown in Figure 8.12, a double-sided slotline rat-race hybrid-ring coupler with a maximum amplitude imbalance of 1 dB and isolation of over 20 dB has been achieved with a bandwidth of more than 26%. The insertion loss at the center frequency of 3 GHz is 3.6 dB. For an ideal 3-dB coupler, the insertion loss should be 3 dB. The 0.6 dB extra loss is partly due to the microstrip—slotline transitions. Besides the insertion loss, the measured and calculated results shown in Figures 8.12 and 8.13 agree very well.

Compared with the microstrip rat-race hybrid-ring coupler, which can be implemented with a typical bandwidth of 20%, the double-sided slotline rat-race hybrid-ring coupler has a bandwidth of more than 26%.

8.2.3 Asymmetrical Coplanar Strip Hybrid-Ring Couplers

The asymmetrical coplanar strip (ACPS) like CPW has the advantages of small dispersion, simple realization of short-circuited ends, easy integration with lumped elements and active devices, and no need for via holes. These attractive characteristics make ACPS important in MIC and MMIC designs [30–33]. However, when used in conjunction with hybrid rings, CPW presents a problem of having to plate the inner circular conducting ground plane, as well as bond to it at T-junction. This incorporates extra time and cost in the fabrication process.

Figure 8.14 shows the physical configuration and equivalent circuit of the 1.5 $\lambda_{g,ACPS}$ circumference uniplanar hybrid-ring coupler that is realized on one side of substrate using CPW to ACPS transmission lines [34]. The circuit consists of four CPW to ACSP T-junctions and a circular ACSP ring that is divided into three $\lambda_{g,ACPS}/4$ sections and one $3\lambda_{g,ACPS}/4$ section. The characteristic impedance of the circular ACPS ring is $Z_R = \sqrt{2}Z_o$, where Z_o is the characteristic impedance of the CPW feed lines. The circuit was designed at a center frequency of 3 GHz and fabricated on RT/Duriod 6010 ($\varepsilon_r = 10.8$) substrate with thickness $h = 0.635$ mm, a characteristic impedance of $Z_o = Z_{cpw} = 50$ ohms

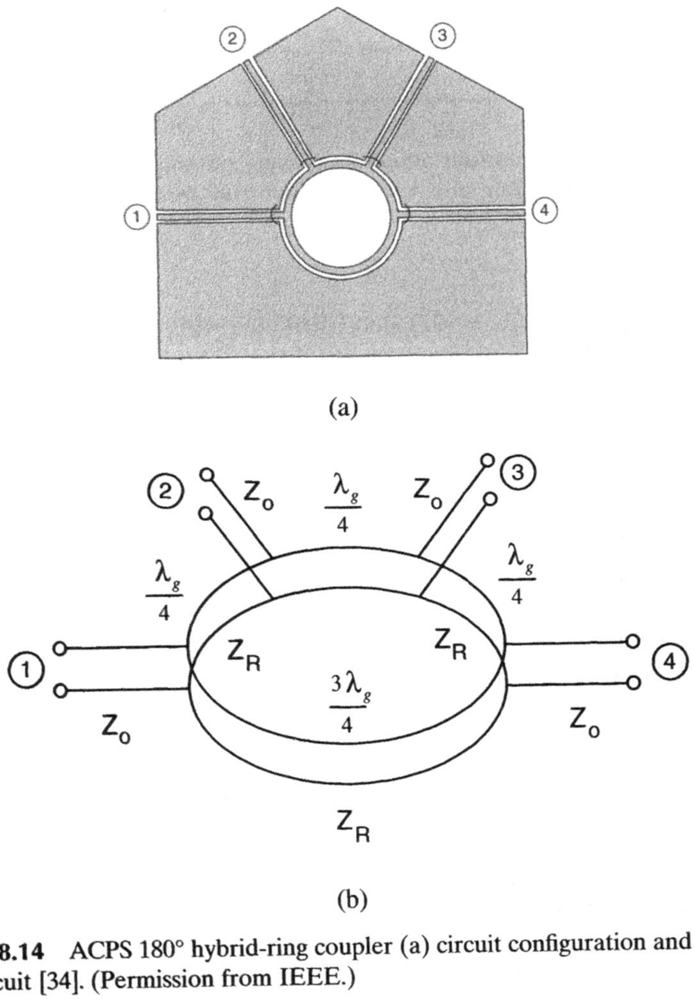

FIGURE 8.14 ACPS 180° hybrid-ring coupler (a) circuit configuration and (b) equivalent circuit [34]. (Permission from IEEE.)

for four CPW feed lines, and Z_R = 71 ohms for the circular ring. The four CPW feed lines have a gap of G = 0.29 mm and a center conductor width of w_{cpw} = 0.6 mm. The 71-ohms ACPS consists of the ground plane, a space of s = 0.27 mm, and a line width of w_{ACPS} = 0.4 mm. The quarter-wavelength sections are 10.73 mm.

The simulated and measured results with good agreement are shown in Figure 8.15. In the measured results, the coupling ($|S_{21}|$ or $|S_{41}|$) is 3.5 ± 0.4 dB (3 dB for ideal coupling, the insertion loss is less than 0.9 dB, which includes two coaxial-to-CPW transitions and 40-mm-long input/output CPW lines), the input return loss ($|S_{11}|$) is greater than 23 dB, and the isolation ($|S_{31}|$) is greater than 21 dB. Compared with the typical microstrip hybrid-ring coupler with a typical bandwidth of 20%, the ACPS coupler has a bandwidth of 25%.

8.3 180° REVERSE-PHASE BACK-TO-BACK BALUNS

The 180° reverse-phase back-to-back baluns are the alternatives to microstrip baluns [25, 35]. Microstrip/parallel-plate-line tapered transition is the well-known microstrip balun used in many microwave balanced circuits. It is frequently incorporated in microwave mixers to connect the coaxial input ports to a balanced bridge of mixer diodes. The concept of the microstrip tapered-balun was first proposed by Duncan and Minerva [36] in 1960. They used the tapered-baluns to drive wide-band balanced aerials. Figure 8.16*a* shows the well-known 180° reverse-phase microstrip back-to-back tapered-balun that is commonly used in balanced mixers. The circuit consists of two microstrip tapered-baluns that are connected in twisted form. The ground plane of tapered-balun 1 is on the bottom side of the substrate; the ground plane of tapered-balun 2 is on the top side of the substrate. In the middle of the circuit, the metal strips on both sides of the substrate have equal widths and are symmetric. This symmetric transmission line is called a parallel-plate line. In contrast to the microstrip transmission line, which has a ground plane and is unsymmetric and unbalanced, the parallel-plate is a balanced line. The mixer diodes are inserted in the balanced parallel-plate line. The signal excited from the microstrip line of tapered-balun 1 and the signal excited from the microstrip line of tapered-balun 2 have a 180° phase difference in the middle of the parallel-plate line, because the ground planes of tapered-baluns 1 and 2 are on opposite sides of the substrate. Figure 8.17*a* shows the equivalent circuit of the microstrip back-to-back balun. The twisted parallel-plate line connects two ground planes on the opposite sides of the substrate and results in the 180° phase reversal. The 180° phase reversal is essential for balanced mixer circuits.

Although the 180° reverse-phase microstrip back-to-back tapered-balun has a very wide bandwidth, the use of double-sided ground planes results in very complicated fabrication and packaging processes. To overcome this

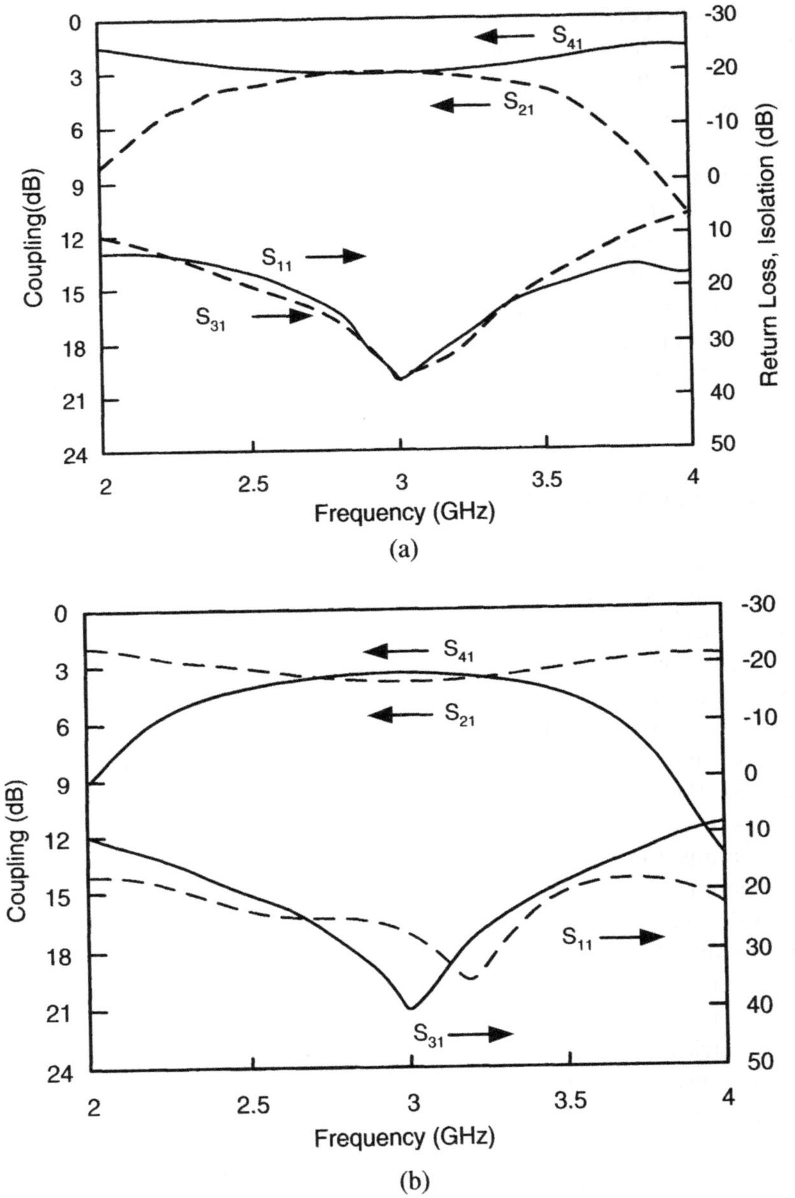

FIGURE 8.15 ACPS 180° hybrid-ring coupler (a) simulation and (b) measurement [34]. (Permission from IEEE.)

problem, a new uniplanar 180° reverse-phase back-to-back balun was developed using the broad-band CPW-slotline transition [25, 35].

Figure 8.16*b* shows the circuit configuration of the 180° reverse-phase CPW back-to-back balun [35, 37]. The two CPW-slotline transitions in Figure 8.16*b*

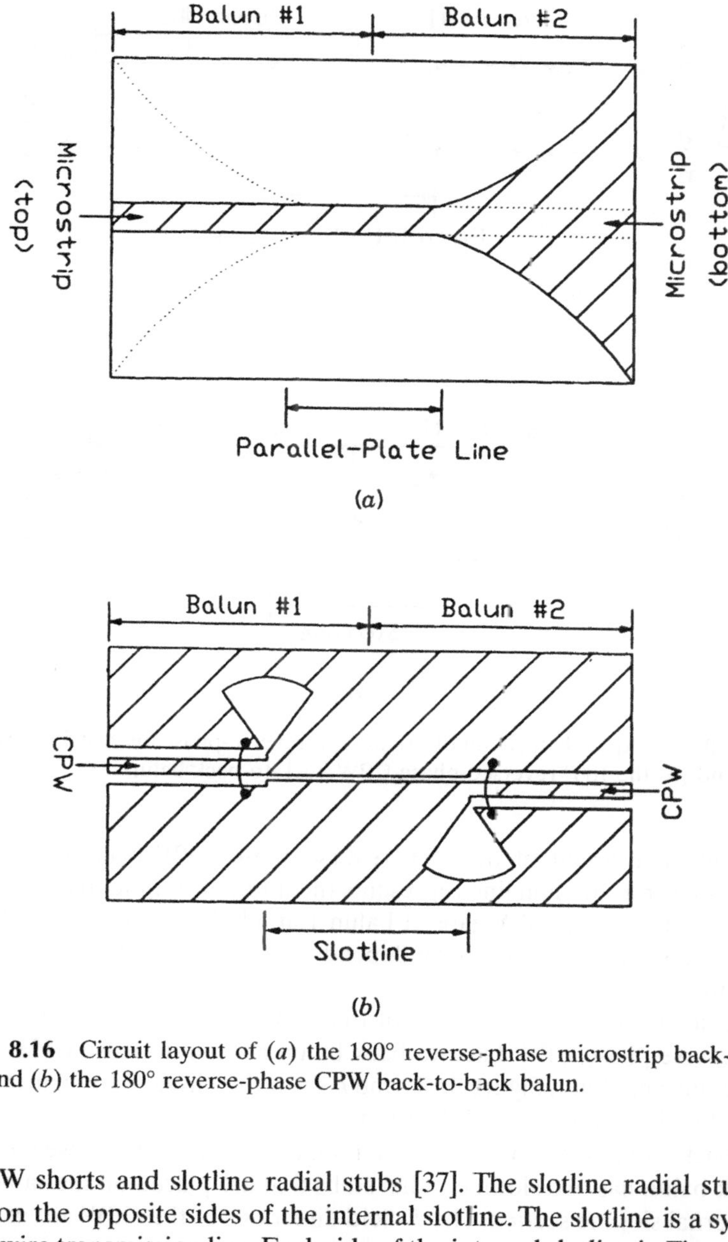

FIGURE 8.16 Circuit layout of (*a*) the 180° reverse-phase microstrip back-to-back balun, and (*b*) the 180° reverse-phase CPW back-to-back balun.

use CPW shorts and slotline radial stubs [37]. The slotline radial stubs are placed on the opposite sides of the internal slotline. The slotline is a symmetric two-wire transmission line. Each side of the internal slotline in Figure 8.16*b* connects the center conductor (or ground plane) of the CPW in balun 1 and the ground plane (or center conductor) of the CPW in balun 2. Referring to Figure 8.16*b*, a 180° phase shift of the *E*-field is introduced into the output signal at balun 2 when the input signal is excited from balun 1. The change of the *E*-field direction is caused by the inserted slotline section that connects the opposite sides of the CPW gaps at balun 1 and balun 2. Figure 8.17*b* shows

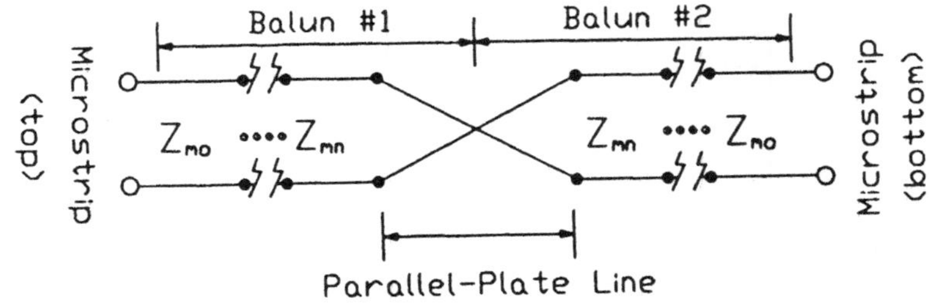

(a)

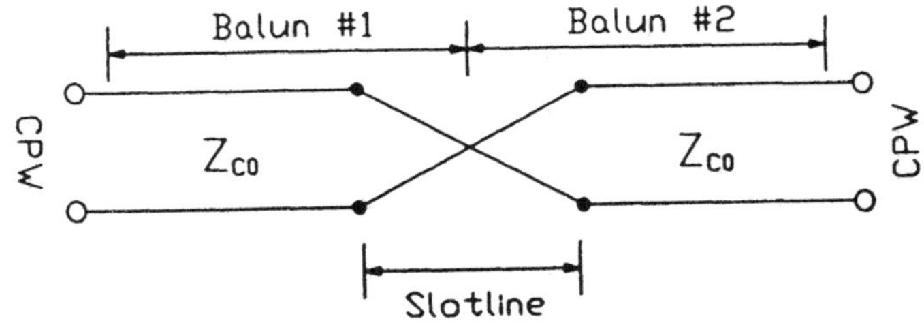

(b)

FIGURE 8.17 Equivalent circuit of (*a*) the 180° reverse-phase microstrip back-to-back balun, and (*b*) the 180° reversed-phase CPW back-to-back balun.

the equivalent circuit of the 180° reverse-phase CPW back-to-back balun. The twisted transmission line represents the internal slotline that connects the opposite sides of the CPW gaps at balun 1 and balun 2. The phase change of the twisted slotline is frequency independent and can thus be applied to wise-band circuits.

To test the 180° phase reversal of the twisted CPW back-to-back balun, a 180° reverse-phase CPW back-to-back balun and an in-phase CPW back-to-back balun were built. Figure 8.18*a* and *b* show the physical configurations and schematic diagram of *E*-field distribution for the in-phase and reverse-phase CPW back-to-back baluns. As shown in Figure 8.18*a*, the in-phase CPW back-to-back balun has two slotline radial stubs that are placed on the same side of the internal slotline section. The *E*-field directions of the CPW are in phase at balun 1 and balun 2.

The test circuits were built on a RT/Duroid 6010.8 ($\varepsilon_r = 10.8$) substrate with the following dimensions: substrate thickness $h = 1.27$ mm, characteristic impedance of the CPW $Z_c = 50\,\Omega$, CPW center conductor width $S_C = 0.51$ mm, CPW gap size $G_C = 0.25$ mm, characteristic impedance of the slotline $Z_S = 54.39$, slotline line width $W_s = 0.1$ mm, radius of the slotline radial stub $r = 6$ mm, and angle of the slotline radial stubs $\theta = 90°$. The measurements were made using

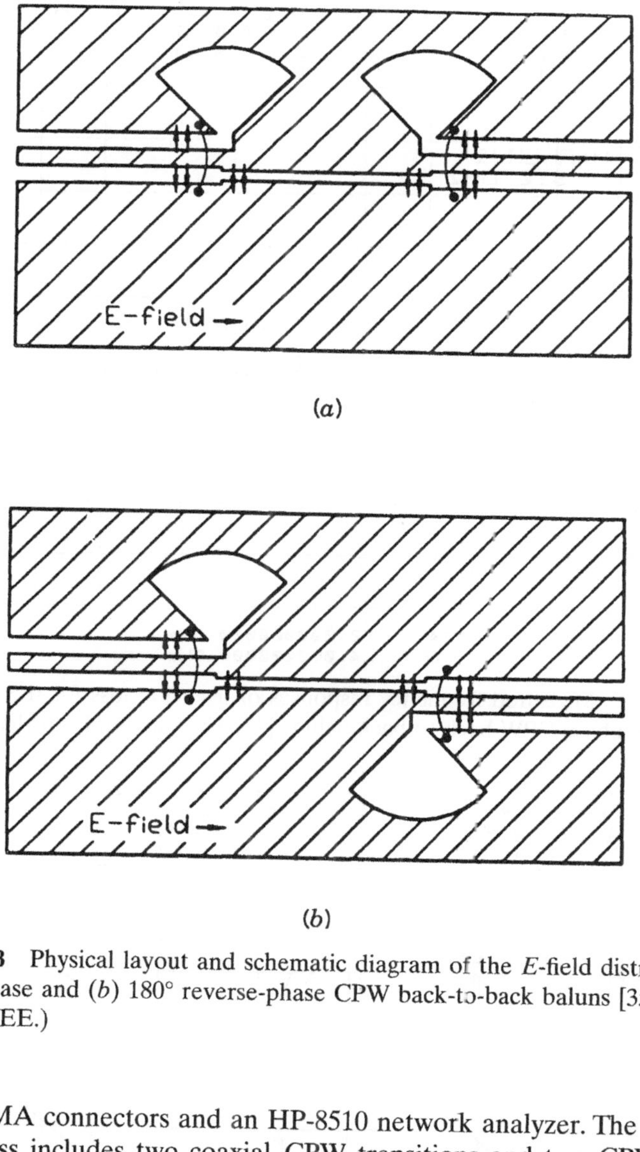

FIGURE 8.18 Physical layout and schematic diagram of the *E*-field distribution for the (*a*) in-phase and (*b*) 180° reverse-phase CPW back-to-back baluns [35]. (Permission from IEEE.)

standard SMA connectors and an HP-8510 network analyzer. The measured insertion loss includes two coaxial–CPW transitions and two CPW–slotline transitions.

Figures 8.19 and 8.20 show the measured frequency responses of insertion loss and phase angle for the 180° reverse-phase and in-phase CPW back-to-back baluns. Figure 8.21 shows the amplitude and phase differences between the in-phase and 180° reverse-phase CPW back-to-back baluns. The amplitude difference is within 0.3 dB from 2 GHz to 4 GHz. Over the same range, maximum phase difference is maintained within 5° as shown in Figure 8.21. The 5° phase error is due to the mechanical tolerances, misalignments, connectors, and discontinuities.

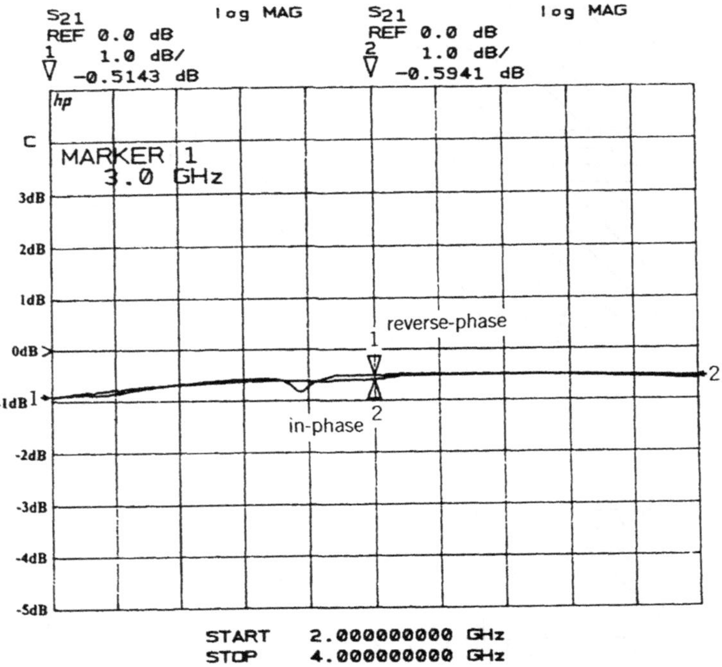

FIGURE 8.19 Measured frequency responses of insertion loss for the in-phase and 180° reverse-phase CPW back-to-back baluns [35]. (Permission from IEEE.)

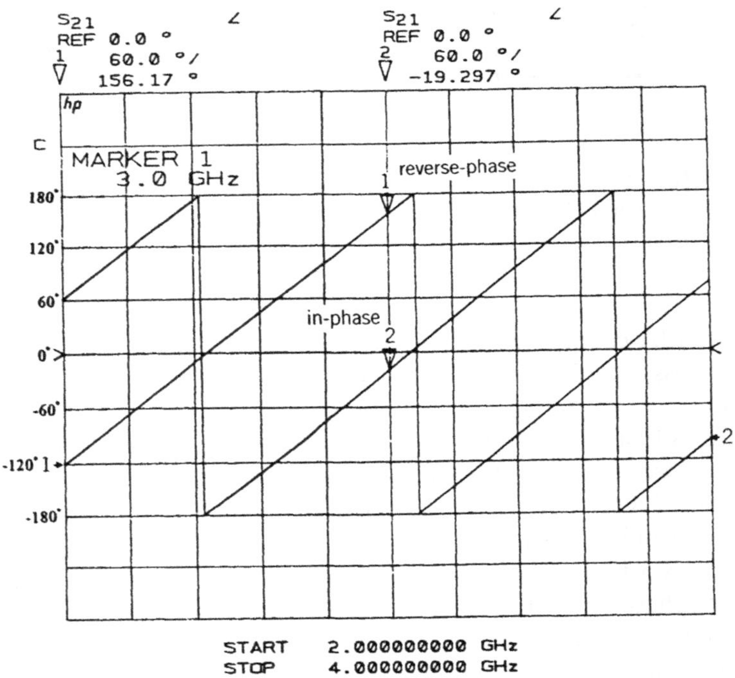

FIGURE 8.20 Measured frequency responses of phase angles for the in-phase and 180° reverse-phase CPW back-to-back baluns [35]. (Permission from IEEE.)

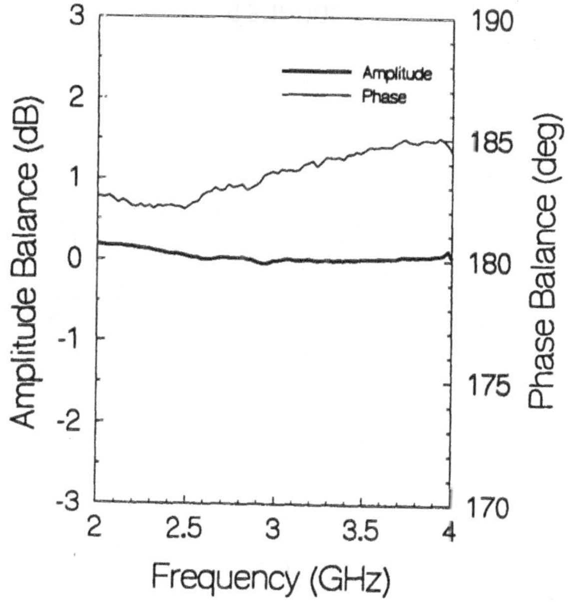

FIGURE 8.21 Amplitude and phase differences between the in-phase and 180° reverse-phase CPW back-to-back baluns.

8.4 180° REVERSE-PHASE HYBRID-RING COUPLERS

8.4.1 CPW-Slotline 180° Reverse-Phase Hybrid-Ring Couplers

The 20 to 26% bandwidth of the rat-race coupler limits its applications to narrow-band circuits. Several design techniques have been developed to extend the bandwidth. One technique proposed by March [3] in 1968 used a $\lambda_g/4$ coupled-line section to replace the $3\lambda_g/4$ section of the conventional $3\lambda_g/2$ microstrip rat-race hybrid-ring coupler. Figures 8.22 and 8.23 show the physical configuration and equivalent circuit of the microstrip reverse-phase hybrid-ring coupler with a shorted parallel coupled-line section. The shorted parallel coupled-line section provides a 180° phase delay, as shown in Figure 8.23. The even- and odd-mode admittances of the coupled-line section vary more slowly with frequency than those of the conventional three quarter-wavelength phase-delay sections [3]. Consequently the coupling and other parameters of the reverse-phase coupler are less frequency dependent. Although the bandwidth of the reverse-phase hybrid-ring coupler has been improved up to more than one octave, the difficulty of constructing the shorted coupled-line section limits its use at low frequencies.

As mentioned before, the uniplanar structure does not use the backside of the substrate and circumvents the need of via holes for short circuits.

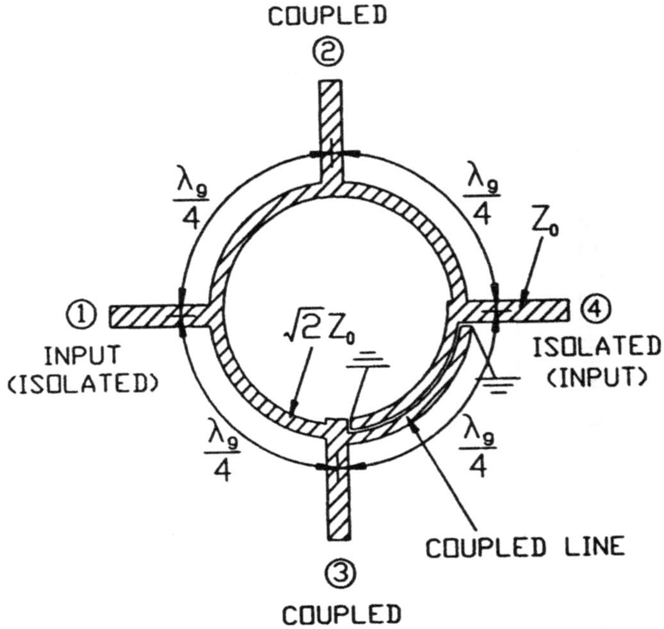

FIGURE 8.22 Circuit layout of the microstrip reverse-phase hybrid-ring coupler.

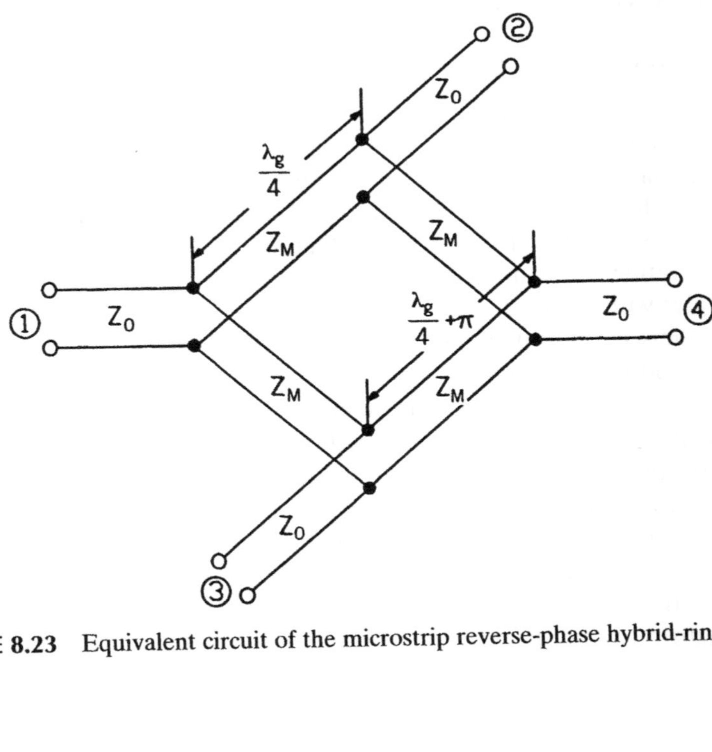

FIGURE 8.23 Equivalent circuit of the microstrip reverse-phase hybrid-ring coupler.

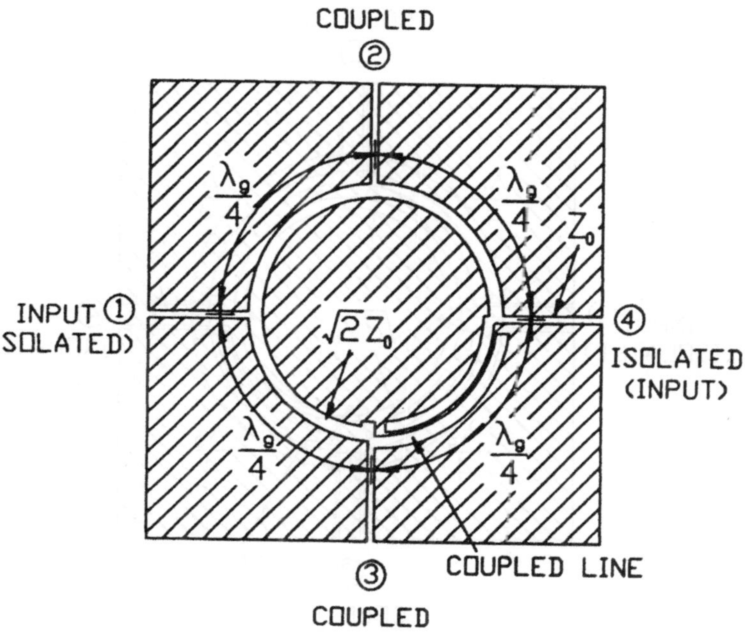

FIGURE 8.24 Circuit layout of the slotline reverse-phase hybrid-ring coupler [37]. (Permission from IEEE.)

Figures 8.24 and 8.25 show uniplanar implementations of the reverse-phase hybrid-ring coupler using slotline and CPW rings, respectively [25, 37]. The coupled-slotline section in Figure 8.24 and the coupled-CPW section in Figure 8.25 require no via holes for the short terminations. Although the uniplanar reverse-phase hybrid-ring couplers are easier to fabricate than the microstrip couplers, they still demand a very small gap within the coupled-slotline or coupled-CPW section when 3-dB coupling is required.

In 1970 Rehnmark [2] proposed a modified reverse-phase hybrid-ring coupler using a balanced twin-wire line. The $3\lambda_g/4$ phase delay section is replaced by a $\lambda_g/4$ section plus a phase reversal obtained by twisting the pair of lines. However, this circuit is only possible in a twin-wire configuration that seriously restricts its applications.

Another modified microstrip reverse-phase hybrid-ring coupler was proposed by Chua [4] in 1971. He substituted a $\lambda_g/4$ slotline for the $3\lambda_g/4$ phase-delay section of the conventional microstrip rat-race hybrid-ring coupler. The microstrip-slotline transition provides a 180° phase delay. Since the phase change of the microstrip-slotline transition is frequency independent, the resulting microstrip reverse-phase hybrid-ring coupler has a wider bandwidth than the conventional rat-race hybrid-ring coupler. Although the modified version has a symmetric geometry, an excellent coupling bandwidth and fairly good isolation, the double-sided implementation of a curved $3\lambda_g/4$ microstrip

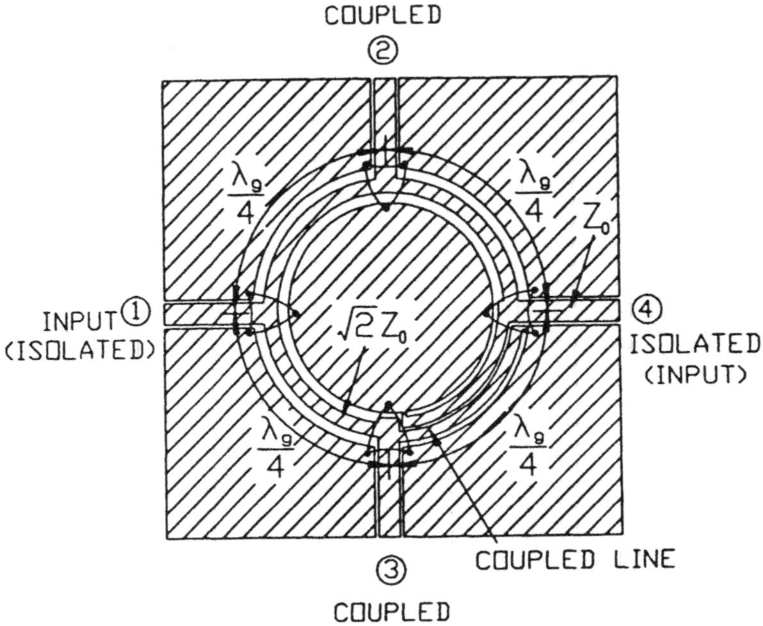

FIGURE 8.25 Circuit layout of the CPW reverse-phase hybrid-ring coupler.

line with an inserted $\lambda_g/4$ slotline is very difficult to realize in a photolithographic process. Also, the unity of the ring structure is destroyed, and the inserted slotline section may cause some discontinuity problems.

To overcome these problems, this section presents a new uniplanar reverse-phase CPW hybrid-ring coupler using a 180° reverse-phase CPW back-to-back balun [25, 35]. As mentioned in Section 8.3, the 180° reverse-phase CPW-slotline back-to-back transition produces a phase shift that is 180° longer than that of the in-phase CPW-slotline back-to-back transition. The 180° phase shift is frequency independent. Figure 8.26 shows the circuit layout of the uniplanar reverse-phase CPW hybrid-ring coupler. The circuit consists of four CPW-slotline T-junctions, three quarter-wavelength CPW sections, and one 180° reverse-phase CPW back-to-back balun. As shown in Figure 8.26, the new hybrid-ring coupler substitutes a 180° CPW-slotline phase shifter for the phase delay section used in the conventional rat-race hybrid-ring coupler. The resulting uniplanar reverse-phase hybrid-ring coupler has a broad bandwidth, because the phase change of the 180° reverse-phase CPW-slotline back-to-back balun is frequency independent. Figure 8.27 shows the equivalent transmission-line model. The twisted transmission line represents the 180° phase reversal of the CPW-slotline back-to-back balun.

To test the circuit, a truly uniplanar reverse-phase hybrid-ring coupler was built on a RT/Duroid 6010.8 ($\varepsilon_r = 10.8$) substrate with the following dimensions: substrate thickness $h = 1.524$ mm, characteristic impedance of the CPW

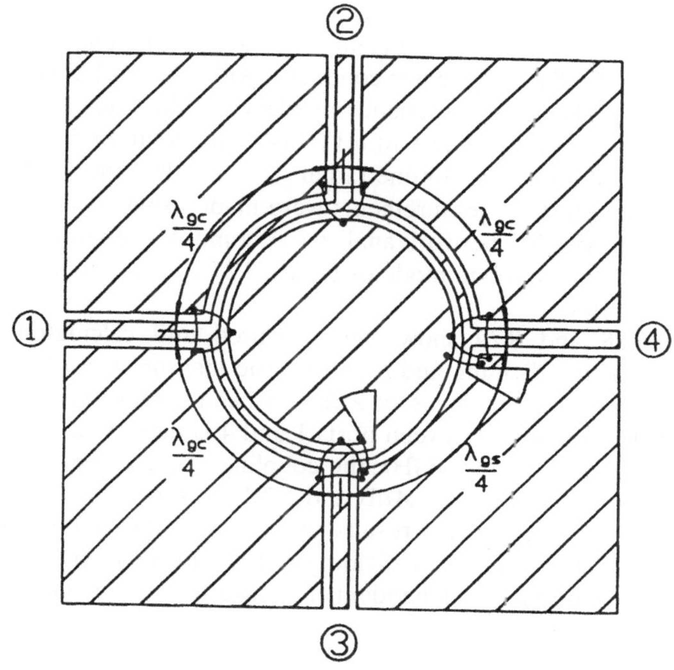

FIGURE 8.26 Circuit layout of the CPW/slotline reverse-phase hybrid-ring coupler [35]. (Permission from IEEE.)

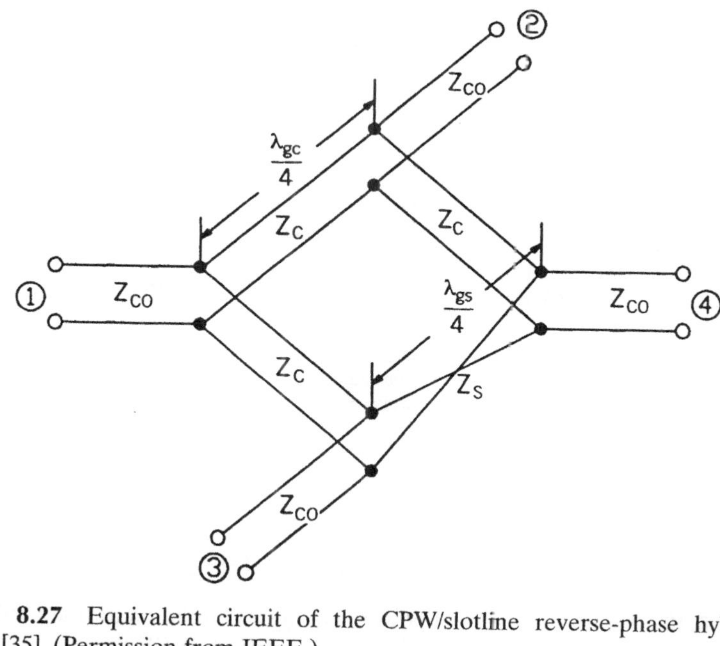

FIGURE 8.27 Equivalent circuit of the CPW/slotline reverse-phase hybrid-ring coupler [35]. (Permission from IEEE.)

feed lines $Z_{c0} = 50\,\Omega$, CPW feed lines center conductor width $S_{c0} = 0.53$ mm, CPW feed lines gap size $G_{c0} = 0.34$ mm, characteristic impedance of the CPW ring $Z_C = 70.7\,\Omega$, CPW ring center conductor width $S_C = 0.205$ mm, CPW ring gap size $G_C = 0.47$ mm, characteristic impedance of the reverse-phase slotline section $Z_S = 70.7\,\Omega$, slotline line width $W_S = 0.47$ mm, radius of the slotline radial stub $r = 5$ mm, angle of the slotline radial stubs $\theta = 30°$, and CPW ring mean radius $r = 6.88$ mm. The measurements were made using standard SMA connectors and an HP-8510 network analyzer. A computer program based on the equivalent transmission-line mode of Figure 8.27 was developed and used to analyze the circuit.

Figures 8.28 and 8.29 show the measured and calculated frequency responses of power dividing and isolation. The measured results show that a maximum amplitude imbalance of 2 dB has been achieved over a 2–4-GHz bandwidth. The isolation between ports 1 and 4 is more than 17 dB over the same octave bandwidth. At 3 GHz, the coupling of the power from port 1 to the balanced arms 2 and 3 is 3.4 dB and 3.7 dB, respectively. The isolation is 25 dB at 3 GHz. The calculated results agree very well with the experimental results. As expected, the power dividing characteristics of the reverse-phase coupler are less frequency dependent. The insertion loss is mainly from the CPW-slotline transition.

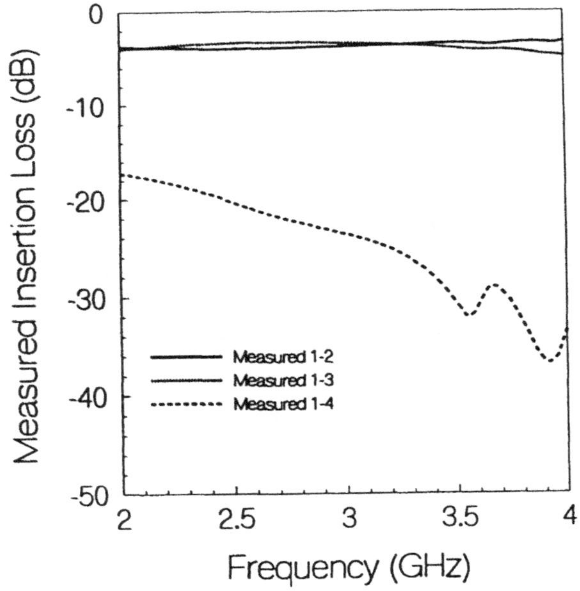

FIGURE 8.28 Measured results of power dividing and isolation for the CPW/slotline reverse-phase hybrid-ring coupler [35]. (Permission from IEEE.)

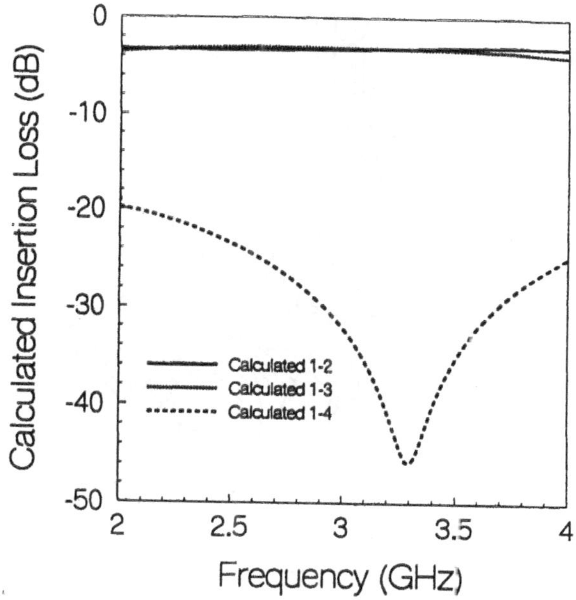

FIGURE 8.29 Calculated results of power dividing and isolation for the CPW/slotline reverse-phase hybrid-ring coupler [35]. (Permission from IEEE.)

8.4.2 Reduced-Size Uniplanar 180° Reverse-Phase Hybrid-Ring Couplers

The conventional hybrid-ring couplers constructed with quarter-wavelength lines occupy large areas in MMICs, and the bandwidths are limited due to the electrical line lengths. Although several designs have been developed to extend the bandwidth [38–40], some disadvantages of using microstrip include the precision fabrication for constructing the shorted coupled-line section [38, 39] and the difficulty of inserting the ground pins for the microstrip shorts [40].

The 180° reverse-phase CPW-slotline back-to-back balun with the advantage of frequency independence makes the coupler achieve a small size of $\lambda_g/5$ for each section and a wide band operation [41]. Figure 8.30 shows the circuit layout of the wideband reduced-size uniplanar hybrid-ring coupler. The circuit consists of four CPW-slotline tee junctions and one 180° reverse-phase CPW-slotline back-to-back balun that is formed using a pair of CPW-to-slotline transitions as shown in Figure 8.30*a*. Figure 8.30*b* shows the equivalent circuit. The twisted transmission line presents the 180° reverse-phase CPW-slotline back-to-back balun.

The hybrid-ring coupler in Figure 8.30 was fabricated on RT/Duroid 6010.5 ($\varepsilon_r = 10.5$) with a thickness of 1.524 mm. The dimensions of the circuit are as follows:

Coupler's circumference: $C = 0.8\lambda_g$ (mean radius $R = 4.52$ mm)
slotline ring: $Z_s = 66.9$ ohms (slotlline width $w_s = 0.31$ mm)

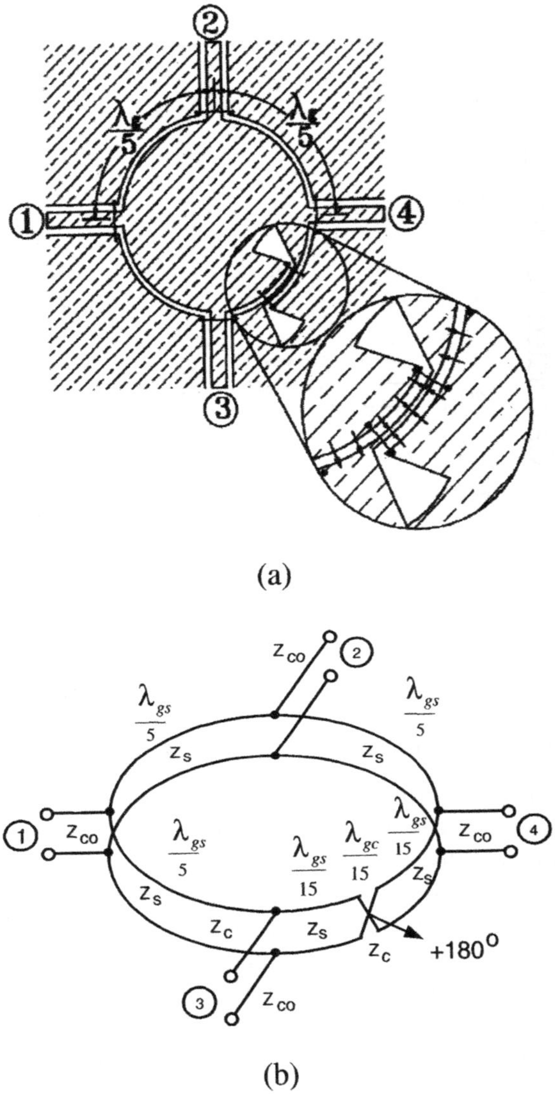

FIGURE 8.30 Reduced-size reverse-phase hybrid-ring coupler (a) layout and (b) equivalent circuit [41]. (Permission from IEEE.)

CPW section: $Z_c = 66.9$ ohms (center strip $S_c = 0.2$ mm and gap size $G_c = 0.31$ mm)

CPW feed lines: $Z_{co} = 50$ ohms (center strip $S_{co} = 0.6$ mm and gap size $G_{co} = 0.31$ mm)

Slotline radial stub radius: $r_s = 5$ mm

Slotline radial stub angle: $\varphi_s = 45°$

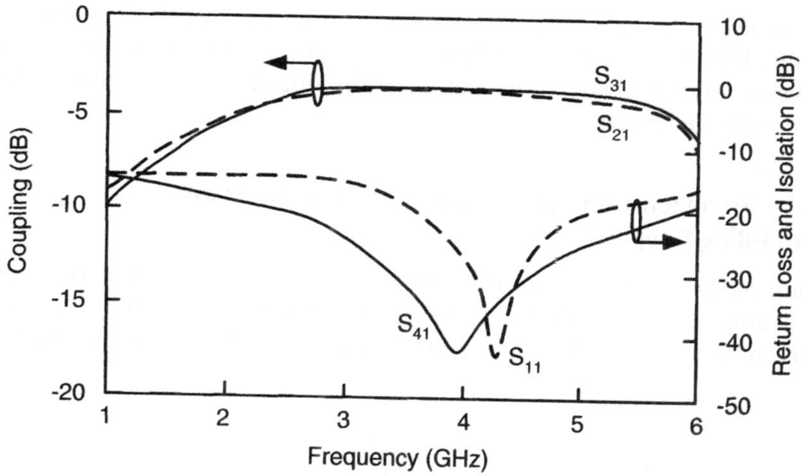

(a)

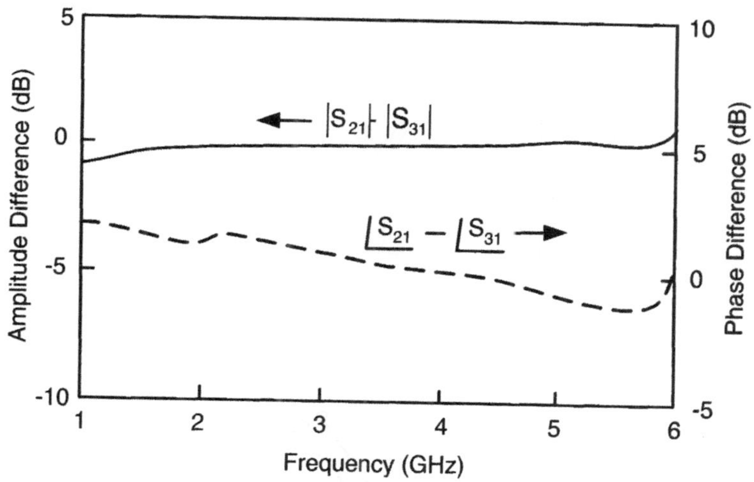

(b)

FIGURE 8.31 Measured results for (a) coupling, return loss, and isolation and (b) amplitude imbalance and phase imbalance [41]. (Permission from IEEE.)

To eliminate the coupled slotline mode propagating on the CPW lines, bonding wires have been used at the coupler's CPW-slotline discontinuities.

Figure 8.31 shows the hybrid-ring coupler's measured frequency responses of coupling, isolation, return loss, amplitude, and phase imbalance, respectively. The measured results show that the couplings of power from port 1 to ports 2 and 3 are 3.6 and 3.7 dB at 4 GHz, respectively. The isolation between ports 1 and 4 is greater than 19 dB, and return loss is more than 15 dB both over a frequency range from 2.7 to 6 GHz. The amplitude and phase imbalance between ports 2 and 3 are excellent over a broad bandwidth. The reduction of the line length to 72° has no deleterious effect on performance of the circuit. However, the radial stub in the center of the ring can cause a problem for the smaller circumference.

8.4.3 Asymmetrical Coplanar Strip 180° Reverse-Phase Hybrid-Ring Couplers

Figure 8.32*a* shows the circuit configuration of the new hybrid-ring coupler consisting of four CPW to ACPS T-junctions and four ACPS arms (one of them with a 180° phase reversal) [34]. Figure 8.32*b* shows the equivalent

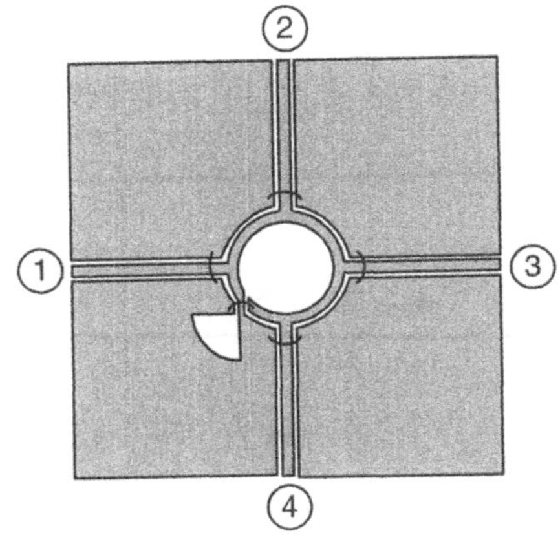

(a)

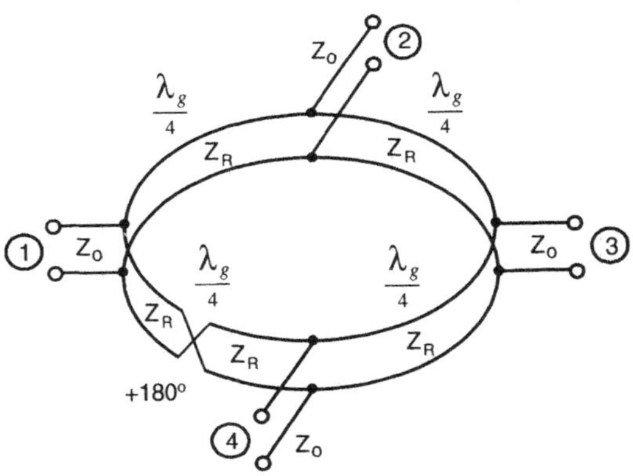

(b)

FIGURE 8.32 ACPS 180° reverse-phase hybrid-ring coupler (a) configuration and (b) equivalent circuit [34]. (Permission from IEEE.)

transmission-line model of the coupler. The twisted transmission line represents the phase reversal of the ACPS crossover. When the signal is fed to port 1, it splits into two equal components that arrive at ports 2 and 4 in phase, but are canceled out at port three.

The 180° reverse-phase hybrid-ring coupler was fabricated on an $h = 0.635$-mm-thick RT/Duroid 6010.8 ($\varepsilon_r = 10.8$) substrate. The coupler was designed at the center frequency of 3 GHz. The circuit's CPW feed lines have a characteristic impedance of $Z_o = Z_{cpw} = 50$ ohms (strip width $w_{cpw} = 0.6$ mm, gap size $G = 0.29$ mm), and the ACPS lines have a characteristic impedance of $Z_R = \sqrt{2}\ Z_o = 71$ ohms (strip width $w_{ACPS} = 0.4$ mm, spacing size $s = 0.27$ mm). The four ACPS arms each have a length of $\lambda_{g,ACPS}/4 = 10.73$ mm. The slotline radial stub's radius is $r = 6$ mm with an angle of 90°. Adding air bridges at the circuit's discontinuities is important to prevent the coupled slotline mode from propagating on the CPW and ACPS lines.

The measured data of the reverse-phase hybrid coupler are shown in Figure 8.33*a*. Over an octave bandwidth from 2 to 4 GHz, Figure 8.33*a* shows that the coupling ($|S_{21}|$ or $|S_{41}|$) is 3.95 ± 0.45 dB (3 dB for ideal coupling) and the isolation ($|S_{31}|$) is greater than 23 dB. The input return loss ($|S_{11}|$) is greater than 15 dB from 2.2 to 4 GHz, and it is greater than 13.5 dB from 2 to 4 GHz. Figure 8.33*b* illustrates an important feature of the coupler. The output amplitude imbalance (±0.4 dB) and phase difference (±4°) are excellent over a bandwidth from 2 to 4 GHz because the ACPS crossover provides an almost perfect 180° phase shift over the entire frequency range. This is an advantage with respect to the microstrip implementations of the 180° hybridring coupler, where the $\lambda_g/2$ delay line gives a 180° phase shift only at the center frequency.

8.5 90° BRANCH-LINE COUPLERS

8.5.1 Microstrip Branch-Line Couplers

The microstrip branch-line coupler [25, 37] is a basic component in applications such as power dividers, balanced mixers, frequency discriminators, and phase shifters. Figure 8.34 shows the commonly used microstrip branch-line coupler. To analyze the branch-line coupler, an even-odd mode method is used. When a unit amplitude wave is incident at port 1 of the branch-line coupler, this wave divides into two components at the junction of the coupler. The two component waves arrive at ports 2 and 3 with a net phase difference of 90°. The component waves are 180° out of phase at port 4 and cancel each other. This case can be decomposed into a superposition of two simpler circuits and excitations, as shown in Figures 8.35 and 8.36. The amplitudes of the scattered waves are [26]

$$B_1 = \frac{1}{2}\Gamma_e + \frac{1}{2}\Gamma_o \tag{8.16a}$$

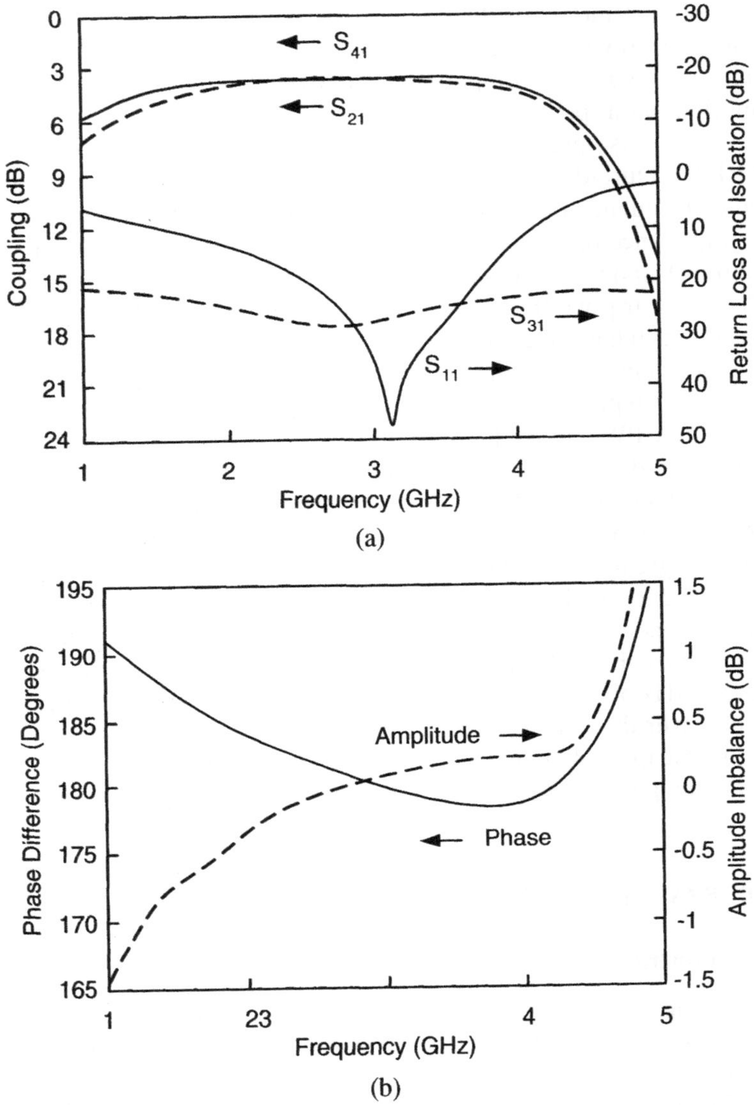

FIGURE 8.33 Measured results for ACPS 180° reverse-phase hybrid-ring coupler (a) coupling, return loss and isolation and (b) amplitude and phase difference [34]. (Permission from IEEE.)

$$B_2 = \frac{1}{2}T_e + \frac{1}{2}T_o \tag{8.16b}$$

$$B_3 = \frac{1}{2}T_e - \frac{1}{2}T_o \tag{8.16c}$$

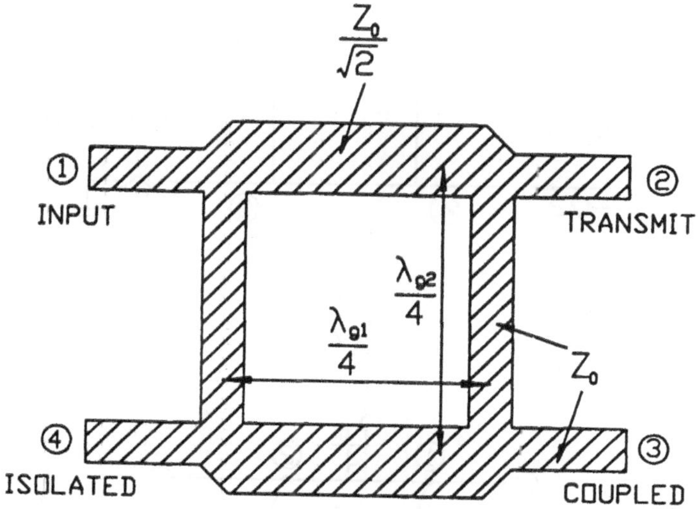

FIGURE 8.34 Physical configuration of the microstrip 2-branch coupler.

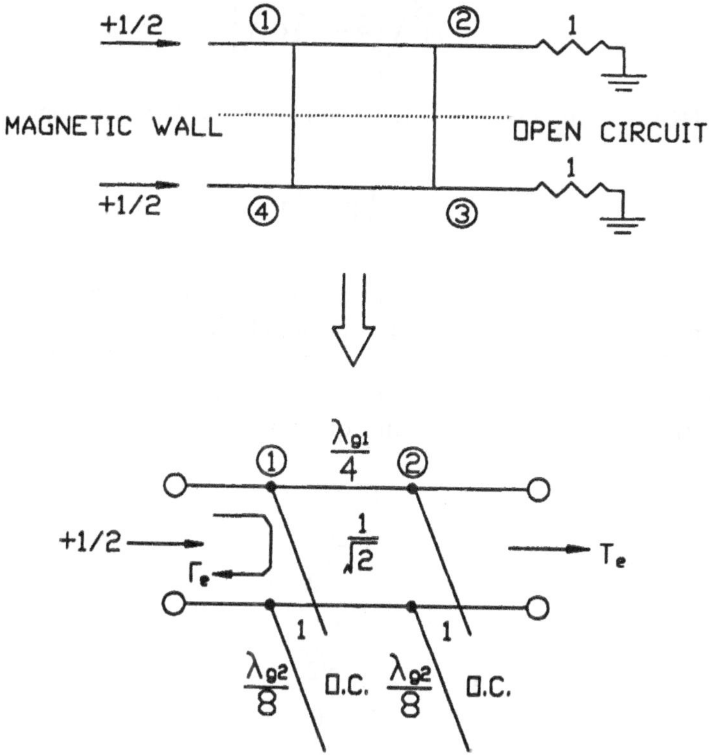

FIGURE 8.35 Even-mode decomposition of the 2-branch coupler.

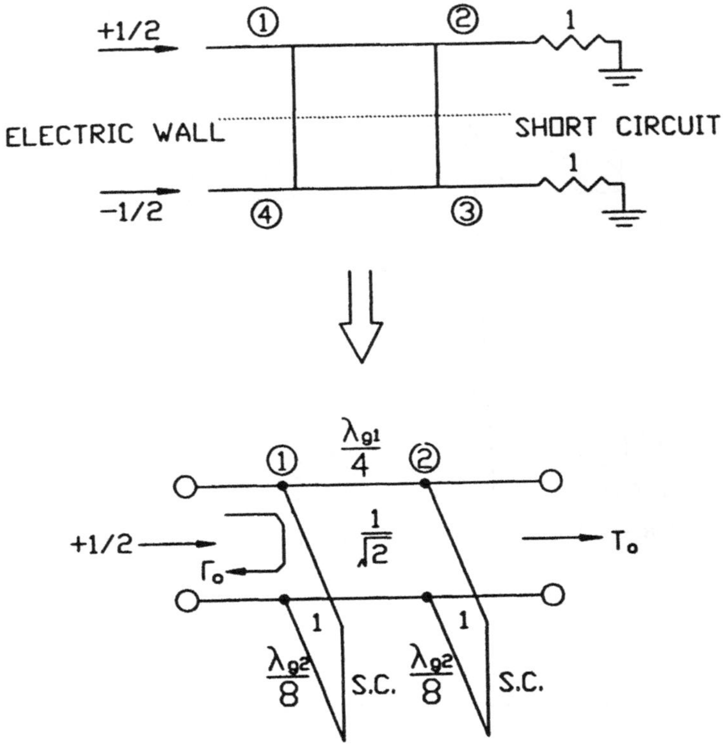

FIGURE 8.36 Odd-mode decomposition of the 2-branch coupler.

$$B_4 = \frac{1}{2}\Gamma_e - \frac{1}{2}\Gamma_o \tag{8.16d}$$

where $\Gamma_{e,o}$ and $T_{e,o}$ are the even- and odd-mode reflection and transmission coefficients, and B_1, B_2, B_3, and B_4 are the amplitudes of the scattered waves at ports 1, 2, 3, and 4, respectively. Using the *ABCD* matrix for the even- and odd-mode two-port circuits shown in Figures 8.35 and 8.36, the required reflection and transmission coefficients in Equation (8.16) are [26]

$$\Gamma_e = 0 \tag{8.17a}$$

$$T_e = \frac{-1-j}{\sqrt{2}} \tag{8.17b}$$

$$T_o = \frac{1-j}{\sqrt{2}} \tag{8.17c}$$

$$\Gamma_o = 0 \tag{8.17d}$$

Using these results in Equation (8.16) gives

$$B_1 = 0 \tag{8.18a}$$

$$B_2 = \frac{-j}{\sqrt{2}} \tag{8.18b}$$

$$B_3 = \frac{-1}{\sqrt{2}} \tag{8.18c}$$

$$B_4 = 0 \tag{8.18d}$$

which shows that the input port is matched, port 4 is isolated from port 1, and the input power is evenly divided at ports 2 and 3 with a 90° phase difference. For impedance matching, the square of the characteristic impedance of the series arms is half of the square of the termination impedance.

8.5.2 CPW-Slotline Branch-Line Couplers

This section presents two uniplanar branch-line couplers using CPW and slotline structures [25, 37]. The design technique for the CPW branch-line couplers uses a shunt connection, while the design technique for the slotline branch-line couplers uses a series connection.

Figure 8.37 shows the physical configuration of the CPW branch-line coupler. When a signal is applied to port 1, outputs appear at ports 2 and 3

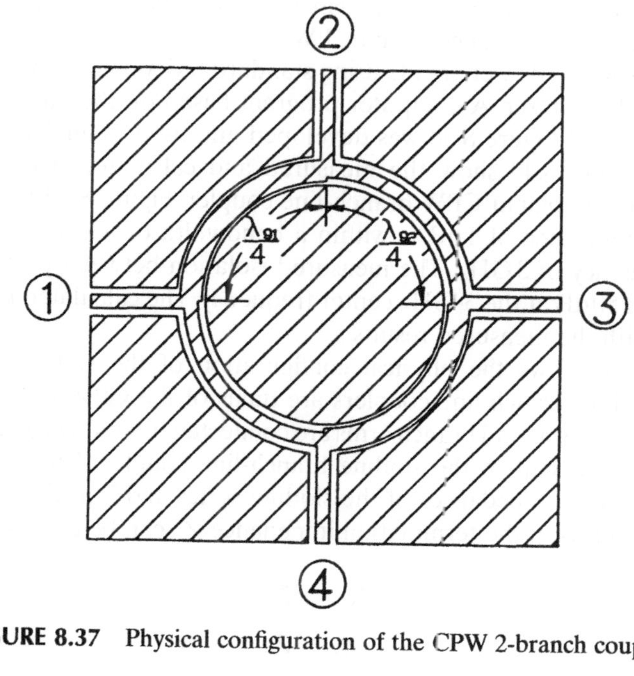

FIGURE 8.37 Physical configuration of the CPW 2-branch coupler.

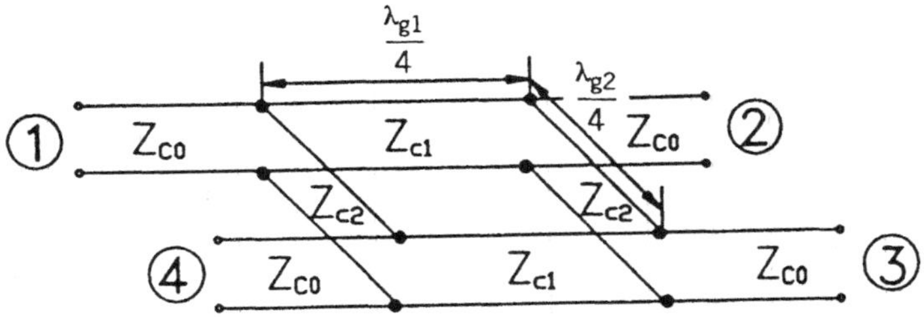

FIGURE 8.38 Equivalent circuit of the CPW 2-branch coupler.

that are equal in amplitude and differ in phase by 90°. Port 4 represents the isolation port. Figure 8.38 shows the equivalent circuit of the uniplanar CPW branch-line coupler. The series arms and branch arms are connected in parallel. The corresponding line characteristic impedances of the CPW series and branch arms for 3-dB coupling, in terms of the termination impedance Z_0, can be expressed as

$$Z_{C1} = \frac{Z_0}{\sqrt{2}} \tag{8.19}$$

$$Z_{C2} = Z_0 \tag{8.20}$$

where Z_{C1} is the characteristic impedance of the CPW series arms, and Z_{C2} is the characteristic impedance of the CPW branch arms.

The measurements were made using standard SMA connectors and an HP-8510 network analyzer. A computer program based on the equivalent transmission model of Figure 8.38 was developed and used to analyze the circuit. Figures 8.39 and 8.40 show the measured and calculated performances of the fabricated uniplanar CPW branch-line coupler. Figure 8.39 shows that the amplitude imbalance of 1 dB is within a bandwidth of less than 20% at the center frequency of 3 GHz. The measured isolation between ports 1 and 4 is greater than 50 dB at the 3-GHz center frequency. The calculated results agree very well with the measured results.

Figure 8.41 shows the physical configuration of the slotline branch-line coupler. Slotline branch-line couplers are duals of the CPW branch-line couplers. The series arms and branch arms are connected in series. Figure 8.42 shows the equivalent circuit of the slotline branch-line coupler. The corresponding line characteristic impedances of the slotline series and branch arms for 3-dB coupling, in terms of the termination impedance Z_0, can be expressed as

$$Z_{S1} = \sqrt{2} Z_0 \tag{8.21}$$

$$Z_{S2} = Z_0 \tag{8.22}$$

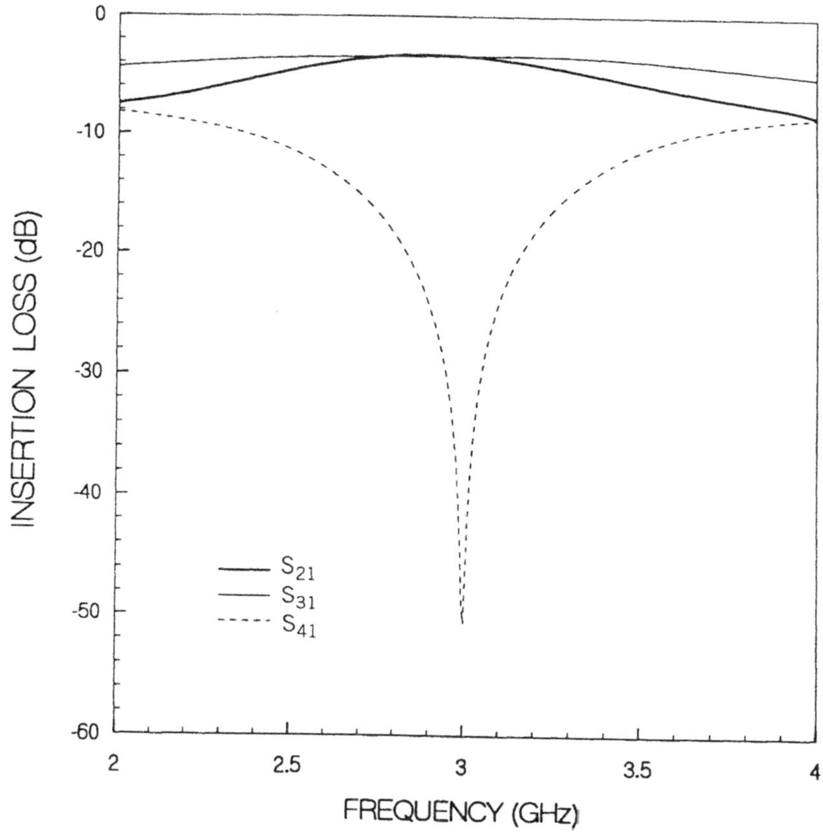

FIGURE 8.39 Measured results of power dividing and isolation for the CPW 2-branch coupler.

where Z_{S1} is the characteristic impedance of the slotline series arms, and Z_{S2} is the characteristic impedance of the slotline branch arms.

Figures 8.43 and 8.44 show the measured and calculated performances of the fabricated uniplanar slotline branch-line coupler. The calculated results were obtained from the equivalent transmission-line model shown in Figure 8.42. Figure 8.43 shows that the amplitude imbalance of 1 dB is within a bandwidth of less than 20% at the 3-GHz center frequency. The measured isolation between ports 1 and 4 is greater than 30 dB at the center frequency 3 GHz.

8.5.3 Asymmetrical Coplanar Strip Branch-Line Couplers

The 90° ACPS branch-line hybrid coupler is shown in Figure 8.45*a*. In a standard branch-line coupler [34], if the port characteristic impedance is Z_o and two of the $\lambda_g/4$ branches have a characteristic impedance of $Z_o/\sqrt{2}$. If $Z_o =$ 50 ohms, then the two $Z_o/\sqrt{2}$ lines would each have a characteristic impedance

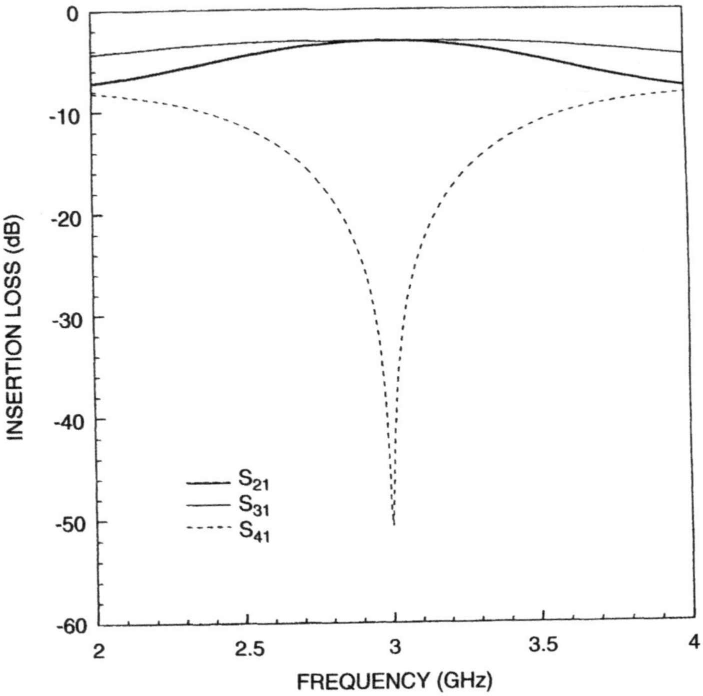

FIGURE 8.40 Calculated results of power dividing and isolation for the CPW 2-branch coupler.

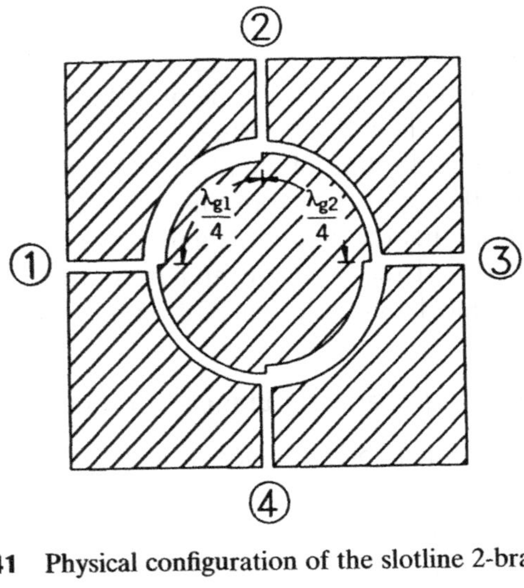

FIGURE 8.41 Physical configuration of the slotline 2-branch coupler.

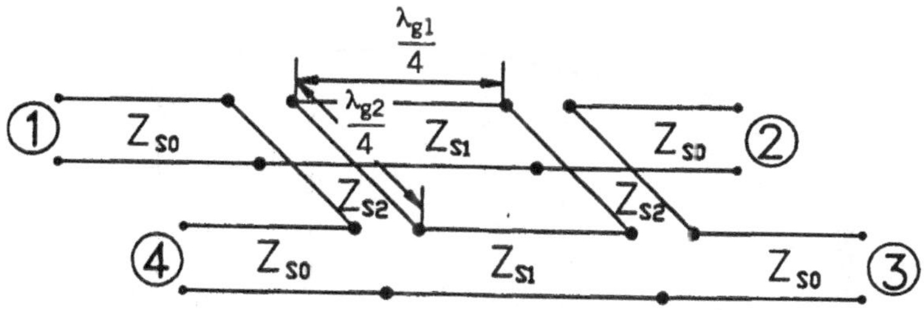

FIGURE 8.42 Equivalent circuit of the slotline 2-branch coupler.

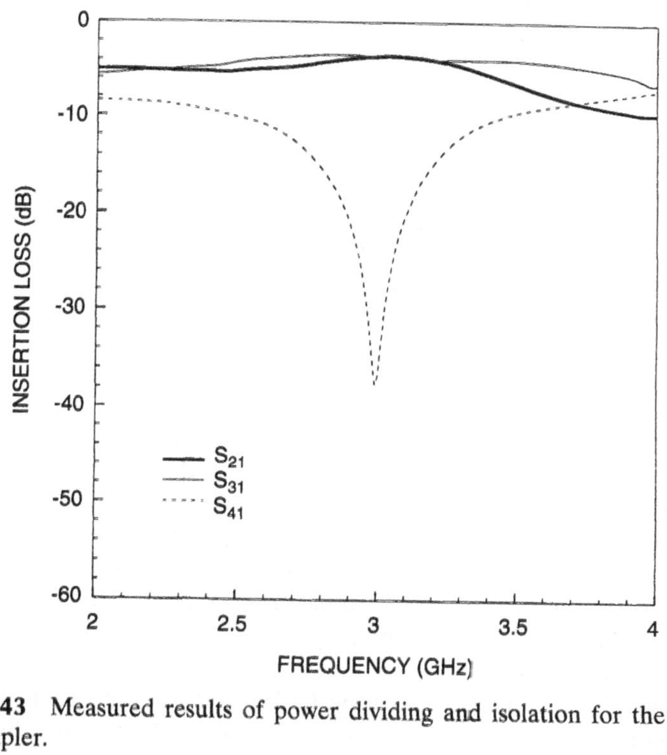

FIGURE 8.43 Measured results of power dividing and isolation for the slotline 2-branch coupler.

of 35.4 ohms. This impedance value is difficult to attain using ACPS. To overcome this problem, the input and output port characteristic impedances were increased to Z_o' (100 ohms). By using a CPW quarter-wavelength transformer, the coupler port impedances ($Z_{cpw}' = Z_o' = 100$ ohms) were matched to the CPW ($Z_{cpw} = Z_o = 50$ ohms), which can be connected to the standard 50-ohms test equipment. Based on the above consideration, two high-impedance branches

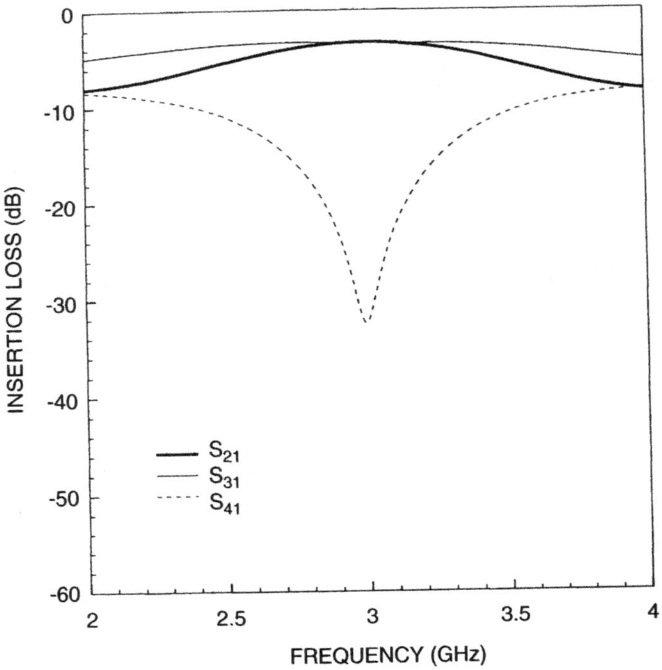

FIGURE 8.44 Calculated results of power dividing and isolation for the slotline 2-branch coupler.

($Z_{100} = Z_o' = 100$ ohms) and two low-impedance branches ($Z_{71} = Z_o'/\sqrt{2} = 71$ ohms) were designed. The equivalent circuit for this branch-line coupler is shown in Figure 8.45*b*. The 71-ohms ACPS branch line ($\lambda_{g,71}/4 = 11.07$ mm) has a spacing of $s = 0.2$ mm and a linewidth of $w_{ACPS} = 0.42$ mm. The 100-ohms ACPS branch line ($\lambda_{g,100}/4 = 10.96$ mm) has a spacing of $s = 0.4$ mm and a linewidth of $w_{ACPS} = 0.18$ mm. For the CPW quarter-wavelength transformer section ($\lambda_{T,cpw}/4 = 10.81$ mm, $Z_{T,cpw} = 71$ ohms), a gap of $G = 0.4$ mm and a linewidth of $w_{T,cpw} = 0.23$ mm are used.

Bond wires were attached over the CPW feed lines at the T-junctions to keep the coupled slotline modes from propagating. The branch-line coupler was fabricated on an $h = 0.635$-mm-thick RT/Duroid 6010 ($\varepsilon_r = 10.8$) substrate. Figure 8.46 shows that the branch-line coupler has attained a 10% bandwidth centered at 3 GHz. The coupling is 3.5 dB at 3 GHz (3 dB for ideal coupling, the insertion loss includes two CPW quarter-wavelength transformers of length 21.8 mm, two CPW input/output sections of length 10 mm, and two coaxial to CPW connectors that were not calibrated out). The input return loss is greater than 17.1 dB, and the isolation is greater than 15.3 dB. The coupler has a worst-case amplitude imbalance of 0.375 dB and a worst-case phase imbalance of 1.9° over the specified bandwidth.

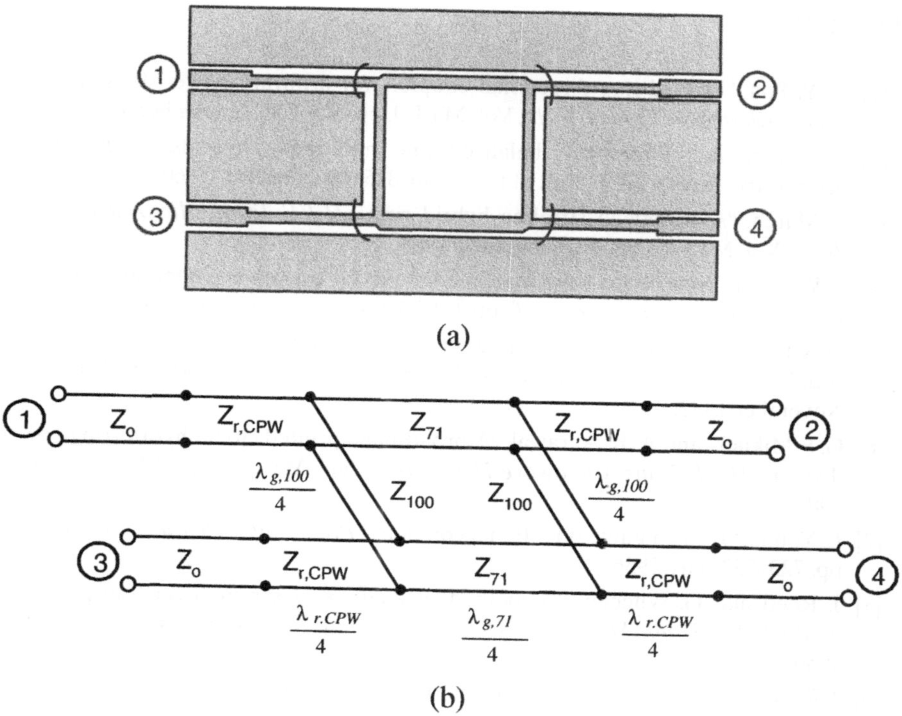

FIGURE 8.45 ACSP 90° branch-line coupler (a) configuration and (b) equivalent circuit [34]. (Permission from IEEE.)

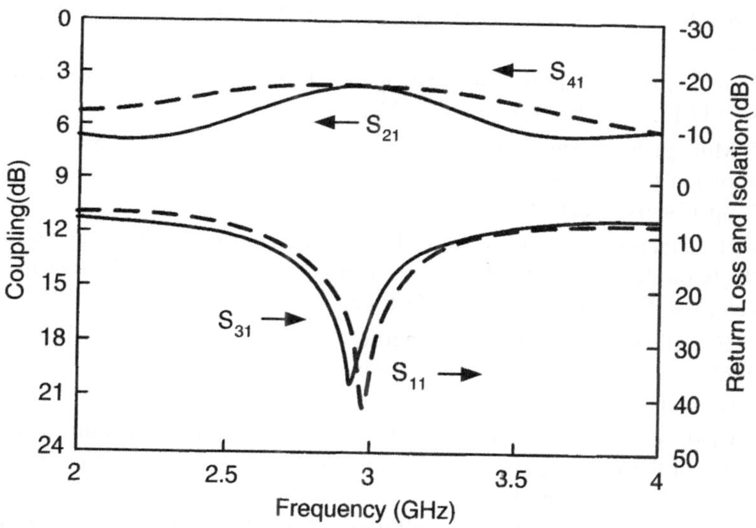

FIGURE 8.46 Measured coupling, return loss, and isolation for the ACSP 90° branch-line coupler [34]. (Permission from IEEE.)

REFERENCES

[1] C. Y. Pon, "Hybrid-ring directional couplers for arbitrary power division," *IRE Trans. Microwave Theory Tech.*, Vol. MTT-9, pp. 529–535, November 1961.

[2] S. Rehnmark, "Wide-band balanced line microwave hybrids," *IEEE Trans. Microwave Theory Tech.*, Vol. MTT-25, pp. 825–830, October 1960.

[3] S. March, "A wideband stripline hybrid ring," *IEEE Trans. Microwave Theory Tech.*, Vol. MTT-16, pp. 361–369, June 1968.

[4] L. W. Chua, "New broad-band matched hybrids for microwave integrated circuits," *Proc. 2nd Eur. Microwave Conf.*, pp. C4/5:1-C4/5:4, September 1971.

[5] D. Kim and Y. Naito, "Broad-band design of improved hybrid-ring 3 dB directional coupler," *IEEE Trans. Microwave Theory Tech.*, Vol. MTT-30, pp. 2040–2046, November 1982.

[6] G. F. Mikucki and A. K. Agrawal, "A broad-band printed circuit hybrid-ring power divider," *IEEE Trans. Microwave Theory Tech.*, Vol. MTT-37, pp. 112–117, January 1989.

[7] L. Young, "Branch guide directional couplers," *Proc. Natl. Electron. Conf.*, Vol. 12, pp. 723–732, July 1956.

[8] J. Reed and G. Wheeler, "A method of analysis of symmetrical four-port networks," *IRE Trans. Microwave Theory Tech.*, Vol. MTT-4, pp. 246–252, October 1956.

[9] J. Reed, "The multiple branch waveguide coupler," *IRE Trans. Microwave Theory Tech.*, Vol. MTT-6, pp. 398–403, October 1958.

[10] L. Young, "Synchronous branch guide directional couplers for low and high power applications," *IRE Trans. Microwave Theory Tech.*, Vol. MTT-10, pp. 459–475, November 1962.

[11] R. Levy and L. Lind, "Synthesis of symmetrical branch-guide directional couplers," *IEEE Trans. Microwave Theory Tech.*, Vol. MTT-16, pp. 80–89, February 1968.

[12] R. Levy, "Zolotarev branch-guide couplers," *IEEE Trans. Microwave Theory Tech.*, Vol. MTT-21, pp. 95–99, February 1973.

[13] M. Muraguchi, T. Yukitake, and Y. Naito, "Optimum design of 3-dB branch-line couplers using microstrip lines," *IEEE Trans. Microwave Theory Tech.*, Vol. MTT-31, pp. 674–678, August 1983.

[14] W. H. Leighton and A. G. Milnes, "Junction reactance and dimensional tolerance effects on X-band –3 dB directional couplers," *IEEE Trans. Microwave Theory Tech.*, Vol. MTT-19, pp. 818–824, October 1971.

[15] A. F. Celliers and J. A. G. Malherbe, "Design curves for –3-dB branch-line couplers," *IEEE Trans. Microwave Theory Tech.*, Vol. MTT-33, pp. 1226–1228, November 1985.

[16] T. Anada and J. P. Hsu, "Analysis and synthesis of triplate branch-line 3 dB coupler based on the planar circuit theory," in *1987 IEEE MTT-S Int. Microwave Symp. Dig.*, pp. 207–210, June 1987.

[17] A. Angelucci and R. Burocco, "Optimized synthesis of microstrip branch-line couplers taking dispersion, attenuation loss and T-junction into account," in *1988 IEEE MTT-S Int. Microwave Symp. Dig.*, pp. 543–546, June 1988.

[18] F. C. de Ronde, "A new class of microstrip directional couplers," in *1970 IEEE MTT-S Int. Microwave Symp. Dig.*, pp. 184–186, June 1970.

[19] J. A. Garcia, "A wide-band quadrature hybrid coupler," *IEEE Trans. Microwave Theory Tech.*, Vol. MTT-19, pp. 660–661, July 1971.

[20] B. Shiek, "Hybrid branch-line couplers—A useful new class of directional couplers," *IEEE Trans. Microwave Theory Tech.*, Vol. MTT-22, pp. 864–869, October 1974.

[21] B. Shiek and J. Koehler, "Improving the isolation of 3-dB couplers in microstrip-slotline technique," *IEEE Trans. Microwave Theory Tech.*, Vol. MTT-26, pp. 5–7, January 1978.

[22] F. C. de Ronde, "Octave-wide matched symmetrical, reciprocal, 4- and 5-ports," in *1982 IEEE MTT-S Int. Microwave Symp. Dig.*, pp. 521–523, June 1982.

[23] R. K. Hoffman and J. Siegl, "Microstrip-slot coupler design," Parts I and II, *IEEE Trans. Microwave Theory Tech.*, Vol. MTT-30, pp. 1205–1216, August 1982.

[24] M. Schoenberger, A. Biswas, A. Mortazawi, and V. K. Tripathi, "Coupled slot-strip coupler in finline," *IEEE MTT-S Int. Microwave Symp. Dig.*, pp. 751–753, June 1991.

[25] C. Ho, "Slotline, CPW ring circuits and waveguide ring cavities for coupler and filter applications," Ph.D. dissertation, Texas A&M University, College Station, May 1994.

[26] D. M. Pozar, *Microwave Engineering*, Addison-Wesley, Reading, Mass., 1990.

[27] C. Ho, L. Fan, and K. Chang, "Ultra wide band slotline ring couplers," in *1992 IEEE MTT-S Int. Microwave Conference Symp. Dig.*, pp. 1175–1178, 1992.

[28] J. B. Knorr, "Slot-line transitions," *IEEE Trans. Microwave Theory Tech.*, Vol. MTT-22, pp. 548–554, June 1974.

[29] S. B. Cohn, "Slotline field components," *IEEE Trans. Microwave Theory Tech.*, Vol. MTT-20, pp. 172–174, January 1972.

[30] I. Kneppo and J. Gotzman, "Basic parameters of nonsymmetrical coplanar lines," *IEEE Trans. Microwave Theory Tech.*, Vol. 25, p. 718, August 1977.

[31] D. Jaisson, "A single-balanced mixer with a coplanar balun," *Microwave J.*, Vol. 35, pp. 87–96, July 1992.

[32] D. Jaisson, "A microwave-coplanar waveguide coupler for use with an attenuator," *Microwave J.*, Vol. 38, No. 9, pp. 120–130, September 1995.

[33] L. Fan and K. Chang, "Uniplanar power dividers using coupled CPW and asymmetrical CPS for MICs and MMICs," *IEEE Trans. Microwave Theory Tech.*, Vol. 44, No. 12, pp. 2411–2420, December 1996.

[34] B. R. Heimer, L. Fan, and K. Chang, "Uniplanar hybrid couplers using asymmetrical coplanar striplines," *IEEE Trans. Microwave Theory Tech.*, Vol. 45, No. 12, pp. 2234–2240, December 1997.

[35] C. Ho, L. Fan, and K. Chang, "New uniplanar coplanar waveguide hybrid-ring couplers and magic-Ts," *IEEE Trans. Microwave Theory Tech.*, Vol. MTT-42, No. 12, pp. 2440–2448, December 1994.

[36] J. W. Duncan and V. P. Minerva, "100:1 bandwidth balun transformer," *Proc. IRE*, Vol. 48, pp. 156–164, January 1960.

[37] C. Ho, L. Fan, and K. Chang, "Broad-band uniplanar hybrid-ring and branch-line couplers," *IEEE Trans. Microwave Theory Tech.*, Vol. MTT-41, No. 12, pp. 2116–2125, December 1993.

[38] S. J. Robinson, "Broad-band hybrid junctions," *IRE Trans. Microwave Theory Tech.*, Vol. 8, pp. 671–672, November 1960.

[39] S. March, "A wide band stripline hybrid ring," *IEEE Trans. Microwave Theory Tech.*, Vol. 16, p. 361, June 1968.

[40] L. W. Chua, "New broad-band matched hybrids for microwave integrated circuits," in *1971 Proc. European Microwave Conf.*, pp. C4/5–C4/5:4, 1971.

[41] L. Fan, C.-H. Ho, S. Karamaluru, and K. Chang, "Wide-band reduced-size uniplanar magic-T. hybrid-ring, and de Ronde's CPW-slot couplers," *IEEE Trans. Microwave Theory Tech.*, Vol. 43, No. 12, pp. 2749–2758, December 1995.

CHAPTER NINE

Ring Magic-T Circuits

9.1 INTRODUCTION

This chapter presents novel ring magic-T circuits in details [1]. Magic-Ts are fundamental components for many microwave circuits such as power combiners and dividers, balanced mixers, and frequency discriminators. The matched waveguide double-T is a well-known and commonly used wave-guide magic-T [2, 3]. Figures 9.1 and 9.2 show the physical configuration and electric field distribution of the waveguide magic-T, respectively. As shown in Figure 9.2*a*, when a TE_{10} mode is incident at port H, the resulting E_y field lines have an even symmetry in port E. This means that there is no coupling between ports H and E. At the T-junction the incident wave will divide into two components, both of which arrive in phase at ports 1 and 2. As shown in Figure 9.2*b*, when a TE_{10} mode is incident at port E, the resulting E_y field lines have an odd symmetry in port H. Again ports E and H are decoupled. At the T-junction the incident wave will divide into two components, both of which arrive at ports 1 and 2 with a 180° phase difference. In practice, tuning posts and irises are used for matching the double-T junction. The tuning posts and irises must be placed symmetrically to maintain proper operation.

In 1964, Kraker [4] first proposed a planar magic-T. The circuit uses an asymmetric coupled transmission-line directional coupler and Shiffman's phase-shift network. In 1965, DuHamel and Armstrong [5] proposed a tapered-line magic-T. The circuit is based on a tapered asymmetrical transformer consisting of two coupled tapered lines. A complete analysis of the tapered-line magic-T was discussed in [6]. Laughlin [7] proposed a planar magic-T using a microstrip balun in 1976. In 1980, Aikawa and Ogawa [8] proposed a double-sided magic-T that is constructed with microstrip–slotline

Microwave Ring Circuits and Related Structures, Second Edition,
by Kai Chang and Lung-Hwa Hsieh
ISBN 0-471-44474-X

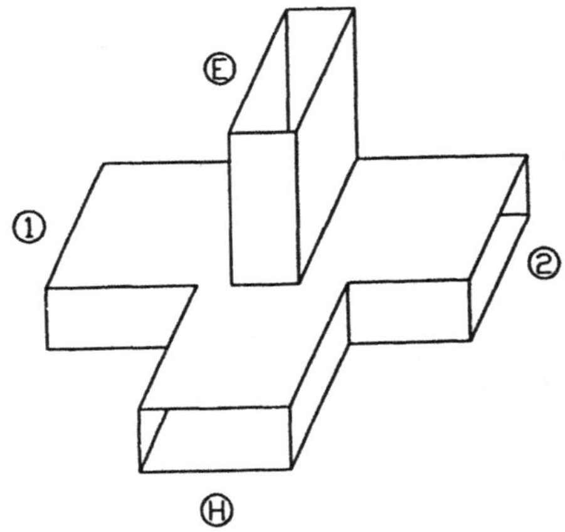

FIGURE 9.1 Physical configuration of the waveguide magic-T.

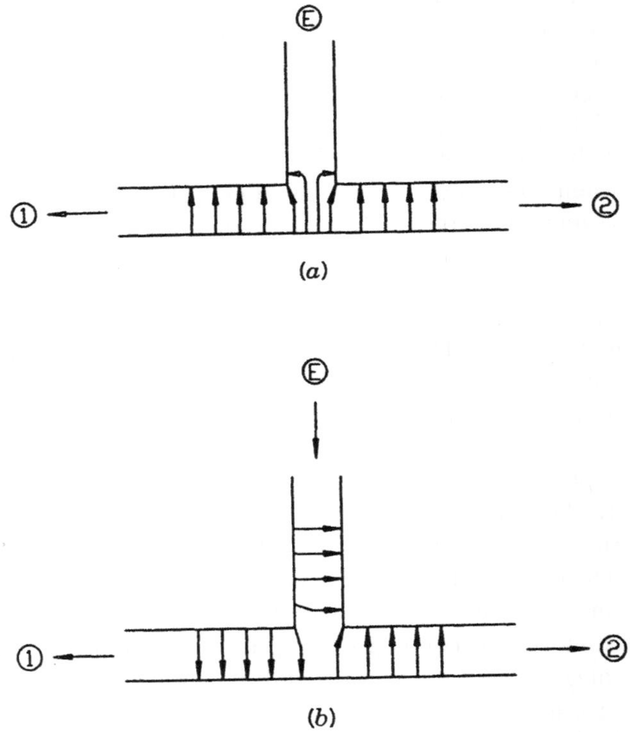

FIGURE 9.2 Schematic diagram of the E-field distribution of the (a) H-arm's excitation and (b) the E-arm's excitation.

T-junctions and coupled slotlines. The double-sided magic-T uses a double-sided structure and has a 2–10-GHz bandwidth. The two balanced arms of the double-sided magic-T are on the same side and they do not need a crossover connection. In recent years, uniplanar transmission lines have emerged as alternatives to microstrip in planar microwave integrated circuits. As mentioned before, the uniplanar microwave integrated circuits do not use the backside of the substrate, and allow easy series and shunt connections of passive and active solid-state devices. The use of uniplanar structures circumvents the need for via holes and reduces processing complexity. In 1987, Hirota et al. [9] proposed a uniplanar magic-T that uses three coplanar waveguide–slotline (CPW) T-junctions and a slotline T-junction. The in-phase CPW excitation is via an air bridge and the slotline T-junction is used as a phase inverter. The uniplanar magic-T has a narrow bandwidth.

This chapter first explains the fundamental characteristics of the 180° reverse-phase CPW–slotline T-junction. The proposed uniplanar T-junction uses a 180° reverse-phase CPW–slotline back-to-back transition as output ports to achieve a 180° phase reversal. The phase shift of the T-junction is frequency independent. The third section presents a new uniplanar CPW magic-T. The circuit consists of a 180° reverse-phase CPW–slotline T-junction and three CPW T-junctions. The fourth section of this chapter discusses the double-sided slotline magic-T. The fifth section discusses the uniplanar slotline magic-T. The circuits discussed in the fourth and fifth sections are based on the 180° phase-reversal of the slotline T-junction.

9.2 180° REVERSE-PHASE CPW–SLOTLINE T-JUNCTIONS

Figure 9.3 shows the circuit configuration and schematic diagram of the *E*-field distribution for a 180° reverse-phase CPW–slotline T-junction [1, 10]. The arrows shown in this figure indicate the schematic expression of the electric field in the CPWs and slotlines. The circuit consists of one CPW–slotline T-junction and two CPW–slotline transitions. As mentioned in Chapter 8, the phase change of the 180° reverse-phase CPW–slotline back-to-back transition is frequency independent and can be applied to wide-band circuits. As shown in Figure 9.3, the *E* field in the input CPW (near the CPW–slotline T-junction) is directed toward the CPW center conductor. This produces two slotline waves with the *E*-field in the $+y$ direction. At the transition of port 1, the $+y$-directed slotline *E*-field causes the *E*-field in the output CPW to be directed toward the CPW center conductor. However, the *E*-field of the output CPW at port 2 is directed toward the CPW ground plane due to the $+y$-directed slotline *E*-field.

According to the preceding principle, a truly uniplanar 180° reverse-phase CPW–slotline T-junction was built on a RT/Duroid 6010.8 ($\varepsilon_r = 10.8$) substrate with the following dimensions: substrate thickness $h = 1.27$ mm, characteristic impedance of the input/output CPW feed lines $Z_{C0} = 50\,\Omega$, input/output CPW

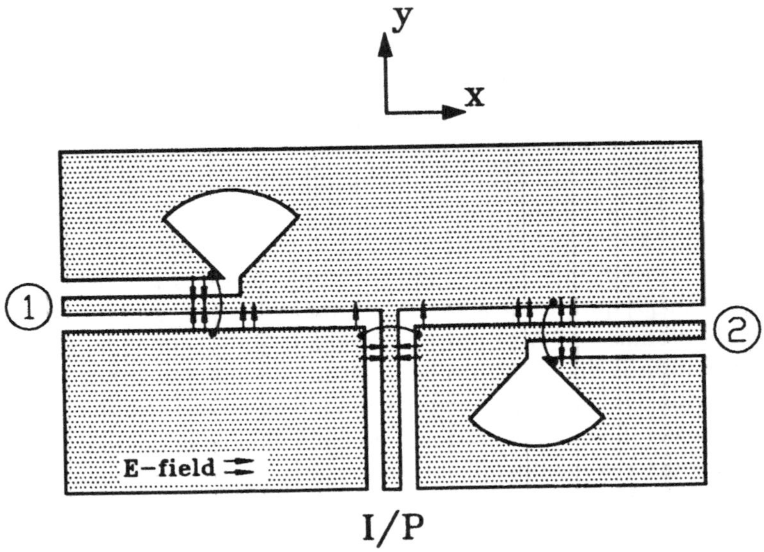

FIGURE 9.3 Physical layout and schematic diagram of the E-field distribution for the 180° reverse-phase CPW–slotline T-junction [10]. (Permission from IEEE.)

feed lines center conductor width $S_{C0} = 0.51\,\text{mm}$, input/output CPW feed lines gap size $G_{C0} = 0.25\,\text{mm}$, characteristic impedance of the slotline $Z_S = 60.6\,\Omega$, slotline line width $W_S = 0.2\,\text{mm}$, radius of the slotline radial stub $r = 6\,\text{mm}$, and angle of the slotline radial stubs $\theta = 90°$. The measurements were made using standard SMA connectors and an HP-8510 network analyzer. The insertion loss includes two coaxial–CPW transitions and one CPW–slotline transition.

Figures 9.4 through 9.6 show the measured performances of the fabricated uniplanar 180° reverse-phase CPW–slotline T-junction. Figure 9.4 shows the measured frequency responses of insertion loss for the output power dividing. Figure 9.5 shows the measured frequency responses of the phase angles at the output ports. Figure 9.6 shows the amplitude and phase differences. The maximum amplitude difference is 0.6 dB from 2 GHz to 4 GHz. Over the same frequency range, the maximum phase difference is 3.5°.

9.3 CPW MAGIC-Ts

Figure 9.7 shows the circuit configuration of the uniplanar CPW magic-T [1, 10]. The uniplanar magic-T consists of a 180° reverse-phase CPW–slotline T-junction and three CPW T-junctions. The 180° reverse-phase CPW–slotline T-junction is used as a phase inverter. In Figure 9.7, ports E and H correspond to the E- and H-arm of the conventional waveguide magic-T, respectively. Ports 1 and 2 are the power-dividing balanced arms. Figure 9.8 shows the

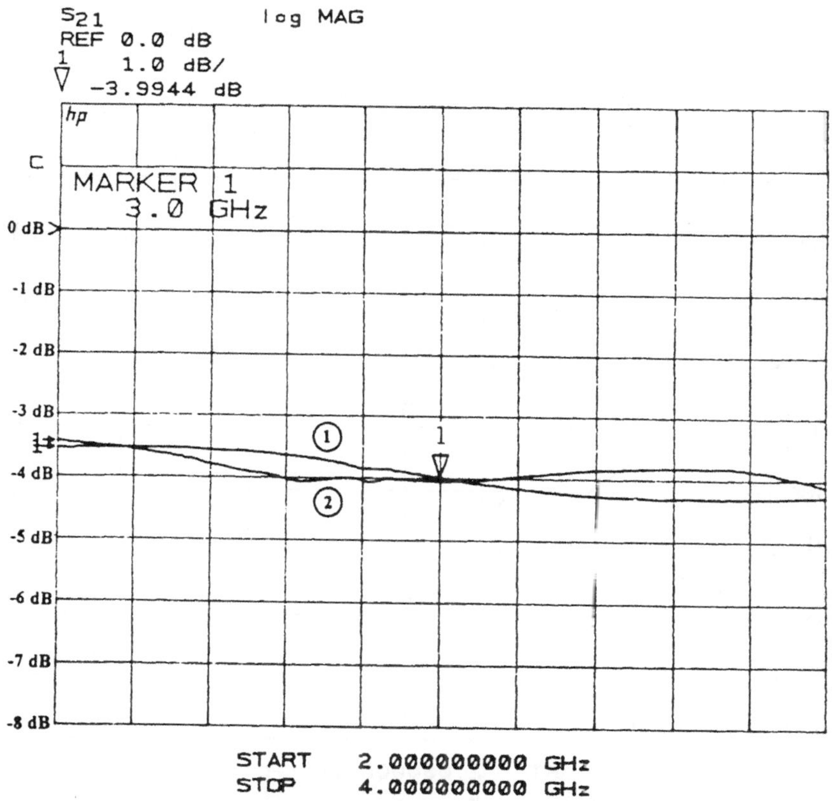

FIGURE 9.4 Measured frequency responses of power dividing for the uniplanar 180° reverse-phase CPW–slotline T-junction.

equivalent transmission-line model of the uniplanar CPW magic-T. The twisted transmission line in Figure 9.8 represents the phase reversal of the CPW–slotline T-junction.

Figures 9.9 and 9.10 show the schematic expressions of the *E*-field distribution and equivalent circuit for the in-phase and 180° out-of-phase couplings, respectively. The arrows shown in Figure 9.9 and 9.10 indicate the schematic expression of the electric field in the CPWs and slotlines. In Figure 9.9, the signal is fed to port H, and then divides into two components, both of which arrive in-phase at ports 1 and 2. The two component waves arrive at port E 180° out of phase and cancel each other. In this case, the symmetry plane at port H corresponds to an open circuit (magnetic wall), while the symmetry plane at port E corresponds to a short circuit (electric wall). In Figure 9.10, the signal is fed to port E, and then divides into two components, which arrive at ports 1 and 2 with a 180° phase difference. The 180° phase difference between the divided signals at ports 1 and 2 is due to the 180° reverse-phase CPW–slotline T-junction. The two component waves arrive at port H 180° out

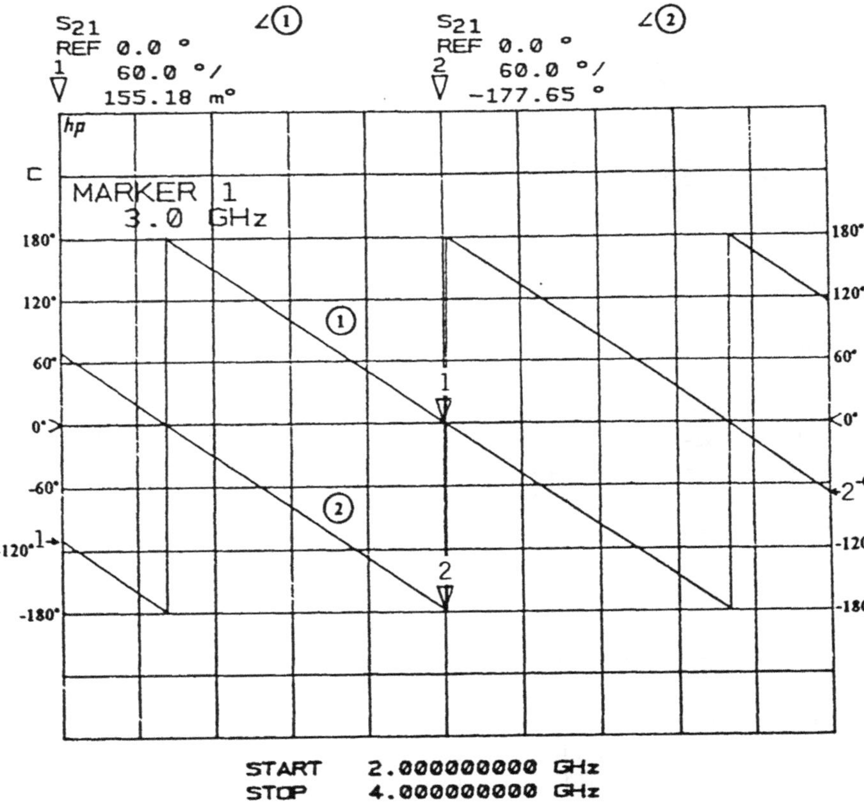

FIGURE 9.5 Measured frequency responses of phase angles for the uniplanar 180° phase-reversed CPW–slotline T-junction.

of phase and cancel each other. The symmetry plane at port E corresponds to an open circuit (magnetic wall); the symmetry plane at port H corresponds to a short circuit (electric wall). The isolation between ports E and H is perfect as long as the mode conversion in the reverse-phase CPW–slotline T-junction is ideal.

The in-phase equivalent circuit in Figure 9.9 is obtained when ports 1 and 2 are excited by two in-phase input signals with the same amplitude. In this case, the symmetry plane at port H corresponds to a short circuit, and the symmetry plane at port H corresponds to an open circuit. The out-of-phase equivalent circuit in Figure 9.10 is obtained when ports 1 and 2 are excited by two 180° out-of-phase input signals with the same amplitude. In this case, the symmetry plane at port B corresponds to an open circuit, and the symmetry plane at port H corresponds to a short circuit. A two-port circuit calculation is used to analyze the isolation and impedance matching instead of the symmetric four-port networks discussed in Chapter 8, because the circuit is

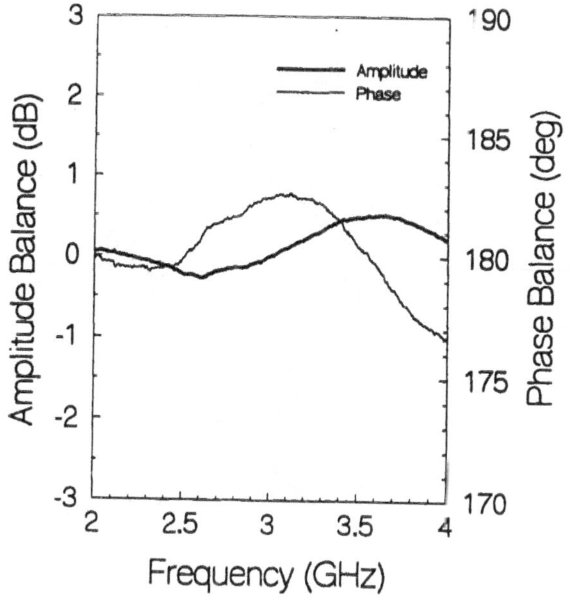

FIGURE 9.6 Amplitude and phase differences for the uniplanar 180° phase-reversed CPW–slotline T-junction.

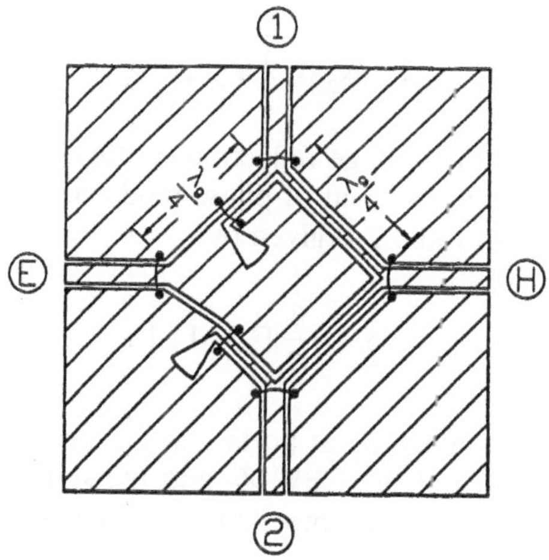

FIGURE 9.7 Physical configuration of the uniplanar CPW magic-T using a 180° reverse-phase CPW–slotline T-junction [10]. (Permission from IEEE.)

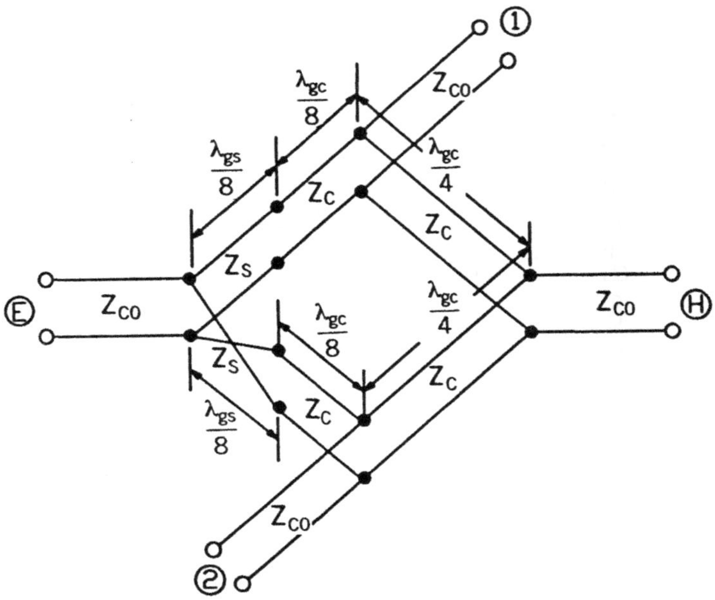

FIGURE 9.8 Equivalent circuit of the uniplanar CPW magic-T in Figure 9.7 [10]. (Permission from IEEE.)

symmetric with respect to ports E and H [8]. The return loss at ports 1 and 2 is given by

$$|S_{11}|, |S_{22}| = \frac{1}{2}|\Gamma_{++} + \Gamma_{+-}| \tag{9.1}$$

where Γ_{++} and Γ_{+-} are the voltage reflection coefficients at port 1 for the in-phase mode coupling and 180° out-of-phase mode coupling, respectively. The isolation between ports 1 and 2 is given by

$$|S_{12}| = \frac{1}{2}|\Gamma_{++} - \Gamma_{+-}| \tag{9.2}$$

To achieve impedance matching at ports 1 and 2, that is, $|S_{11}| = |S_{22}| = 0$, the characteristic impedance of the CPW Z_C and the slotline Z_S in terms of the input/output CPW characteristic impedance Z_{C0} is given by

$$Z_S = Z_C = \sqrt{2}Z_{C0} \tag{9.3}$$

According to Equations (9.1) to (9.3), a truly uniplanar magic-T was built on a RT/Duroid 6010.8 ($\varepsilon_r = 10.8$) substrate with the following dimensions:

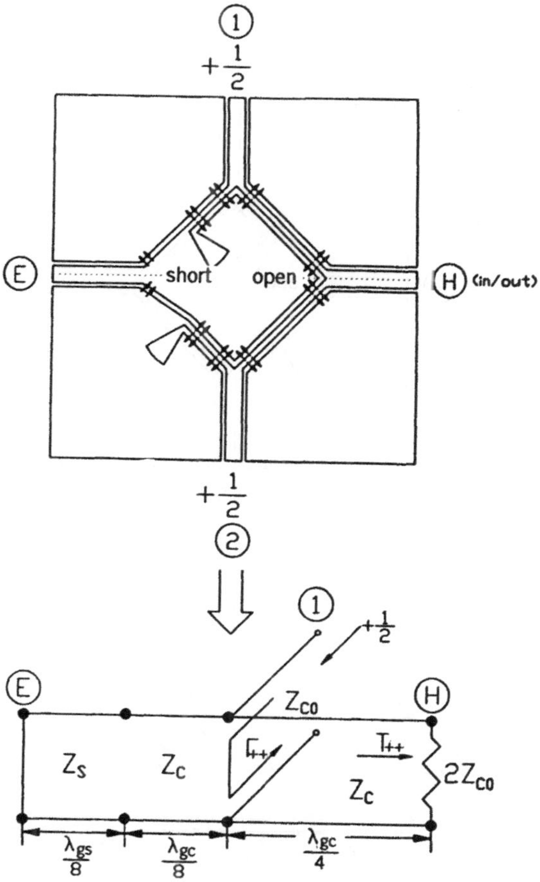

FIGURE 9.9 Schematic expression of the *E*-field distribution and equivalent transmission-line model for the in-phase coupling mode [10]. (Permission from IEEE.)

substrate thickness $h = 1.27\,\text{mm}$, characteristic impedance of the input/output CPW feed lines $Z_{C0} = 50\,\Omega$, input/output CPW feed lines center conductor width $S_{0C} = 0.51\,\text{mm}$, input/output CPW feed lines gap size $G_{C0} = 0.25\,\text{mm}$, characteristic impedance of the CPW in the magic-T $Z_C = 70.7\,\Omega$, magic-T CPW center conductor width $S_C = 0.51\,\text{mm}$, magic-T CPW gap size $G_C = 0.25\,\text{mm}$, characteristic impedance of the slotline in the magic-T $Z_S = 70.7\,\Omega$, magic-T slotline line width $W_S = 0.25\,\text{mm}$, slotline radial stub angle $\theta = 30°$, and slotline radial stub radius $r = 5\,\text{mm}$. The measurements were made using standard SMA connectors and an HP-8510 network analyzer. A computer program based on the equivalent transmission model of Figure 9.8 was developed and used to analyze the circuit.

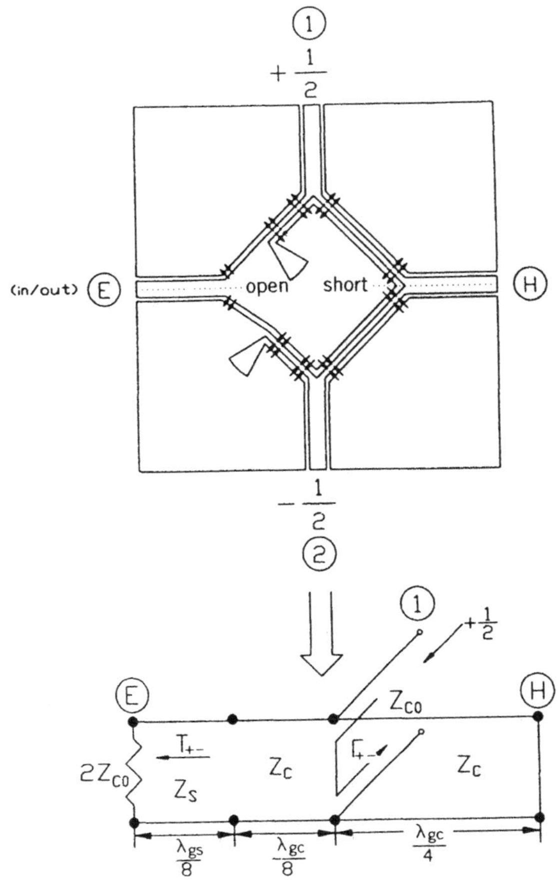

FIGURE 9.10 Schematic expression of the *E*-field distribution and equivalent transmission-line model for the 180° out-of-phase coupling mode [10]. (Permission from IEEE.)

Figure 9.11 shows the measured and calculated frequency responses of insertion loss for the H-arm's power dividing, that is, in-phase mode coupling. The extra insertion loss is less than 0.7 dB at the center frequency of 3 GHz. The maximum amplitude imbalance of the H-arm is 0.3 dB in the frequency range of 2–4 GHz. Figure 9.12 shows the measured and calculated frequency responses of insertion loss for the E-arm's power dividing, that is, 180° out-of-phase mode coupling. The extra insertion loss is less than 1.1 dB at the center frequency of 3 GHz. The maximum amplitude imbalance of the E-arm is 0.5 dB in the frequency range of 2–4 GHz. As shown in Figures 9.11 and 9.12, the calculated results agree in general with the measured results except the insertion loss. The additional insertion loss of the CPW magic-T in the measurement is mainly due to the CPW–slotline transition in the reverse-phase T-junction.

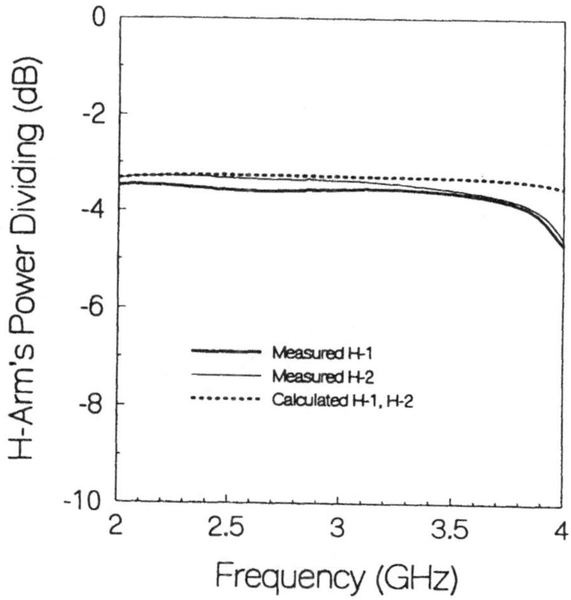

FIGURE 9.11 Measured and calculated frequency responses of the H-arm's power dividing for the uniplanar CPW magic-T [10]. (Permission from IEEE.)

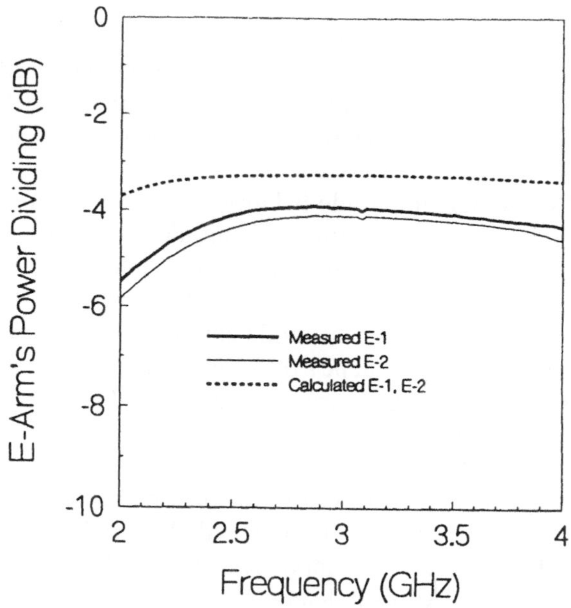

FIGURE 9.12 Measured and calculated frequency responses of the E-arm's power dividing for the uniplanar CPW magic-T [10]. (Permission from IEEE.)

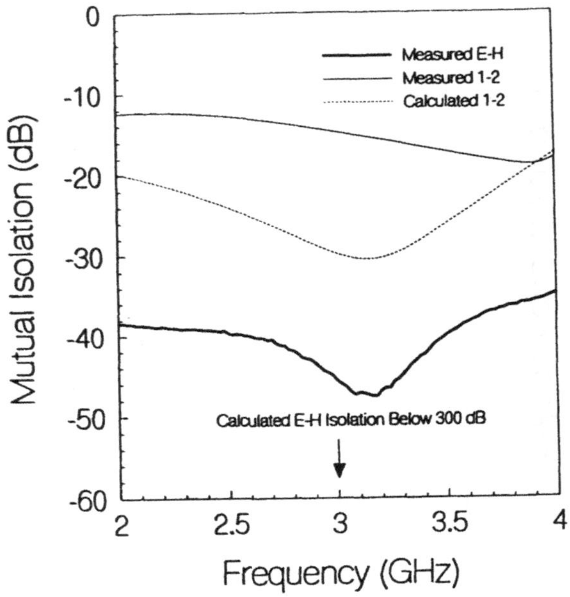

FIGURE 9.13 Measured and calculated frequency responses of the mutual isolation for the uniplanar CPW magic-T [10]. (Permission from IEEE.)

Figure 9.13 shows the measured and calculated frequency responses of mutual isolation between the E- and H-arms and the balanced arms 1 and 2. The isolation between the E- and H-arms is greater than 30 dB from 2 GHz to 4 GHz. Over the same frequency range, the mutual isolation between the two balanced arms is greater than 12 dB.

Figure 9.14 shows the amplitude balance for the 180° out-of-phase and in-phase mode coupling. The maximum amplitude imbalance of the E-arm is 0.5 dB from 2–4 GHz. The maximum amplitude imbalance of the H-arm is 0.3 dB in the same frequency range. Figure 9.15 shows the phase balance for the 180° out-of-phase and in-phase mode coupling. The phase error of the E-arm is 0.8° at the center frequency of 3 GHz. The E-arm's maximum phase imbalance is 3.5° over the frequency range of 2–4 GHz. The phase error of the H-arm is 0.7° at the center frequency of 3 GHz. The H-arm's maximum phase imbalance is 2° from 2–4 GHz.

The experimental and theoretical results just presented show that the uniplanar magic-T has fairly good amplitude and phase balances. With the advantages of broad-band operation, simple design procedure, uniplanar structure, and ease of integrating with solid-state devices, the uniplanar CPW magic-T should have many applications in microwave and millimeter-wave hybrid and monolithic integrated circuits.

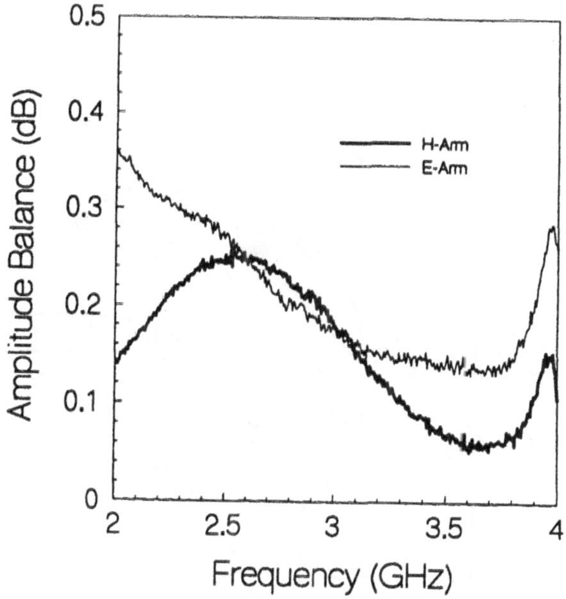

FIGURE 9.14 H- and E-arms' amplitude balances for the uniplanar CPW magic-T [10]. (Permission from IEEE.)

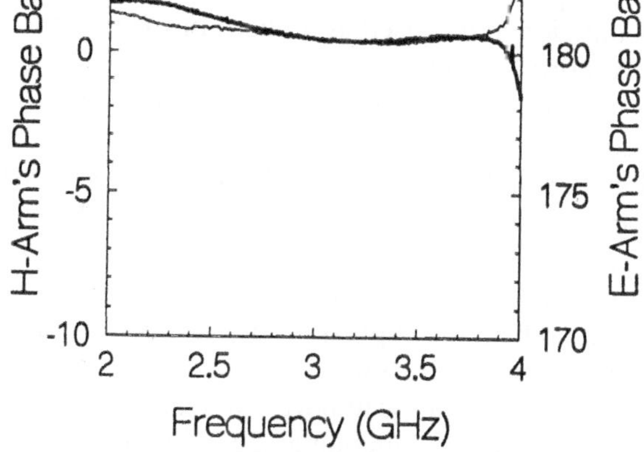

FIGURE 9.15 H- and E-arms' phase balances for the uniplanar CPW magic-T [10]. (Permission from IEEE.)

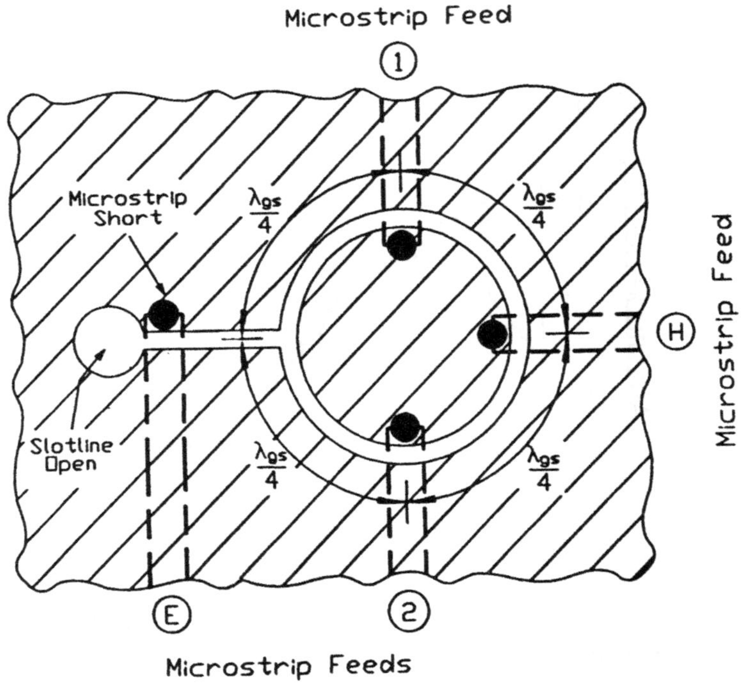

FIGURE 9.16 Physical layout of the double-sided slotline ring magic-T with microstrip feeds.

9.4 180° DOUBLE-SIDED SLOTLINE RING MAGIC-Ts

Figure 9.16 shows the circuit layout of the double-sided slotline ring magic-T [1, 11, 12]. The circuit simply consists of a slotline T-junction and a slotline ring with three microstrip feeds. The slotline T-junction is a well-known 180° reverse-phase T-junction and is used as a phase inverter in the slotline magic-T. In Figure 9.16, ports E and H correspond to the E- and H-arms of the conventional waveguide magic-T, respectively. Ports 1 and 2 are the power-dividing balanced arms. Figure 9.17 shows the equivalent transmission-line model of the double-sided slotline magic-T. The twisted transmission line in Figure 9.17 represents the phase reversal of the slotline T-junction.

Figures 9.18 and 9.19 show the schematic expressions of the in-phase and 180° out-of-phase couplings, respectively. The arrows shown in Figures 9.18 and 9.19 indicate the schematic expression of the electric field in the slotlines. In Figure 9.18, the signal is fed to port H and then divides into two components, both of which arrive in phase at ports 1 and 2. The two component waves arrive in-phase at the slotline T-junction and cannot be extracted from port E. In Figure 9.19 the signal is fed to port E and then divides into two components, which arrive at ports 1 and 2 with a 180° phase difference. The 180°

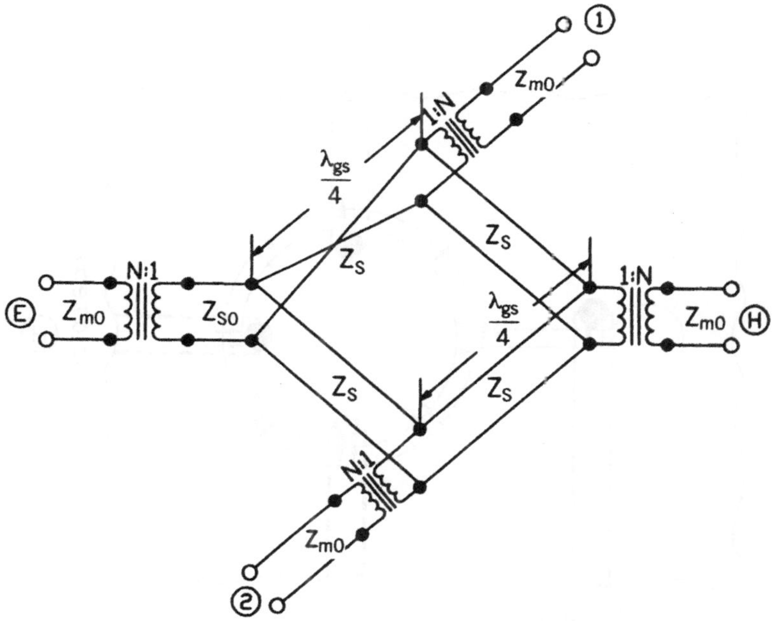

FIGURE 9.17 Equivalent circuit of the double-sided slotline ring magic-T with microstrip feeds.

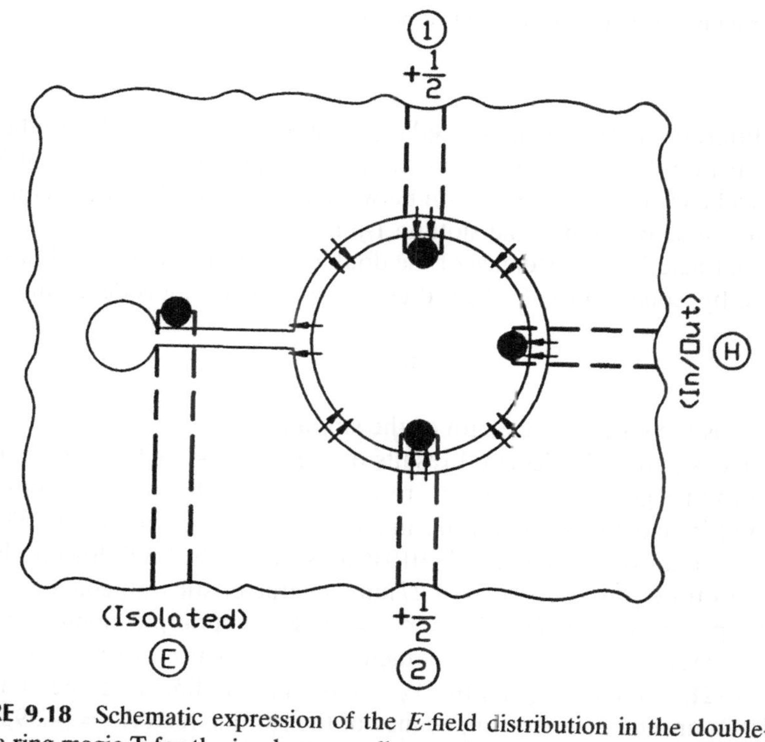

FIGURE 9.18 Schematic expression of the *E*-field distribution in the double-sided slotline ring magic-T for the in-phase coupling mode.

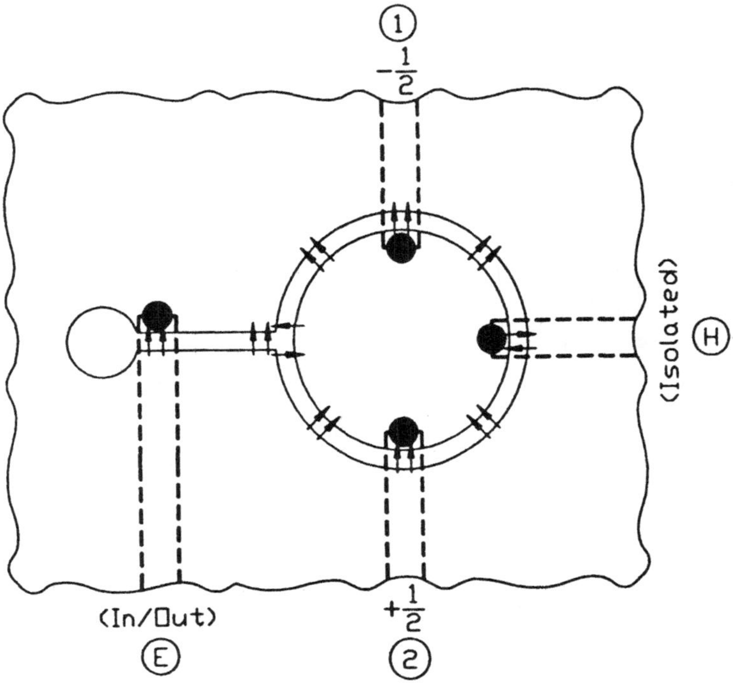

FIGURE 9.19 Schematic expression of the *E*-field distribution in the double-sided slotline ring magic-T for the 180° out-of-phase coupling mode.

phase difference between the divided signals at ports 1 and 2 is due to the slotline T-junction. The two component waves arrive at port H 180° out of phase and cancel each other. The isolation between ports B and H is perfect as long as the mode conversion in the slotline T-junction is ideal.

The characteristic impedance of the double-sided slotline magic-T was also designed by Equation (9.3). The radius of the slotline ring is determined by

$$2\pi r = \lambda_{gs} \tag{9.4}$$

where λ_{gs} is the guide wavelength of the slotline ring.

The measured and calculated results of the double-sided slotline magic-T are shown in Figures 9.20 and 9.21, respectively. The theoretical results were calculated from the equivalent transmission-line model in Figure 9.17. The test circuit was built on a RT/Duroid 6010.8 substrate with the following dimensions: substrate thickness h = 1.27 mm, characteristic impedance of the input/output microstrip feed lines Z_{m0} = 50 Ω, input/output microstrip feed lines line width W_{m0} = 1.09 mm, characteristic impedance of the slotline ring Z_S = 70.7 Ω, slotline ring width W_S = 0.85 mm, slotline ring mean radius r = 8.4193 mm, characteristic impedance of the slotline feed Z_{S0} = 54.39 Ω, and

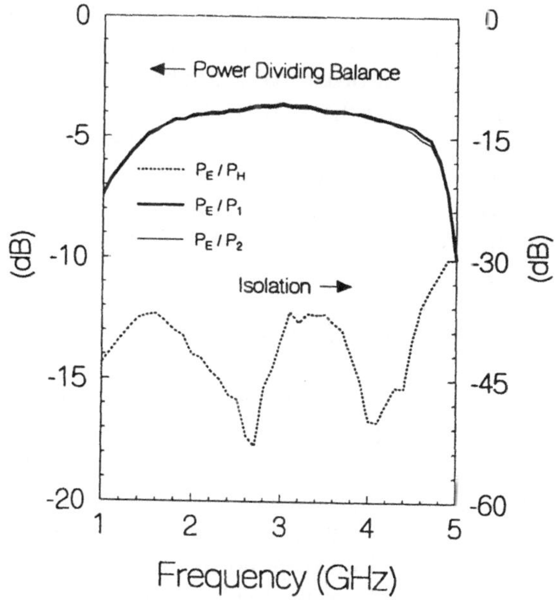

FIGURE 9.20 Measured results of power dividing and isolation for the double-sided slotline ring magic-T with microstrip feeds [12]. (Permission from IEEE.)

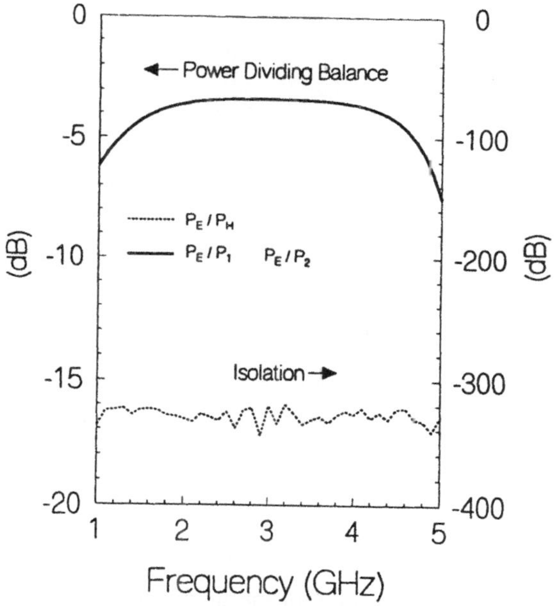

FIGURE 9.21 Calculated results of power dividing and isolation for the double-sided slotline ring magic-T with microstrip feeds [12]. (Permission from IEEE.)

slotline feed line width $W_{S0} = 0.1\,\text{mm}$. The measurements were made using standard SMA connectors and an HP-8510 network analyzer.

As shown in Figure 9.20, a broad-band double-side slotline magic-T with an excellent isolation of greater than 35 dB and a good power-dividing balance of 0.2 dB was achieved over an 80% bandwidth. The 1.3-dB insertion loss at the center frequency of 3 GHz is due to the microstrip–slotline transition and the slotline T-junction. Except for the extra insertion loss, the measured and calculated results shown in Figures 9.20 and 9.21 agree very well.

9.5 180° UNIPLANAR SLOTLINE RING MAGIC-Ts

Figure 9.22 shows the physical configuration of the uniplanar slotline ring magic-T [1, 10, 13, 14]. The E-arm of the uniplanar slotline magic-T is fed through a CPW connected to a broad-band TYPE-AC CPW–slotline transition [13]. The slotline T-junction is used as a phase inverter to achieve the 180° phase reversal. The H-arm and output balanced arms are all fed by CPW lines. In Figure 9.22, ports E and H correspond to the E- and H-arms of the conventional waveguide magic-T, respectively. Ports 1 and 2 are the power-dividing balanced arms. Figure 9.23 shows the equivalent transmission-line

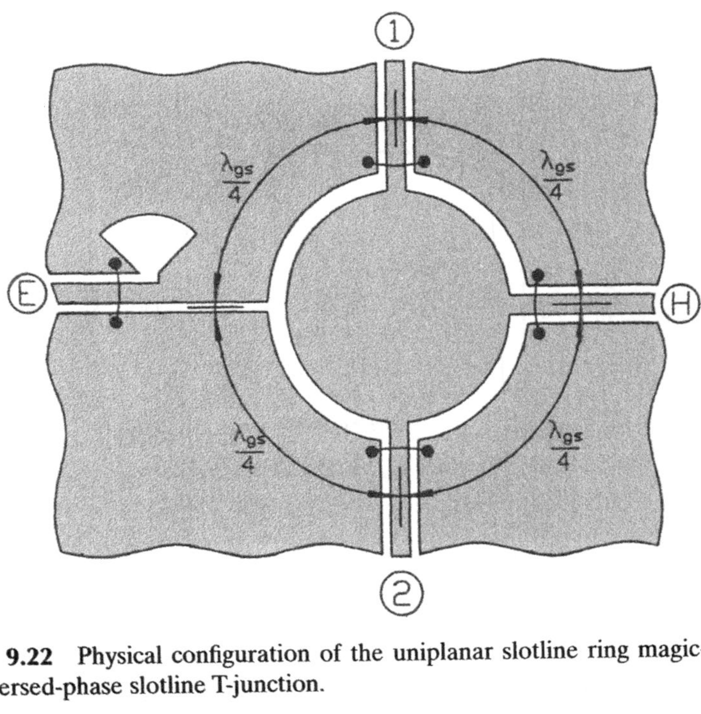

FIGURE 9.22 Physical configuration of the uniplanar slotline ring magic-T using a 180° reversed-phase slotline T-junction.

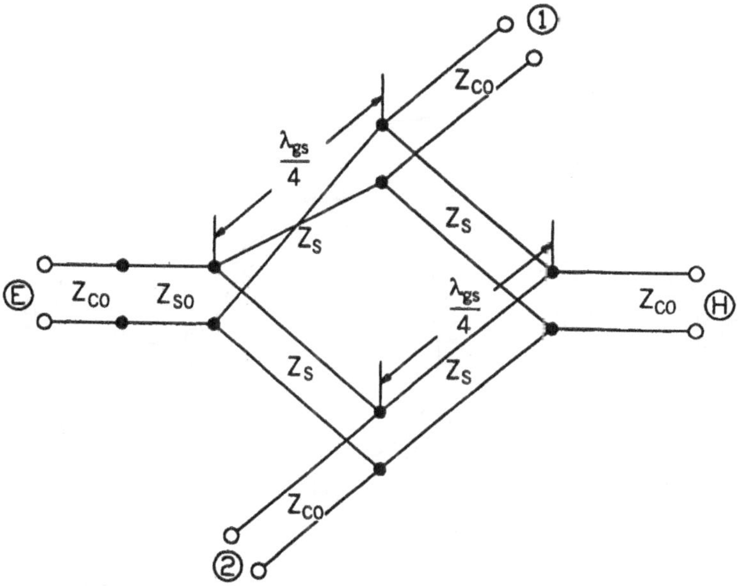

FIGURE 9.23 Equivalent circuit of the uniplanar slotline ring magic-T.

model of the CPW ring magic-T. The twisted transmission line in Figure 9.23 represents the phase reversal of the slotline T-junction.

Figures 9.24 and 9.25 show the schematic expressions of the in-phase and 180° out-of-phase couplings, respectively. The arrows shown in Figures 9.24 and 9.25 indicate the schematic expression of the electric field in the CPWs and slotlines. In Figure 9.24, the signal is fed to port H and then divides into two components, both of which arrive in phase at ports 1 and 2. The two component waves arrive at the slotline T-junction in phase and cannot be extracted from port E. In Figure 9.25, the signal is fed to port E and then divides into two components, which arrive at ports 1 and 2 with a 180° phase difference. The 180° phase difference between the divided signals at ports 1 and 2 is due to the slotline T-junction. The two component waves arrive at port H 180° out of phase and cancel each other. The isolation between ports E and H is perfect as long as the mode conversion in the slotline T-junction is ideal.

The characteristic impedance of the slotline ring and CPW feed lines is determined by Equation (9.3). The mean radius of the slotline ring is given by Equation (9.4). Using Equations (9.3) and (9.4), a truly uniplanar slotline ring magic-T was built on a RT/Duroid 6010.8 ($\varepsilon_r = 10.8$) substrate with the following dimensions: substrate thickness $h = 1.27$ mm, characteristic impedance of the input/output CPW feed lines $Z_{C0} = 50\,\Omega$, input/output CPW feed lines center conductor width $S_{C0} = 0.51$ mm, input/output CPW feed lines gap size $G_{C0} = 0.25$ mm, characteristic impedance of the slotline feed $Z_{S0} = 54.39\,\Omega$, slot-

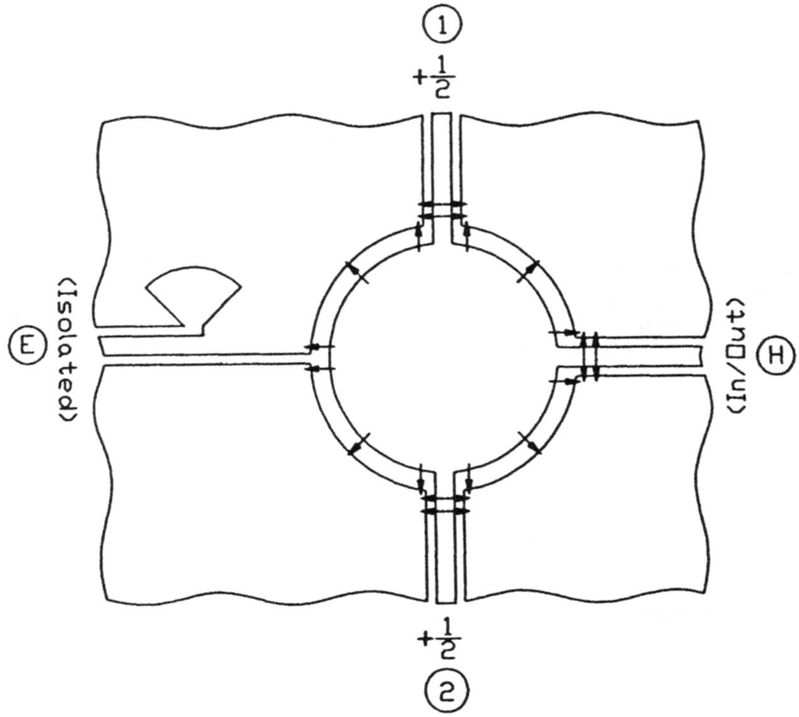

FIGURE 9.24 Schematic expression of the *E*-field distribution in the uniplanar slotline ring magic-T for the in-phase coupling mode.

line feed line width $W_{S1} = 0.1$ mm, characteristic impedance of the slotline ring $Z_S = 70.7\,\Omega$, slotline ring line width $W_{S2} = 0.43$ mm, and slotline ring radius $r = 7.77$ mm. The measurements were made using standard SMA connectors and an HP-8510 network analyzer. A computer program based on the equivalent transmission mode of Figure 9.23 was developed and used to analyzer the circuit.

Figure 9.26 shows the measured and calculated frequency responses of insertion loss for the E-arm's power dividing, that is, 180° out-of-phase mode coupling. The extra insertion loss is less than 1 dB at the center frequency of 3 GHz. The maximum amplitude imbalance of the E-arm is 0.5 dB in the frequency range of 2–4 GHz. Figure 9.27 shows the measured and calculated frequency responses of insertion loss for the H-arm's power dividing, that is, in-phase mode coupling. The extra insertion loss is less than 0.5 dB at the center frequency of 3 GHz. The maximum amplitude imbalance of the H-arm is 0.4 dB in the frequency range of 2–4 GHz. As shown in Figures 9.26 and 9.27, the calculated results are given together with the measured results. The additional insertion loss of the uniplanar slotline ring magic-T is mainly due to the CPW-slotline transition and the slotline T-junction.

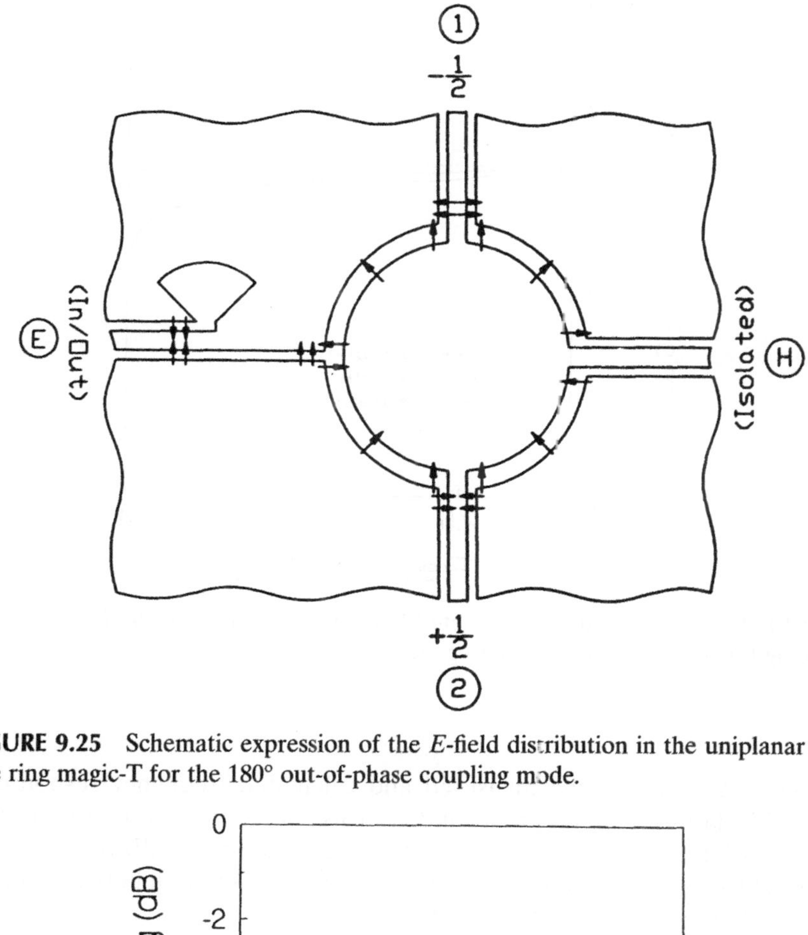

FIGURE 9.25 Schematic expression of the *E*-field distribution in the uniplanar slotline ring magic-T for the 180° out-of-phase coupling mode.

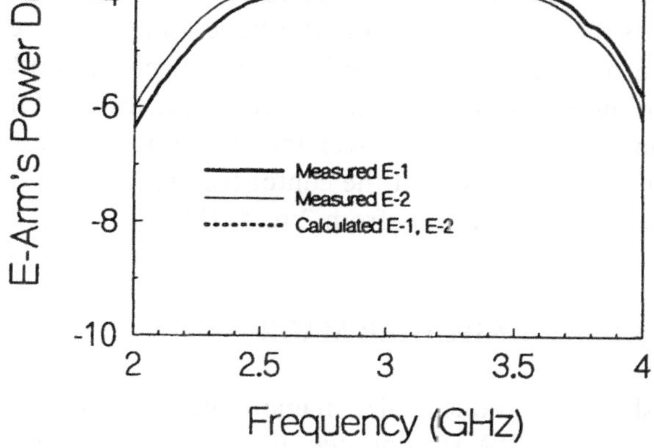

FIGURE 9.26 Measured and calculated frequency responses of the E-arm's power dividing for the uniplanar slotline ring magic-T.

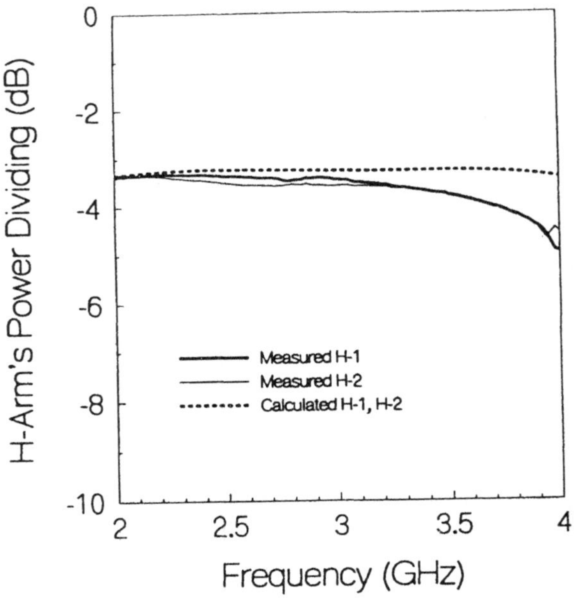

FIGURE 9.27 Measured and calculated frequency responses of the H-arm's power dividing for the uniplanar slotline ring magic-T.

Figure 9.28 shows the measured and calculated frequency responses of mutual isolation between the E- and H-arms and the balanced arms 1 and 2. The isolation between the E- and H-arms is greater than 30 dB from 2 GHz to 4 GHz. Over the same frequency range, the mutual isolation between the two balanced arms is greater than 12 dB.

Figure 9.29 shows the amplitude balance for the 180° out-of-phase and in-phase mode coupling. The maximum amplitude imbalance of the E-arm is 0.5 dB in the frequency range of 2–4 GHz. The maximum amplitude imbalance of the H-arm is 0.4 dB over the same frequency range. Figure 9.30 shows the phase balance for the 180° out-of-phase and in-phase mode coupling. The phase error of the E-arm is 3° at the center frequency of 3 GHz. The E-arm's maximum phase imbalance is 5° over the frequency range of 2–4 GHz. The phase error of the H-arm is 3° at the center frequency of 3 GHz. The H-arm's maximum phase imbalance is 6° from 2 to 4 GHz.

9.6 REDUCED-SIZE UNIPLANAR MAGIC-Ts

Figure 9.31*a* shows the reduced-size magic T that consists of one out-of-phase and three in-phase CPW-slotline tee junctions [15]. The out-of-phase T-junction serves as a phase inverter. In Figure 9.31*a*, ports E and H correspond

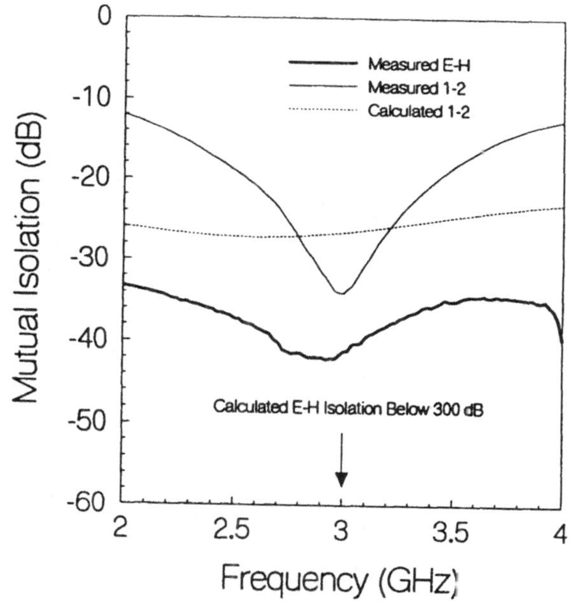

FIGURE 9.28 Measured and calculated frequency responses of the mutual isolation for the uniplanar slotline ring magic-T.

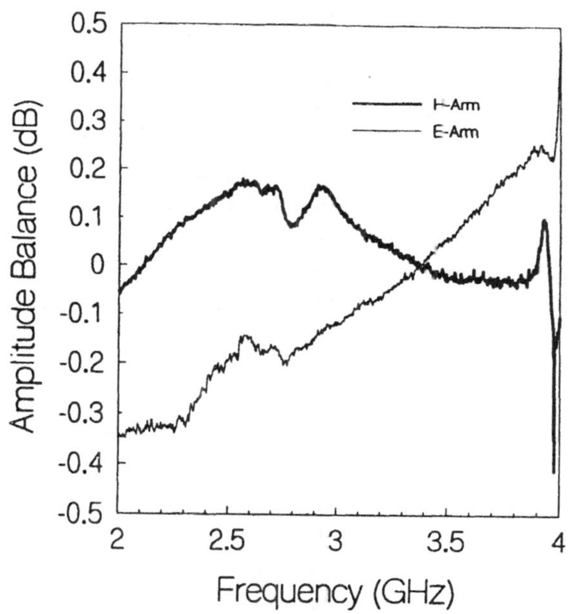

FIGURE 9.29 H- and E-arms' amplitude balances for the uniplanar slotline ring magic-T.

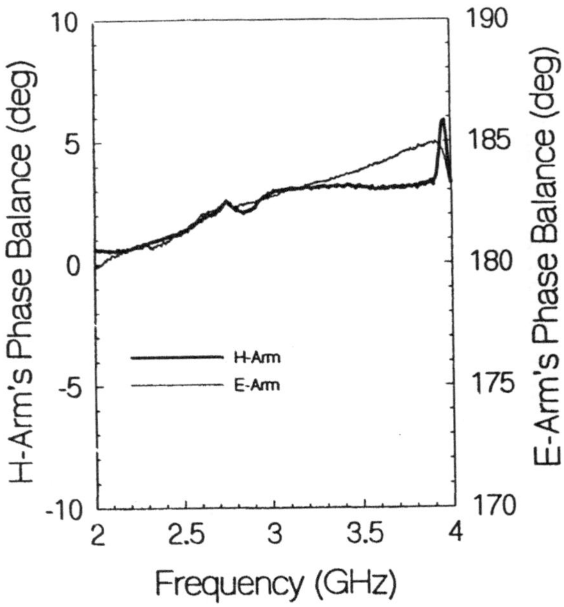

FIGURE 9.30 H- and E-arms' phase balances for the uniplanar slotline ring magic-T.

to the E- and H-arm of the conventional waveguide magic-T, respectively. Ports 1 and 2 are the balanced arms. Figure 9.31*b* shows the equivalent transmission line model of the magic-T. The twisted transmission line represents the reversal of the CPW-slotline T-junction.

Figures 9.32 and 9.33 show the schematic diagram of the E-field distribution and the equivalent circuit for the in-phase and the out-of-phase coupling, respectively. In Figure 9.32*a*, the signal is fed to port H, which then divides into two components, that both arrive in-phase at ports 1 and 2. However, the two components arrive at port E, out-of-phase and cancel out each other. In this case, the symmetry plane at port H corresponds to an open circuit (magnetic wall), whereas the symmetry plane at port E corresponds to a short circuit (electric wall).

In Figure 9.33*a*, the signal is fed to port E, and then divides into two components, which arrive at ports 1 and 2 with a 180° phase difference. The 180° phase difference between the divided signals at ports 1 and 2 is due to the out-of-phase tee junction. The two components waves arrive at port H out-of-phase and cancel out each other. The symmetry plane at port E corresponds to an open circuit (magnetic wall), whereas the symmetry plane at port H corresponds to a short circuit (electric wall). The isolation between ports E and H is perfect as long as the phase reversal in the out-of-phase CPW-soltline T-junction is ideal.

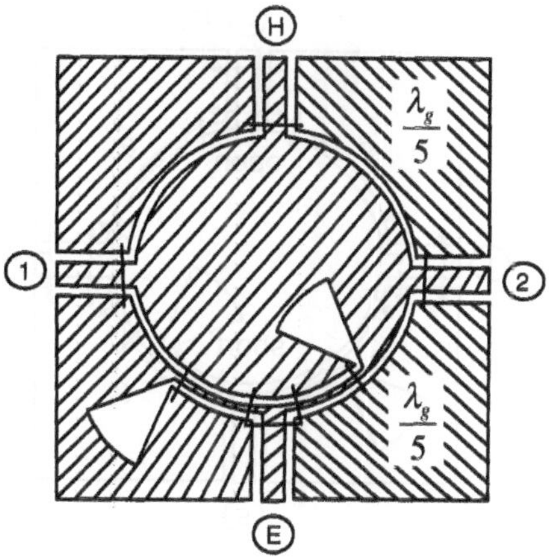

(a)

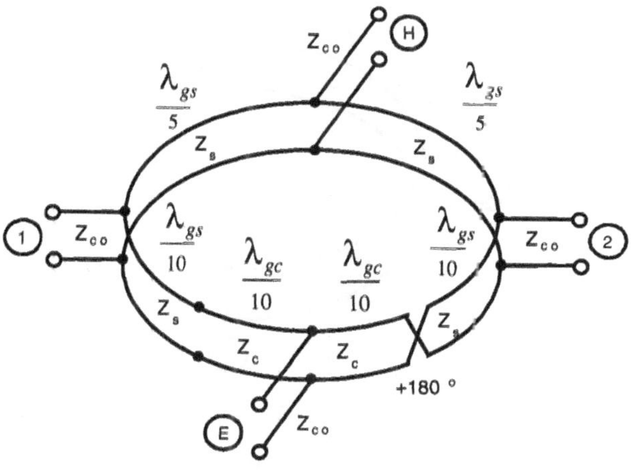

(b)

FIGURE 9.31 Reduced-size uniplanar magic-T (a) layout and (b) equivalent circuit [15].

As shown in Figures 9.32*b* and 9.33*b*, an equivalent circuit was used to analyze the impedance matching. The characteristic impedance of slotline Z_s and CPW Z_c in terms of CPW feed line impedance Z_{co} (usually 50 ohms) and θ (the electric length of a quarter of the slotline ring circumference) are given by [16]

$$Z_s = Z_c = Z_{co}\sqrt{2(1-\cot^2\theta)} \tag{9.5}$$

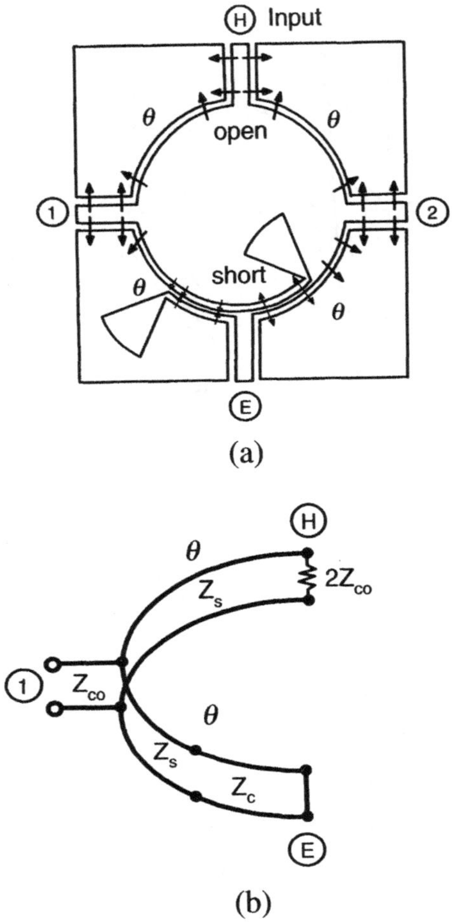

FIGURE 9.32 Out-of-phase coupling mode of the magic-T (a) E-field distribution and (b) equivalent circuit [15]. (Permission from IEEE.)

According to Equation (9.5), the minimum θ is obviously equal to 45°. Simulations indicate that wide band operation is obtained for values of θ, which are smaller in the allowed range. In this design, $\theta = 72°$ (i.e., $\lambda_g/5$) was chosen, resulting in the characteristic impedance Z_s, $Z_c = 66.9$ ohms. The magic-T in Figure 9.33 was designed at the center frequency of 4 GHz and fabricated on a RT/Duroid 6010.5 ($\varepsilon_r = 10.5$) substrate with thickness $h = 1.54$ mm and metal thickness $t = 10\,\mu$m. The radius of the radial stub at CPW-slotline transition is 5 mm. The radial stub angle is 45°. It is important to use air bridges at the magic-T's discontinuities to prevent the coupled slotline mode from propagating on the CPW lines.

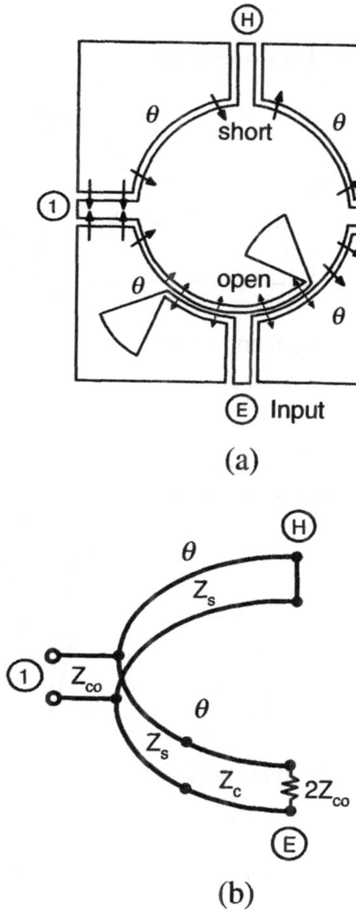

FIGURE 9.33 In-phase coupling mode of the magic-T (a) E-field distribution and (b) equivalent circuit [15]. (Permission from IEEE.)

Figure 9.34 shows the magic-T's measured and calculated transmission, return loss, and isolation, respectively. For the E-port's power division (i.e., out-of-phase mode coupling) shown in Figure 9.34*a*, the insertion loss is less than 0.7 dB at 4 GHz. The return loss for the E-port is greater than 15 dB from 3.1 to 6 GHz. Similarly, Figure 9.34*b* shows the insertion loss of 0.5 dB at 4 GHz for the H port's power division (i.e., in-phase mode coupling). Also, the return loss of for the H-port is greater than 15 dB from 2.7 to 6.2 GHz. The measured and calculated isolations between the E-port and H-port or ports 1 and 2 are shown in Figure 9.34*c*. Figure 9.35 shows that the magic-T has a bandwidth of 1.6 octave from 2 to 6 GHz with maximum power dividing imbalance of 0.4 dB and 2.5° maximum phase imbalance. The measured performances of the various parameters are summarized in Table 9.1.

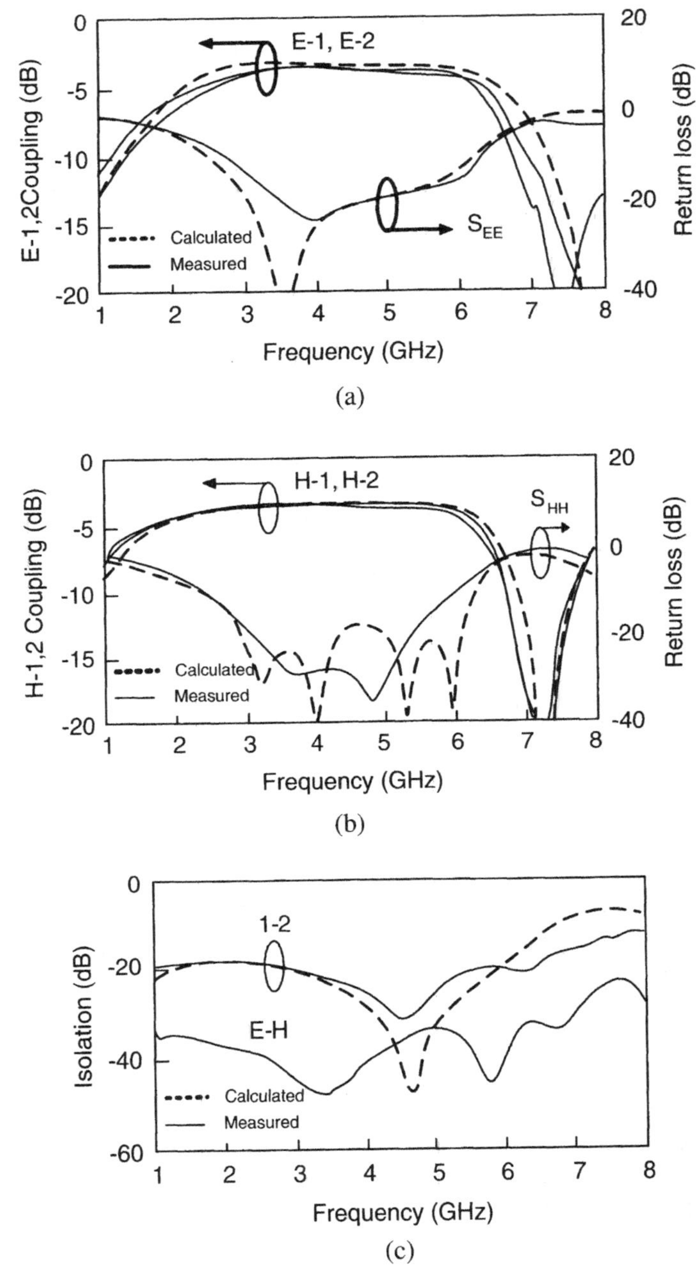

FIGURE 9.34 Measured and calculated frequency responses of the magic-T (a) out-of-phase coupling of E-1, E-2, and E-port's return loss; (b) in-phase coupling of H-1, H-2, and H-port's return loss; and (c) isolations of E-H and 1–2 [15]. (Permission from IEEE.)

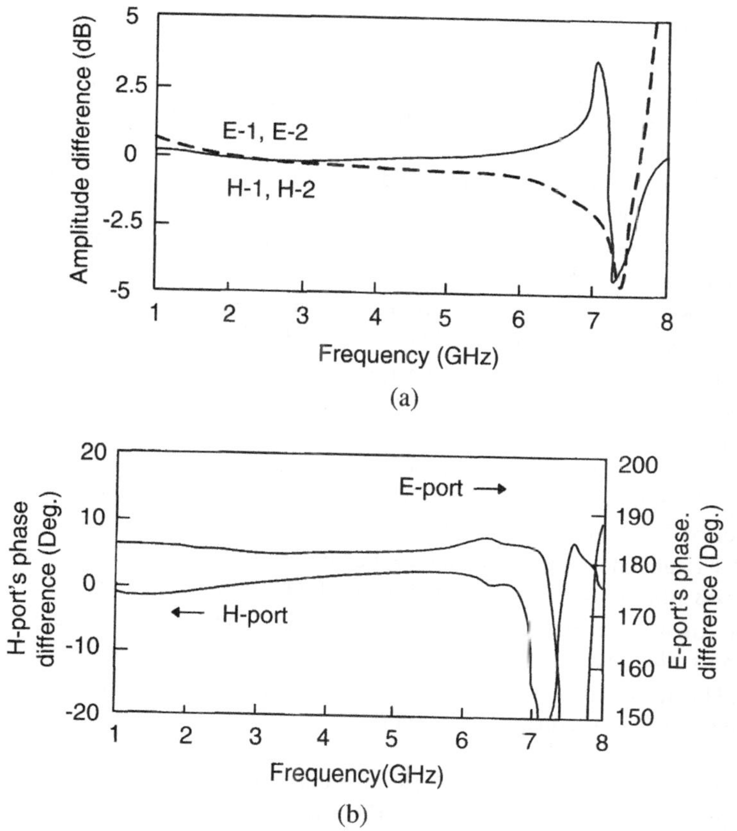

FIGURE 9.35 Measured frequency responses of the magic-T (a) amplitude imbalance and (b) phase imbalance [15].

TABLE 9.1 Summary of Measured Performances of the Magic-T [15]

Parameter		Measured Result	Frequency Range (GHz)	Bandwidth (octave)
Coupling	Fed to port E (S_{1E}, S_{2E})	3.9 ± 0.3 dB	2.8–5.9	>1.075
	Fed to port H (S_{1E}, S_{2E})	3.9 ± 0.3 dB	2.15–6.0	>1.48
Return loss	(S_{11}, S_{22}, S_{EE}, S_{HH})	>15 dB	3.1–6.0	>0.95
Isolation	Port1 and port2	>18 dB	1.0–6.6	>2.5
	Port E and H	>30 dB	1.0–7.7	>2.5
Imbalance	Amplitude E-1/E-2	<0.4 dB	1.8–6.3	>1.8
	Amplitude H-1/H-2	<0.4 dB	1.0–5.9	>2.5
	Phase E-1/E-2	181° ± 1.5°	2.0–7.15	>1.8
	Phase H-1/H-2	<2.5°	1.0–6.4	>2.5
Meeting all the above specifications			3.1–5.9	>0.93

REFERENCES

[1] C. Ho, "Slotline, CPW ring circuits and waveguide ring cavities for coupler and filter applications," Ph.D. dissertation, Texas A&M University, College Station, May 1994.

[2] R. G. Manton, "Hybrid networks and their uses in radio-frequency circuits," *Radio Electron. Eng.*, Vol. 54, pp. 473–489, June 1984.

[3] K. Chang, *Handbook of Microwave and Optical Components*, Vol. 1, Wiley, New York, pp. 145–150, 1990.

[4] D. I. Kraker, "A symmetric coupled-transmission-line magic-T," *IEEE Trans. Microwave Theory Tech.*, Vol. MTT-12, pp. 595–599, November 1964.

[5] R. H. DuHamel and M. E. Armstrong, "The tapered-line magic-T," *Proc. 15th Annu. Symp. Dig. on USAF Antenna Research Program*, Monticello, Ill., pp. 387–388, October 12–14, 1965.

[6] C. P. Tresselt, "Design and computed theoretical performance of three classes of equal-ripple non-uniform line couplers," *IEEE Trans. Microwave Theory Tech.*, Vol. MTT-17, pp. 218–230, April 1972.

[7] G. J. Laughline, "A new impedance-matched wideband balun and magic-T," *IEEE Trans. Microwave Theory Tech.*, Vol. MTT-24, pp. 135–141, March 1976.

[8] M. Aikawa and H. Ogawa, "A new MIC magic-T using coupled slot lines," *IEEE Trans. Microwave Theory Tech.*, Vol. MTT-28, pp. 523–528, June 1980.

[9] T. Hirota, Y. Tarusawa, and H. Ogawa, "Uniplanar MMIC hybrids—A proposed new MMIC structure," *IEEE Trans. Microwave Theory Tech.*, Vol. MTT-35, pp. 576–581, June 1987.

[10] C. Ho, L. Fan, and K. Chang, "New uniplanar coplanar waveguide hybrid-ring couplers and magic-Ts," *IEEE Trans. Microwave Theory Tech.*, Vol. MTT-42, No. 12, pp. 2440–2448, December 1994.

[11] C. Ho, L. Fan, and K. Chang, "Ultra wide band slotline ring couplers," in *1992 IEEE MTT-S Int. Microwave Conf. Dig.*, pp. 1175–1178, 1992.

[12] C. Ho, L. Fan, and K. Chang, "Slotline annular ring elements and their applications to resonator, filter and coupler design," *IEEE Trans. Microwave Theory Tech., Vol.* MTT-41, No. 9, pp. 1648–1650, September 1993.

[13] C. Ho, L. Fan, and K. Chang, "Broad-band uniplanar hybrid-ring and branch-line couplers," *IEEE Trans. Microwave Theory Tech.*, Vol. MTT-41, No. 12, pp. 2116–2125, December 1993.

[14] C. Ho, L. Fan, and K. Chang, "Broadband uniplanar hybrid ring coupler," *Electron. Lett.*, Vol. 29, No. 1, pp. 44–45, January 7, 1993.

[15] L. Fan, C.-H. Ho, and K. Chang, "Wide-band reduced-size uniplanar magic-T, hybrid-ring, and de Ronde's CPW-slot couplers," *IEEE Trans. Microwave Theory Tech.*, Vol. 43, No. 12, pp. 2749–2758, December 1995.

[16] M.-H. Murgulescu, E. Moisan, P. Legaud, E. Penard, and I. Zaquine, "New wideband, 0.67 λ_g circumference 180° hybrid ring couplers," *Electron. Lett.*, Vol. 30, pp. 299–300, Feburary 1994.

CHAPTER TEN

Waveguide Ring Resonators and Filters

10.1 INTRODUCTION

The annular ring structure has been studied thoroughly for the planar transmission structure [1–10]. Many attractive applications for the planar ring circuits have been published [11–23]. This chapter presents a new type of rectangular waveguide ring cavity that can be used as a resonator or a building block for filters or multiplexers [24, 25]. Compared with planar ring circuits, the waveguide ring cavities have higher Q values and can handle higher power. This new type of waveguide component has the flexibility of mechanical and electronic tuning as well as good predictable performance.

The second section of this chapter discusses the single-mode operation of the waveguide ring cavities. Two fundamental structures for the waveguide ring cavities, H- and E-plane waveguide ring cavities, are introduced in this section. Section 10.2 also discusses regular resonant modes, split resonant modes, and forced resonant modes. Mechanically tuned and electronically tuned waveguide ring resonators that are based on the tuning from regular resonant modes to forced resonant modes are also discussed in the second section. The third section discusses the dual-mode operation of the waveguide ring cavities, plus two new dual-mode filters that use the dual resonant modes. A single-cavity dual-mode filter using the H-plane waveguide ring cavity has been developed with a bandwidth of 0.77%, a stopband attenuation of more than 40 dB, and a sharp gain slope transition. The other two-cavity dual-mode filter using two E-plane waveguide ring cavities has been fabricated with a

Microwave Ring Circuits and Related Structures, Second Edition,
by Kai Chang and Lung-Hwa Hsieh
ISBN 0-471-44474-X

bandwidth of 1.12%, a stopband attenuation of 70 dB, and a sharp gain slope transition. The dual-mode index related to the generation of transmission zeros is also discussed in the third section.

10.2 WAVEGUIDE RING RESONATORS

The waveguide ring cavity can be classified as either an *H*-plane waveguide ring cavity or an *E*-plane waveguide ring cavity [24, 25]. Figures 10.1 and 10.2 show the physical configurations of the *H*-plane and *E*-plane waveguide ring cavities, respectively. The *H*-plane waveguide ring cavity is formed by a circle of rectangular waveguide that is curved in the plane of the magnetic field. The *E*-plane waveguide ring cavity consists of a circle of rectangular waveguide that is curved in the plane of the electric field. The differing geometric configurations make the *H*-plane ring cavity more suitable for a pileup design and make the *E*-plane ring cavity more suitable for a cascaded design. Because the electromagnetic field bending in the *E*- and *H*-planes are different, these two structures bear different characteristics and need different excitation methods.

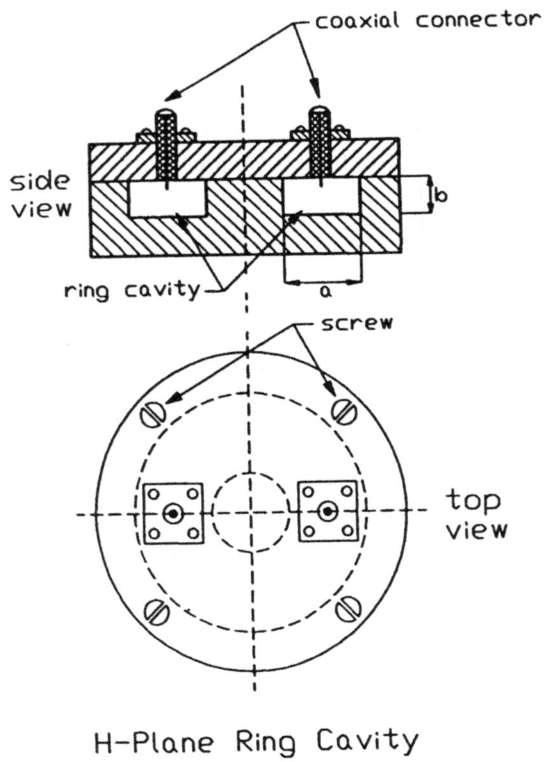

FIGURE 10.1 Physical configuration of the *H*-plane waveguide ring structure.

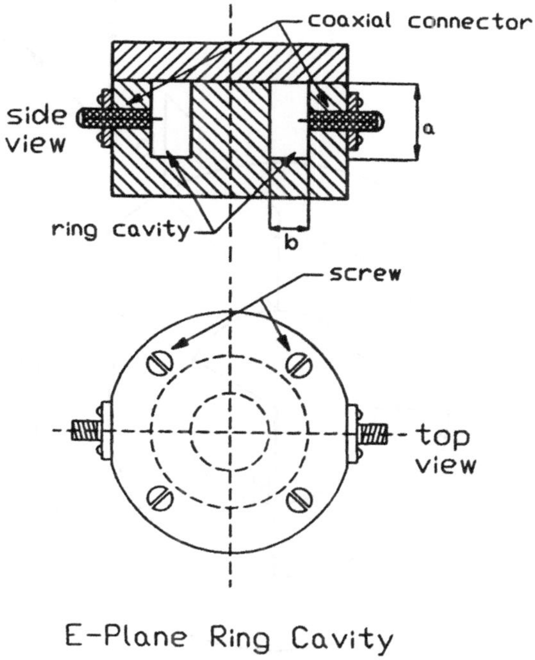

FIGURE 10.2 Physical configuration of the *E*-plane waveguide ring structure.

Both waveguide and coaxial couplings are suitable for exciting the waveguide ring cavities. The external feeds of the waveguide ring cavities use coaxial–waveguide transitions. The *H*-plane waveguide ring cavity has coaxial feeds on the top side of the cavity, whereas the *E*-plane waveguide ring cavity has coaxial feeds on the annular side of the cavity. These coaxial feeds for the *H*-plane and *E*-plane annular ring waveguide cavities are designed to excite the dominant TE_{10n} modes, where n is the mode number of the annular ring resonators.

Figure 10.3 shows the coordinate systems for the *H*-plane ring cavity of cross section $a \times b$ with its axis bent to a curvature of $\chi = 1/R$, where R is the mean radius of the waveguide ring cavity. Figure 10.4 shows the coordinate systems for the *E*-plane ring cavity of cross section $b \times a$ with its axis bent to a curvature of $\chi = 1/R$, where R is the mean radius of the waveguide ring cavity. The second-order correction to the guide wavelength for the dominant mode in the *H*- and *E*-plane ring cavities is given by [26] to be

$$\frac{1}{\lambda_H^2} = \frac{1}{\lambda_g^2} + \frac{\chi^2}{24}\left\{1 - \frac{12+\pi^2}{24}\left(\frac{2a}{\lambda_0}\right)^2 + \frac{15-\pi^2}{2\pi^2}\left(\frac{2a}{\lambda_0}\right)^4\right\} \quad \text{for the H-plane ring cavity} \tag{10.1a}$$

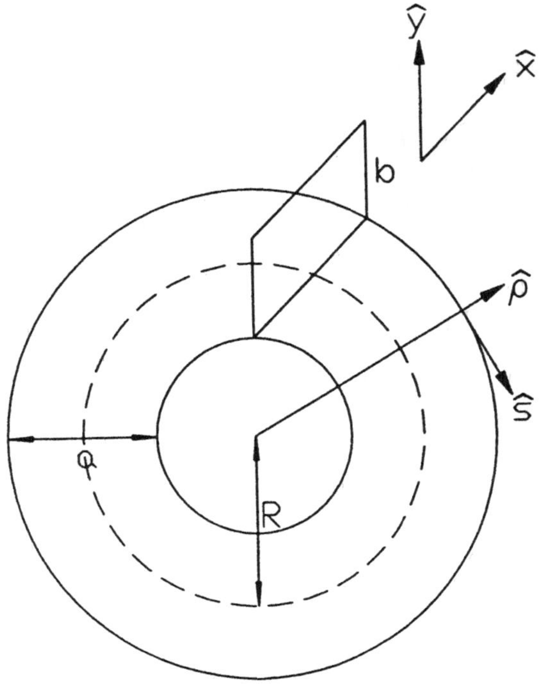

FIGURE 10.3 Coordinate system for the circular *H*-plane bend.

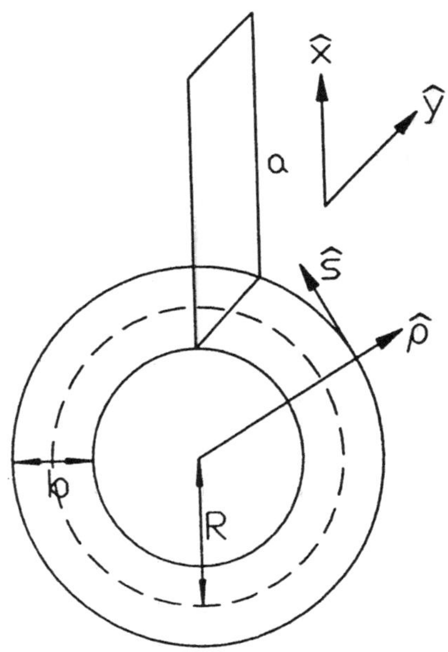

FIGURE 10.4 Coordinate system for the circular *E*-plane bend.

$$\frac{1}{\lambda_E^2}=\frac{1}{\lambda_g^2}\left[1-\frac{\chi^2 b^2}{12}\left(1-\frac{8\pi^2 b^2}{5\lambda_g^2}\right)\right] \quad \text{for the E-plane ring cavity} \tag{10.1b}$$

where a is the broad side of the rectangular waveguide, b is the narrow side of the rectangular waveguide, χ is the curvature of the waveguide ring cavity, λ_0 is the wavelength in free space, and λ_g is the guide wavelength in the rectangular waveguide.

The waveguide ring cavity can be treated as a closed rectangular waveguide. Figure 10.5*a–c* show the equivalent waveguide circuits for the waveguide ring cavities. According to the equivalent circuits shown in Figure 10.5, the wave functions of the dominant mode in the waveguide ring cavity are given by

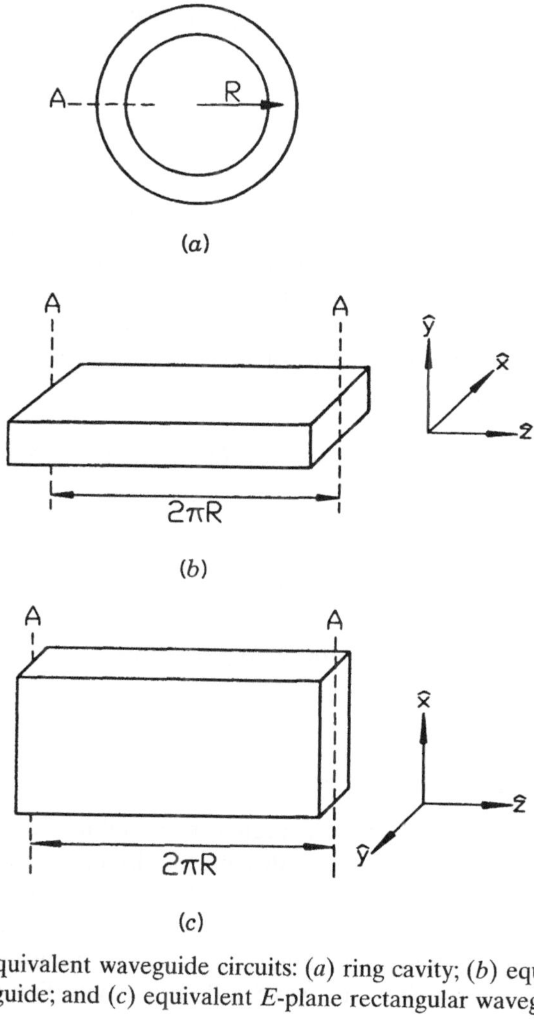

FIGURE 10.5 Equivalent waveguide circuits: (*a*) ring cavity; (*b*) equivalent *H*-plane rectangular waveguide; and (*c*) equivalent *E*-plane rectangular waveguide.

$$H_z = H_0 \cos\left(\frac{\pi x}{a}\right)\cos(k_z' z) \quad \text{or} \quad H_z = H_0 \cos\left(\frac{\pi x}{a}\right)\sin(k_z' z) \tag{10.2}$$

$$E_y = \frac{j\omega\mu}{k_c^2}\frac{\partial H_z}{\partial x} \qquad H_x = \frac{-jk_z'}{k_c^2}\frac{\partial H_z}{\partial x} \tag{10.3}$$

where H_0 is the amplitude constant, k_z' is defined by

$$k_z' = \frac{2\pi}{\lambda_H} \qquad \text{for the } H\text{-plane ring cavity} \tag{10.4a}$$

and

$$k_z' = \frac{2\pi}{\lambda_E} \qquad \text{for the } E\text{-plane ring cavity} \tag{10.4b}$$

and

$$k_c^2 = \omega^2 \mu_0 \varepsilon_0 - k_z'^2 \tag{10.5}$$

As shown in Equation (10.2), both of the sine (odd) and cosine (even) solutions can satisfy the boundary conditions of the waveguide ring cavities. This means that by applying appropriate perturbations it is possible to excite dual resonant modes in a single waveguide ring cavity. This phenomenon is discussed later in Section 10.3. The resonant conditions for the waveguide ring cavities are determined by

$$2\pi R = n\lambda_H \qquad \text{for the } H\text{-plane ring cavity} \tag{10.6a}$$

and

$$2\pi R = n\lambda_E \qquad \text{for the } E\text{-plane ring cavity} \tag{10.6b}$$

where R is the mean radius of the waveguide ring cavity and n is the mode number.

10.2.1 Regular Resonant Modes

Symmetric external feeds excite the regular resonant modes in waveguide ring resonators. The regular resonant modes are the dominant TE_{10n} modes, where n is the mode number of the ring structure. Figure 10.6 shows the mode chart of the E-field for the regular resonant modes of a symmetrically coupled waveguide ring cavity. As shown in Figure 10.6, the symmetric feeds generate the single-mode operation of the waveguide ring cavity. Figure 10.7 shows the

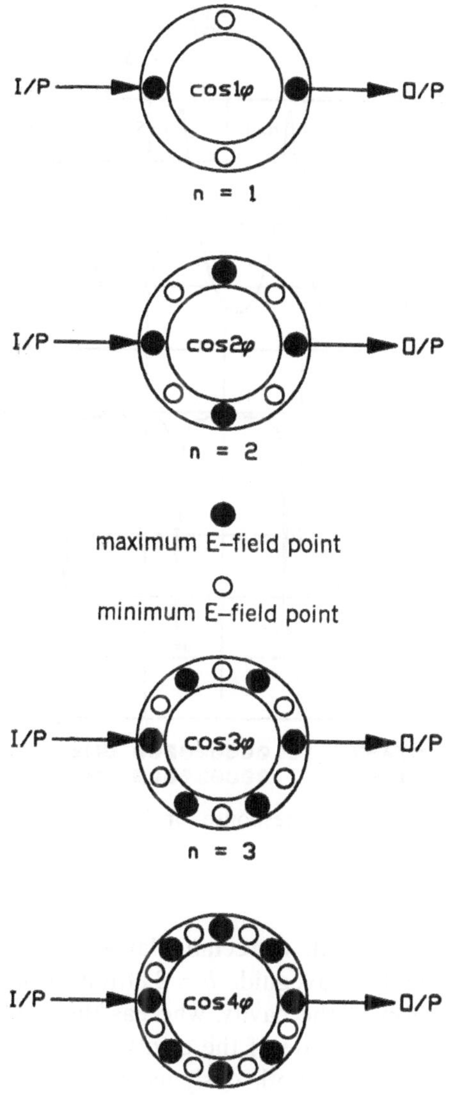

FIGURE 10.6 Mode chart of the *E*-field for the regular resonant modes.

measured frequency responses of insertion loss and return loss for an *H*-plane ring cavity, and Figure 10.8 illustrates the measured frequency responses of insertion loss and return loss for an *E*-plane ring cavity. The test *H*-plane ring cavity was designed to operate in *K*-band with the following dimensions: mean radius $R = 16.185\,\text{mm}$, broad side of rectangular waveguide $a = 10.73\,\text{mm}$, and narrow side of rectangular waveguide $b = 4.44\,\text{mm}$. The test *E*-plane ring cavity was also designed as a *K*-band cavity with the following dimensions: mean

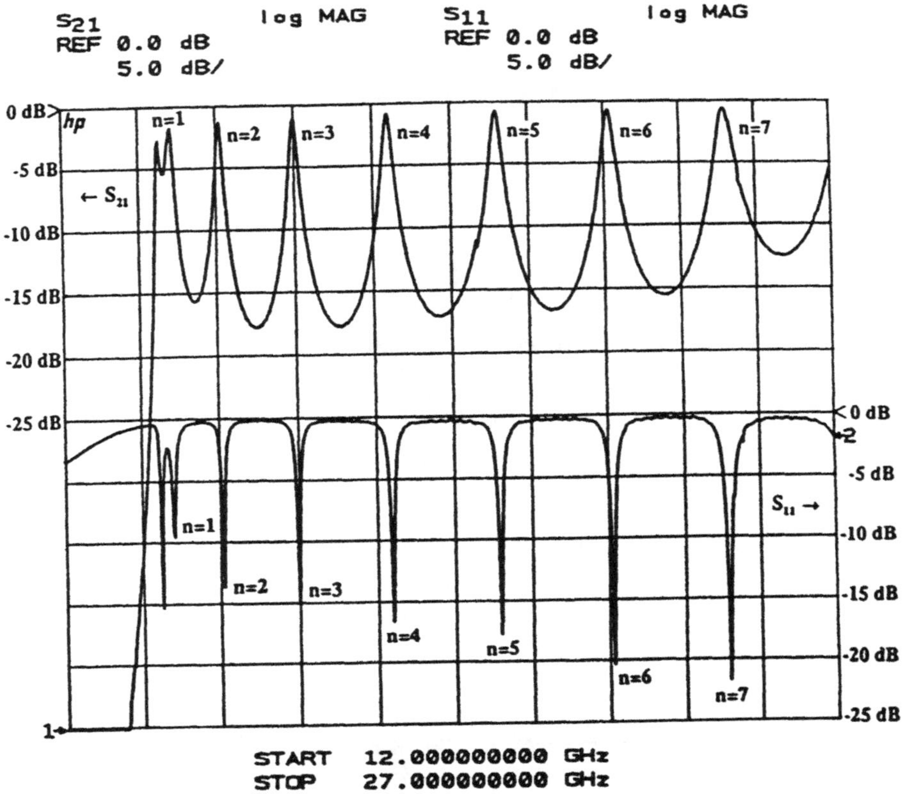

FIGURE 10.7 Measured frequency response for the regular resonant modes of the *K*-band *H*-plane ring cavity.

radius $R = 10.11$ mm, broad side of rectangular waveguide $a = 10.20$ mm, and narrow side of rectangular waveguide $b = 3.88$ mm. The *H*-plane ring cavity has coaxial feeds on top of the cavity, whereas the *E*-plane ring cavity has coaxial feeds on the annular side of the cavity. The coaxial feeds for the *H*- and *E*-plane ring cavities are both designed to excite the dominant TE_{10n} modes.

Figures 10.9 and 10.10 show the theoretical and experimental results for the regular resonant frequencies of the *H*-plane and *E*-plane ring cavities, respectively. The theoretical results shown in Figures 10.9 and 10.10 are calculated from Equations (10.1) and (10.6). As shown in Figure 10.9, the regular resonant frequencies of the *H*-plane ring cavity can be predicted correctly within an error or less than 0.32%. The regular resonant frequencies of the *E*-plane ring cavity can be predicted within an error of less than 0.23%. Easy and correct prediction of resonant frequencies and a simple design procedure make the waveguide ring cavity a good candidate for many waveguide circuits.

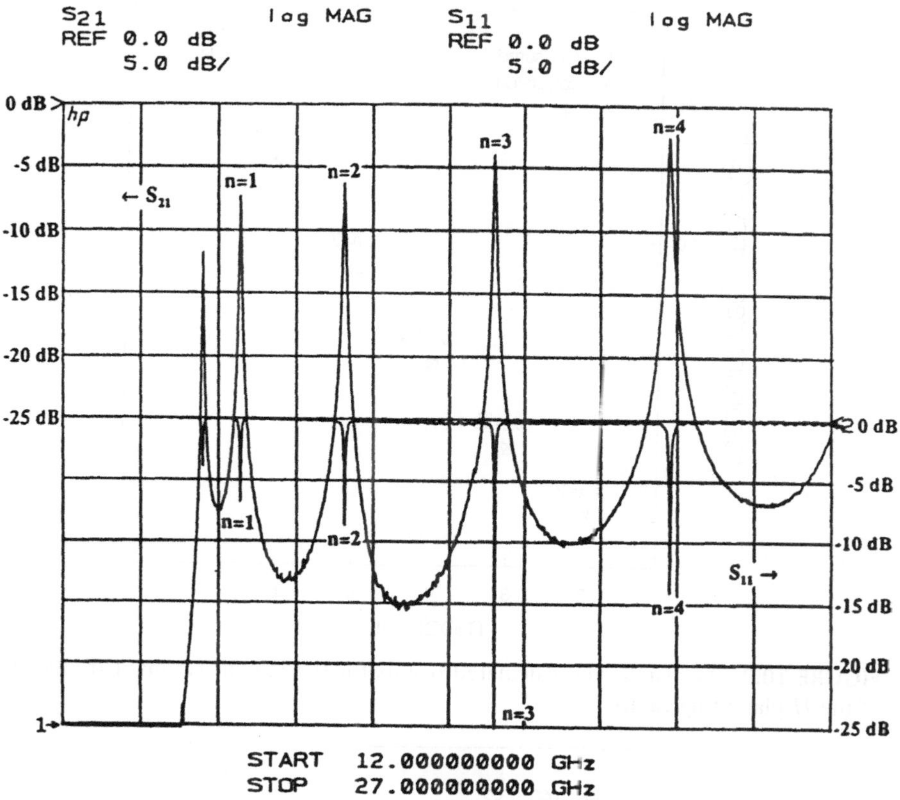

FIGURE 10.8 Measured frequency response for the regular resonant modes of the *K*-band *E*-plane ring cavity.

Tables 10.1 and 10.2 list the loaded and unloaded Q's for each type of waveguide ring cavity at various resonant frequencies. The loaded Q, Q_L, shown in Tables 10.1 and 10.2 is measured using the following equation [27]

$$Q_L = \frac{f_0}{2\delta f} \cdot \frac{\delta X}{R_0} \tag{10.7}$$

where R_0 is the resistance at the resonant frequency f_0, and δX is the actual change in reactance between the f_0 and $f_0 + \delta f$ points. Both the R_0 and δX values were read from the Smith chart using an HP-8510 network analyzer. The relationship between loaded Q and unloaded Q for the waveguide ring cavities, which are transmission forms of resonators, is given by [27]

$$\frac{Q_u}{Q_L} = \frac{1 + \gamma_1 + \gamma_2 + \gamma_1\gamma_2}{\gamma_1\gamma_2 - 1} \tag{10.8}$$

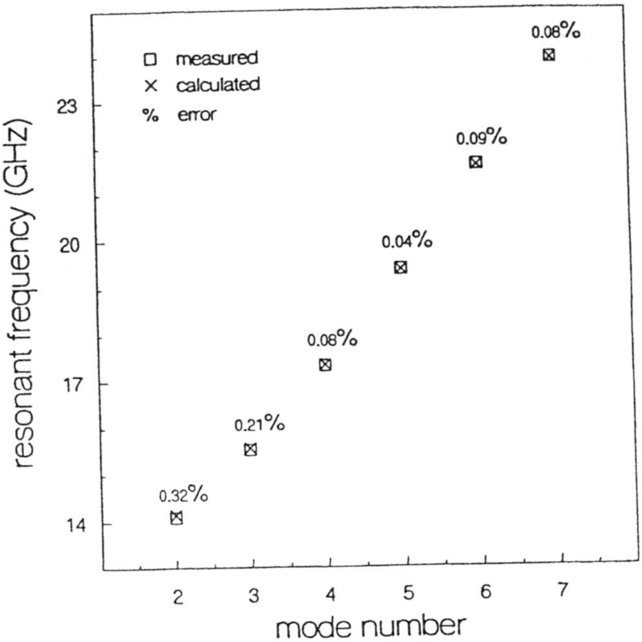

FIGURE 10.9 Measured and calculated results for the regular resonant frequencies of the *H*-plane ring cavity.

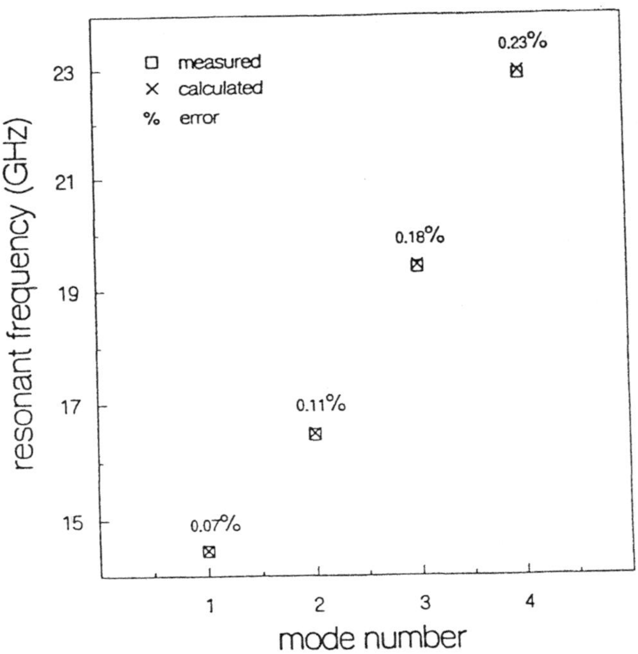

FIGURE 10.10 Measured and calculated results for the regular resonant frequencies of the *E*-plane ring cavity.

TABLE 10.1 Measured Q_L and Q_u for the *H*-plane Ring Cavity

	Calculated f_0 (GHz)	Measured f_0 (GHz)	Error (%)	Measured Q_L	Measured Q_u
$n = 2$	15.146	15.098	0.320	198.200	1,190.9
$n = 3$	16.564	16.530	0.210	222.100	2,038
$n = 4$	18.348	18.333	0.080	194.600	3,570.6
$n = 5$	20.386	20.395	0.040	183.300	3,823.4
$n = 6$	22.593	22.615	0.090	163.600	3,157.2
$n = 7$	24.907	24.928	0.080	151.700	6,272.4

TABLE 10.2 Measured Q_L and Q_u for the *E*-plane Ring Cavity

	Calculated f_0 (GHz)	Measured f_0 (GHz)	Error (%)	Measured Q_L	Measured Q_u
$n = 1$	15.454	15.443	0.070	714.900	1,483.13
$n = 2$	17.502	17.482	0.110	684.900	1,559.69
$n = 3$	20.449	20.412	0.180	775.900	2,505.17
$n = 4$	23.945	23.889	0.230	604.100	2,187.48

where γ_1 is the voltage standing-wave ratio (VSWR) of the input coupling circuit with a matched output load and γ_2 is the VSWR of the output coupling circuit with a matched input load. Table 10.1 shows that *H*-plane ring cavity has an average unloaded Q of 3342.06, whereas Table 10.2 shows that the *E*-plane ring cavity has an average unloaded Q of 1933.87.

The cutoff frequency of the waveguide ring cavity is determined by

$$f_c = \frac{c}{2a} \tag{10.9}$$

where a is the broad side of the rectangular waveguide and c is the speed of light in free space. The designed cutoff frequency of the *H*-plane ring cavity is 14.02 GHz; the measured cutoff frequency of the *H*-plane ring cavity is 13.91 GHz. The designed and measured cutoff frequencies for the *E*-plane ring cavity are 14.93 GHz and 14.72 GHz, respectively.

10.2.2 Split Resonant Modes

The split resonant modes of waveguide ring cavities are generated by using a tuning post at a specific angle of the circle. Figure 10.11 shows the mode chart of *E*-field for the symmetrically coupled waveguide ring cavity with a tuning post at 45°. According to the mode chart shown in the figure, the resonant modes with mode numbers

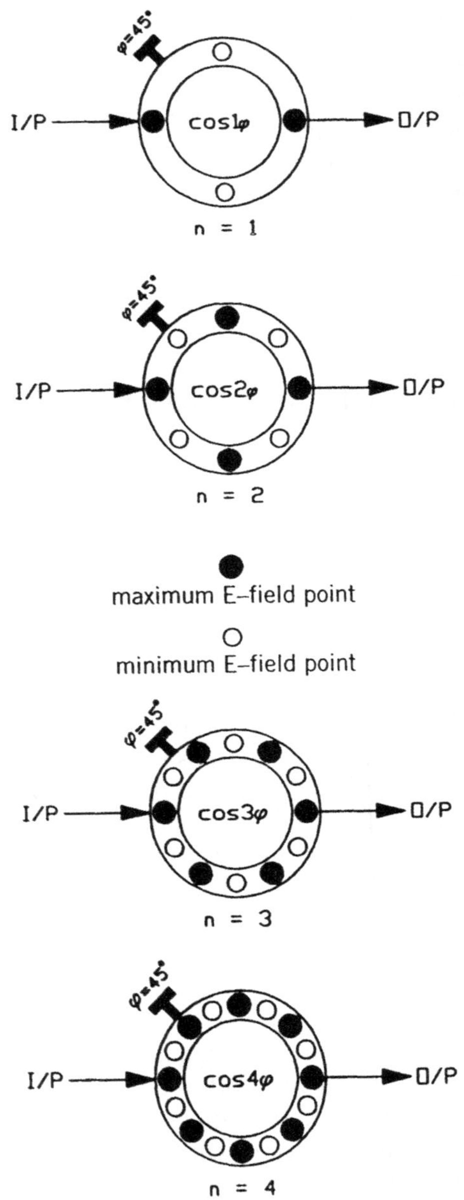

FIGURE 10.11 Mode chart of the *E*-field for the split resonant modes.

$$n = m\frac{90^\circ}{\theta} \tag{10.10}$$

where $\theta = 45^\circ$ and $m = 1, 2, 3, \ldots$, have a maximum or minimum *E*-field point that corresponds to a magnetic or electric wall at the position of the tuning post and will not be split. Figure 10.12 shows the measured frequency response

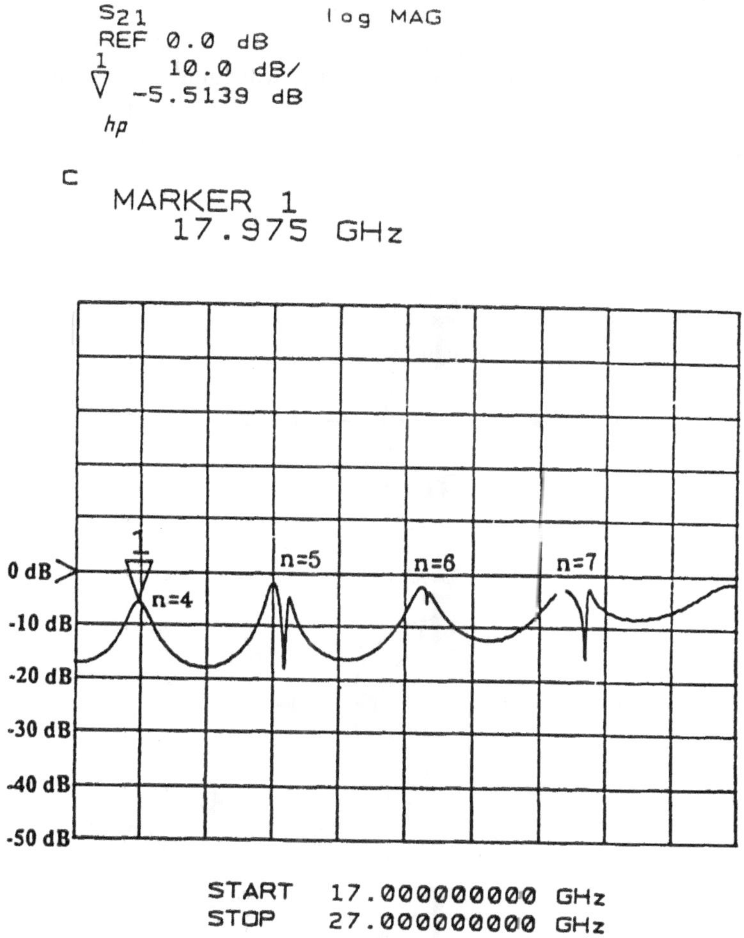

FIGURE 10.12 Measured frequency response for the split resonant modes of the *H*-plane ring cavity with a tuning post at 45°.

of insertion loss for an *H*-plane ring cavity that has a tuning post at 45°. As expected, the fourth resonant mode is not split.

10.2.3 Forced Resonant Modes

A tunable–switchable resonator has been introduced by Martin et al. [17] on microstrip ring circuits. However, the tunable and switchable conditions have not been studied thoroughly. This section discusses the forced resonant modes of the ring cavities, which are caused by the short or open boundary conditions on the ring structure. The tunable and switchable conditions can be derived using the concept of forced resonant modes.

Forced resonant modes are excited by forced boundary conditions, that is, open or short circuits, on waveguide ring cavities. The short boundary can be obtained by inserting a tuning post across the waveguide inside the ring cavities. Figure 10.13 shows the mode chart of the E-field for a symmetrically coupled waveguide ring cavity with a fully inserted tuning post at 90°. As shown in Figure 10.13, the tuning post forces minima of the E-field to occur

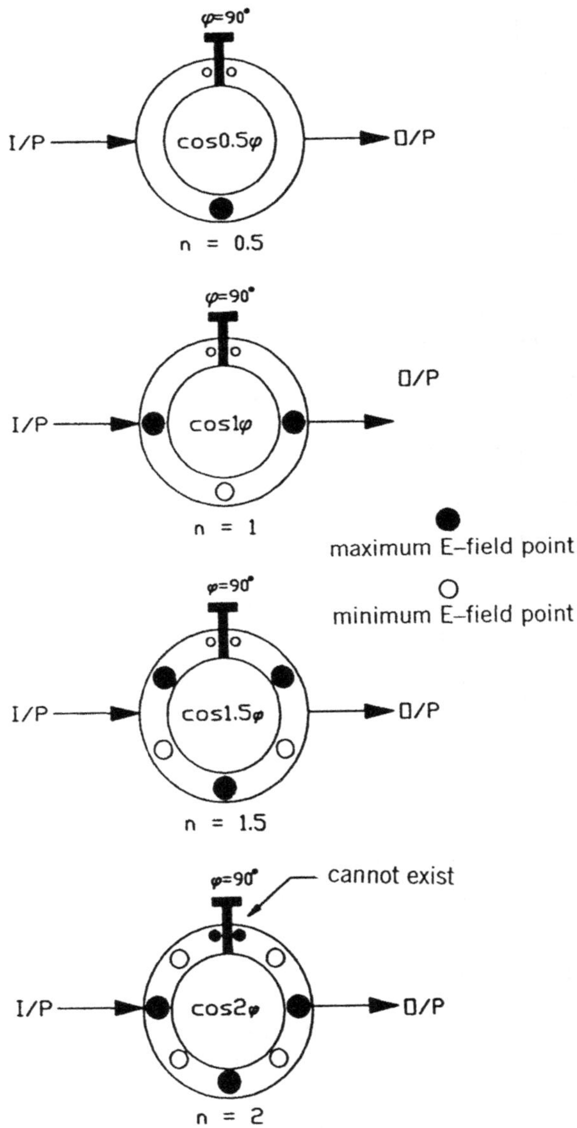

FIGURE 10.13 Mode chart of the E-field for the forced resonant modes.

on both sides of the short plane. Therefore, the resonant modes with even mode numbers cannot exist. On the other hand, the half-wavelength resonant modes with mode numbers $\nu = m/2$, where $m = 1, 3, 5, \ldots$, will be excited due to the short boundary condition.

By inserting the post from zero-depth to full-depth across the waveguide inside the ring cavities, the resonant modes will change from regular resonant modes to forced resonant modes. When the post is fully inserted across the waveguide, the even resonant modes will disappear and the half-wavelength resonant modes will be excited. The maximum tuning range of the forced resonant modes is determined by

$$\delta f_n = |f_n - f_\nu| \tag{10.11}$$

where f_n is the resonant frequency of the full-wavelength resonant mode with even mode number n, and f_ν is the resonant frequency of the excited half-wavelength resonant mode with mode number $\nu = n \pm 1/2$. The general design rule of Equation (10.11) is applied in the following discussion of the mechanically tuned and varactor-tuned waveguide ring cavities.

The mechanically tuned ring cavity was designed as a waveguide ring cavity with symmetrical feed lines and a tuning post at 90°. According to the mean radius of the mechanically tuned ring cavity, a maximum tuning range of 993 MHz for the fourth resonant mode is obtained from Equations (10.1), (10.6), and (10.11). Figure 10.14 shows the measured frequency response of insertion loss for the mechanically tuned ring cavity. As shown in the figure, the capacitance of the tuning post increases with the insertion depth. The variance of the capacitance changes the resonant modes from n to $n - 1/2$. For the fourth resonant mode, the tuning range from $n = 4$ to $\nu = 3.5$ is 677 MHz, which is within the maximum tuning range predicted by Equation (10.11).

The varactor-tuned ring cavity has a tuning post mounted with a varactor. By applying different bias voltages, the capacitance of the mounted varactor will change. The capacitance of the varactor decreases with increasing bias voltage. The variance of the capacitance changes the resonant modes from $n - 1/2$ to n. The measured frequency response of insertion loss for the vactor-tuned ring cavity is shown in Figure 10.15. The tuning range shown in the figure is 190 MHz, which is limited by the tunable capacitance of the varactor. The tunable capacitance of the varactor is varied from 2.64 to 0.75 pF, which is controlled by varying the bias voltage from 0 to 25 V. A varactor with a larger tunable capacitance can be used to achieve better tuning range.

10.3 WAVEGUIDE RING FILTERS

Dual-mode filters have been reported which use circular and rectangular waveguide linear cavities [28–30]. The excitation of dual resonant modes in linear waveguide cavities uses a tuning post inserted at the corner of a square

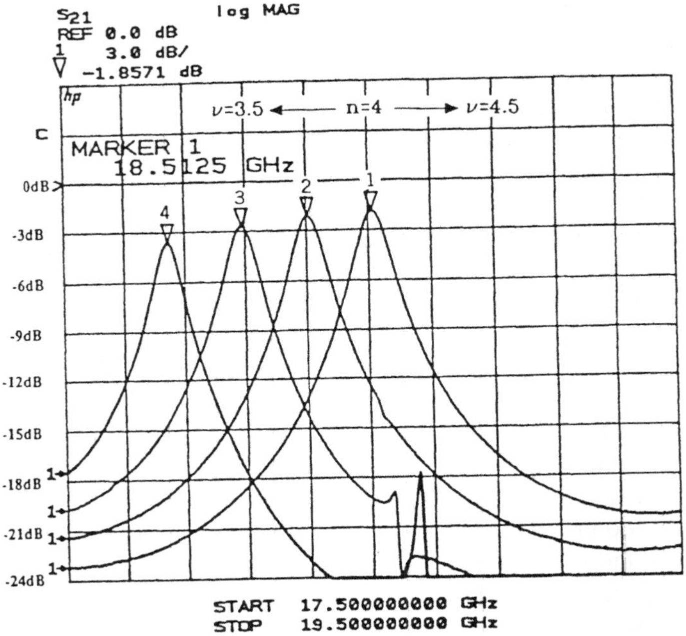

FIGURE 10.14 Measured frequency response for the mechanically tuned *H*-plane ring cavity.

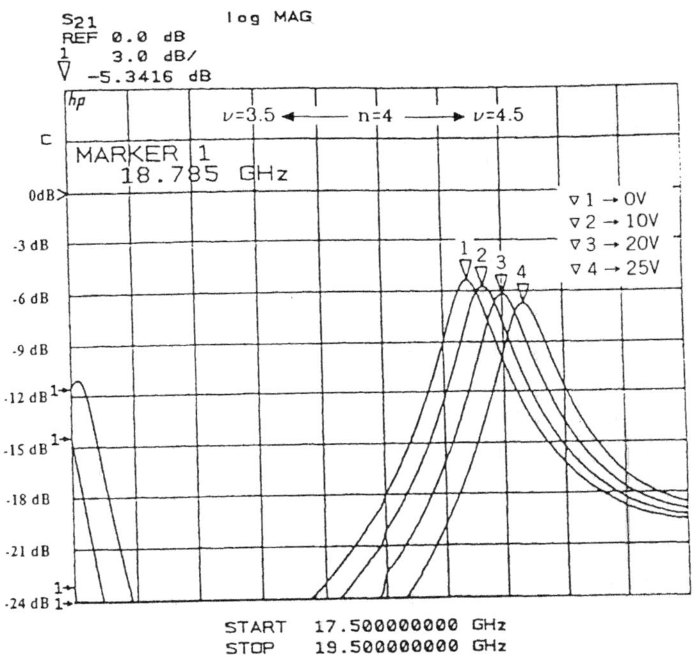

FIGURE 10.15 Measured frequency response for the electronically tuned *H*-plane ring cavity.

waveguide [30]. The sign of the mutual coupling coefficient between the dual resonant modes depends on the position of the inserted post. As shown in Equations (10.2) and (10.3), both sine (odd) and cosine (even) wave functions can exist in waveguide ring cavities. The sine and cosine wave functions have 90° phase differences and are orthogonal to each other. This section discusses the excitation of dual resonant modes, that is, sine (odd) and cosine (even) resonant modes, in waveguide ring cavities and the applications of dual resonant modes in filter design [24, 25].

10.3.1 Decoupled Resonant Modes

Figure 10.16 shows the mode chart of the *E*-field for an asymmetrically coupled waveguide ring cavity whose external feeds are 90° apart. According

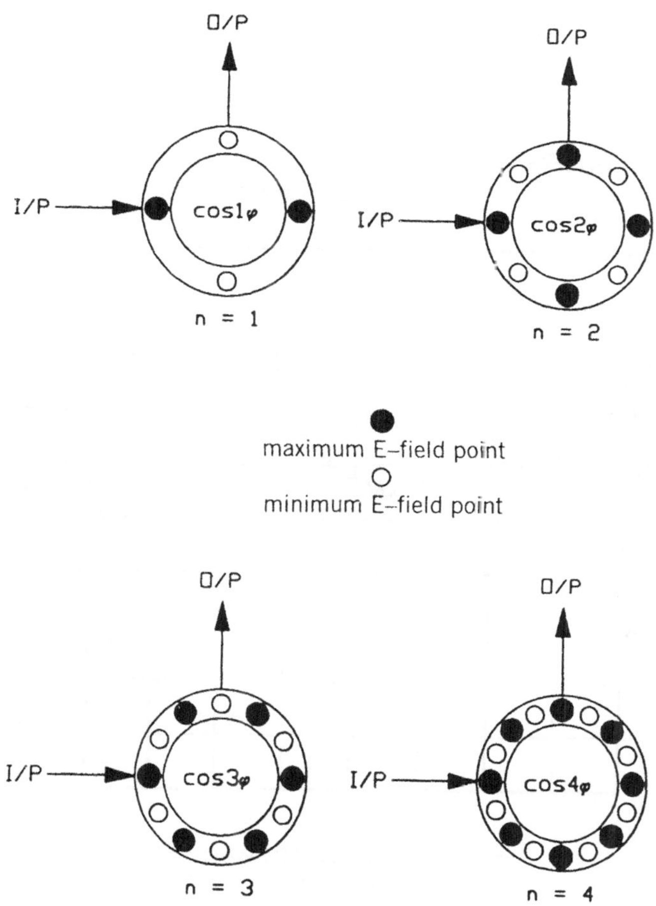

FIGURE 10.16 Mode chart of the *E*-field for the asymmetrically coupled resonant modes.

to the mode chart of the *E*-field in the figure, the resonant modes with odd mode numbers have minimum *E*-field at the output coupling point. Thus the electromagnetic energy cannot be coupled through the output port. These resonant modes are called decoupled resonant modes. Figure 10.17 shows the measured frequency response of insertion loss for an asymmetrically coupled *H*-plane ring cavity. As shown in Figure 10.17, the resonant modes of the single *H*-plane ring cavity, which has a 90° split of the external feed lines without any tuning post, display maximum insertion loss for those decoupled odd modes. The measured results shown in the figure agree with the prediction of the mode-chart analysis.

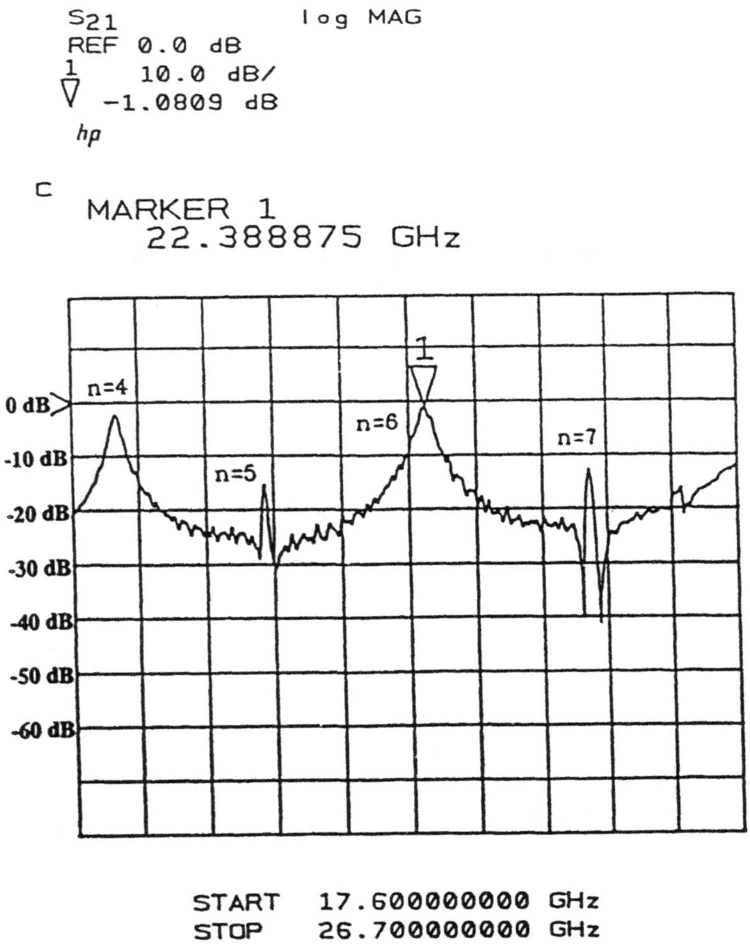

FIGURE 10.17 Measured frequency response for the asymmetrically coupled *H*-plane ring cavity.

10.3.2 Single-Cavity Dual-Mode Filters

By inserting a tuning post at 45° or 135°, the dual resonant modes can be excited from the decoupled resonant modes. Figures 10.18 and 10.19 show the mode chart of E-field and the phase relationship between the input-coupled cosine (even) and post-excited sine (odd) resonance for dual resonant modes with mode number $n = 1$, respectively. Figures 10.20 and 10.21 show the mode chart of E-field and the phase relationship between the input-coupled cosine (even) and post-excited sine (odd) resonance for dual resonant modes with mode number $n = 3$, respectively. As shown in Figures 10.19 and 10.21, the inserted post forces zero E-field on its metal surface. According to this inserted boundary condition, for the dual resonant mode $n = 1$ shown in Figure 10.19, an inverted-sine (odd) resonance is excited to cancel out the E-field of the input-coupled cosine (even) resonance on the post metal surface. For the dual resonant mode $n = 3$ shown in Figure 10.21, however, a sine (odd) resonance is generated to meet the inserted boundary condition.

Figure 10.22*a* and *b* show the equivalent circuits for the dual resonant modes $n = 1$ and $n = 3$. As shown in Figure 10.22*a*, the polarizations of even and odd resonances for the dual resonant mode $n = 1$ are opposite. This means

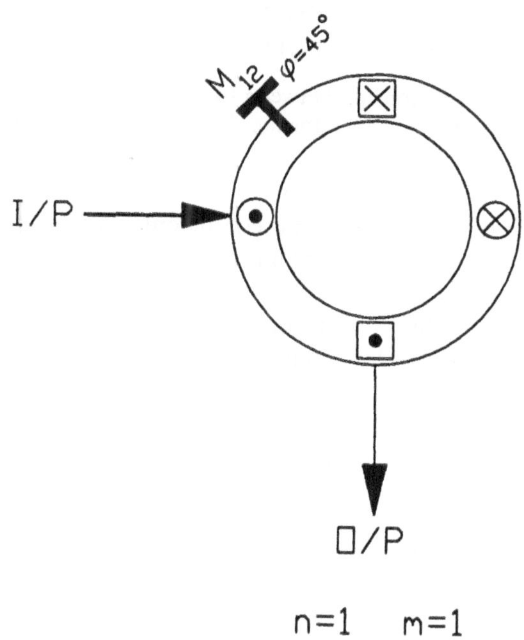

FIGURE 10.18 Mode chart of the E-field for the dual resonant mode $n = 1$. Key: ⊙ = maximum $\mathbf{E}_y$ point for cosine wave; ⊗ = maximum $-\mathbf{E}_y$ point for cosine wave; ⊡ = maximum $\mathbf{E}_y$ point for sine wave; ⊠ = maximum $-\mathbf{E}_y$ point for sine wave.

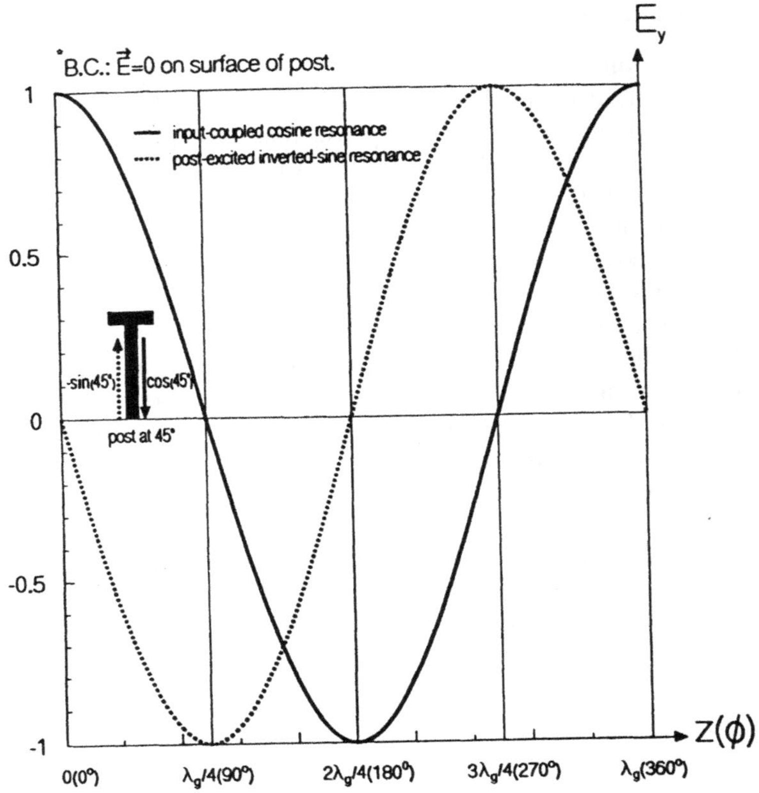

FIGURE 10.19 Phase relationship between the input-coupled cosine resonance and post-excited inverted-sine resonance for the dual resonant mode $n = 1$.

that the sign of the mutual coupling coefficient between the dual resonant mode $n = 1$ is negative. For the implementation of a two-pole canonical dual-mode filter, the transfer function has two transmission zeros when the mutual coupling coefficient of the dual resonant modes is negative [31]. The polarizations of the even and odd resonances for the dual resonant mode $n = 3$, as shown in Figure 10.22*b*, are in the same direction. This means that the sign of the mutual coupling coefficient between the dual resonant mode $n = 3$ is positive. It is not possible to generate transmission zeros using the dual resonant mode $n = 3$.

A dual-mode index is used for the prediction of the polarization of the even and odd resonance and defined by

$$m = \frac{n}{2} + \frac{\theta}{90^\circ} \tag{10.12}$$

where n is the resonant mode number and θ is the location of the tuning post that can be 45°, 135°, 225°, or 315°. For the dual resonant mode $n = 1$ with a

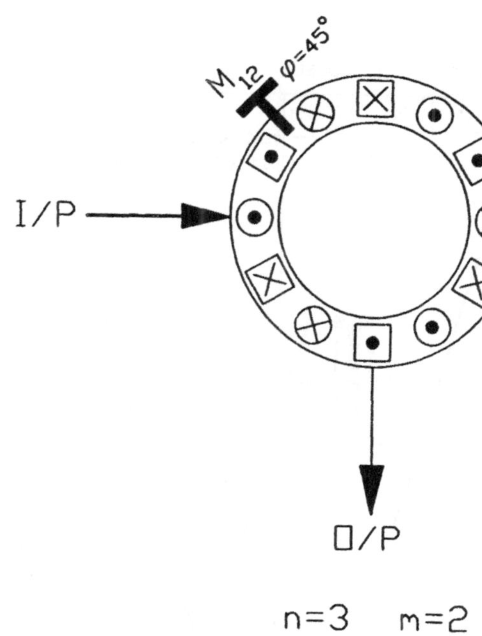

FIGURE 10.20 Mode chart of the E-field for the dual resonant mode $n = 3$. Key: ⊙ = maximum $\mathbf{E}_y$ point for cosine wave; ⊗ = maximum $-\mathbf{E}_y$ point for cosine wave; ⊡ = maximum $\mathbf{E}_y$ point for sine wave; ⊠ = maximum $-\mathbf{E}_y$ point for sine wave.

45° perturbation, the dual-mode index of $m = 1$ implies that the perturbation post is within the first phase domain of a full guide wavelength, as shown in Figure 10.19, and the polarizations of even and odd resonances are opposite, as shown in Figure 10.22*a*. Similarly, for the dual resonant mode $n = 3$ with a 45° perturbation, the dual-mode index of $m = 2$ predicts the second phase domain perturbation of a full guide wavelength, as shown in Figure 10.21, and the same polarizations for the even and odd resonances, as shown in Figure 10.22*b*.

Figures 10.23 and 10.24 show the measured frequency responses of insertion loss for a 90° asymmetrically coupled H-plane ring cavity with tuning posts at 45° and 135°, respectively. As shown in the figures, the dual-mode filter with an odd dual-mode index has two transmission zeros and bears a sharp gain slope transition. However, the dual-mode filter with an even dual-mode index does not have any transmission zero. The measured results agree with the prediction of Equation (10.12). The single-cavity dual-mode filter shown in Figure 10.23, using a single H-plane ring cavity with a tuning post at 45°, has been achieved for the $n = 5$ mode with the following results: (1) center frequency $f_0 = 20.28$ GHz; (2) bandwidth $BW = 250$ MHz; (3) midband insertion loss $IL = 2.69$ dB; and (4) stopband attenuation $A = 40$ dB. The other dual-mode filter caused by a tuning post at 135°, as shown in Figure 10.24, was obtained for the $n = 7$ mode with the following results: (1) center frequency

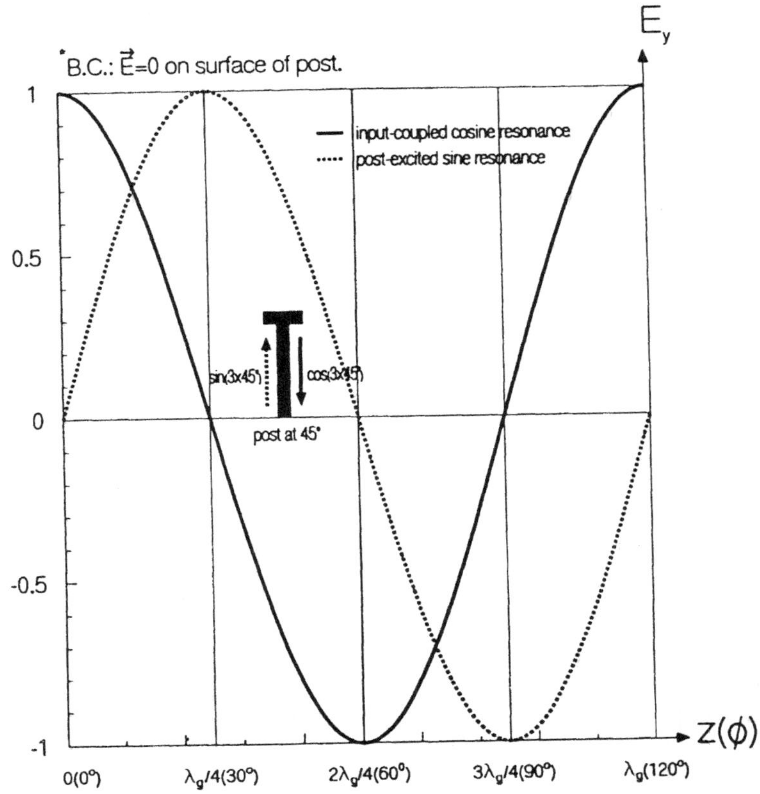

FIGURE 10.21 Phase relationship between the input-coupled cosine resonance and post-excited sine resonance for the dual resonant mode $n = 3$.

$f_0 = 24.62\,\text{GHz}$; (2) bandwidth $BW = 190\,\text{MHz}$; (3) midband insertion loss $IL = 1.5\,\text{dB}$; and (4) stopband attenuation $A = 48\,\text{dB}$.

Though an excellent in-band performance has been achieved with a single waveguide ring cavity, the out-band behavior cannot meet the requirements in some system applications. The following illustrates a new type of two-cavity dual-mode filter that uses two E-plane ring cavities to achieve better in-band and out-band performance.

10.3.3 Two-Cavity Dual-Mode Filters

A two-cavity dual-mode filter using two E-plane ring cavities with 90° splits of the external feed lines was designed to improve the stopband attenuation. The two-cavity dual-mode filter was built by cascading two identical E-plane ring cavities. Two tuning posts located at 45° and 135° were used in each E-plane ring cavity. The measured frequency response of insertion loss for the two-cavity dual-mode filter is shown in Figure 10.25. As shown in the

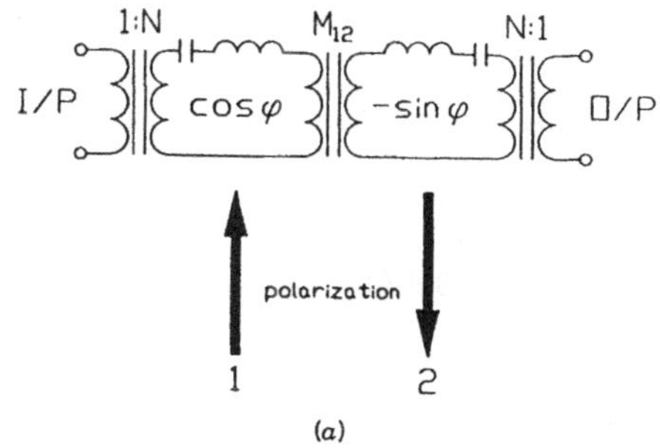

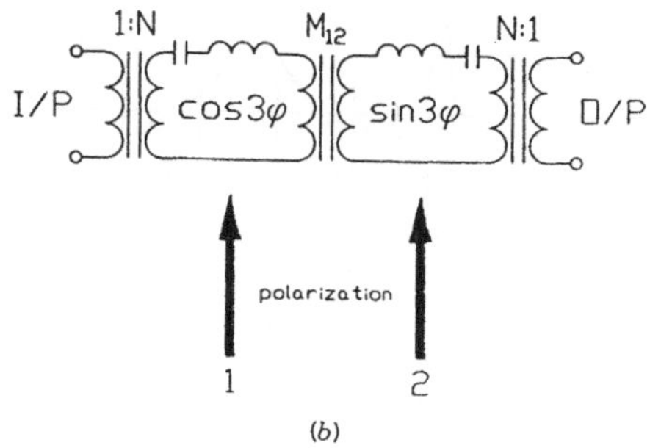

FIGURE 10.22 Equivalent circuits for the dual-mode filters with mode number (*a*) $n = 1$ and (*b*) $n = 3$.

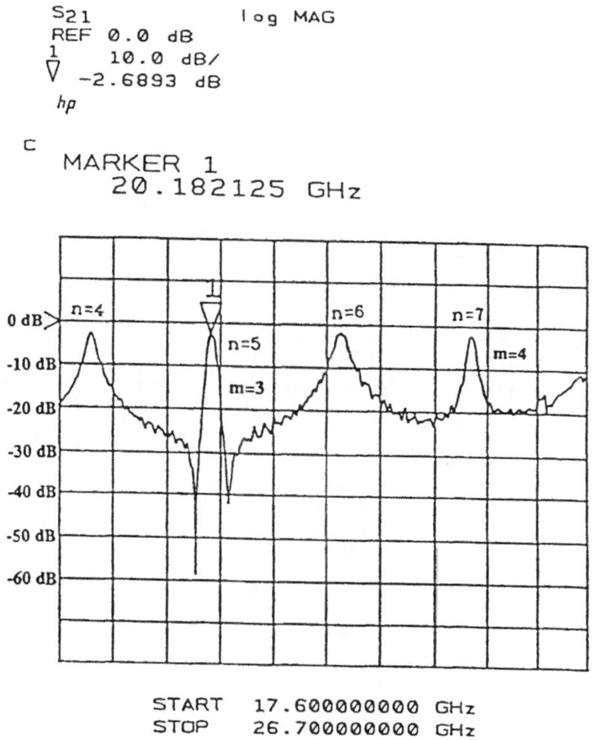

FIGURE 10.23 Measured frequency response for the single-cavity dual-mode filter with a tuning post at 45°.

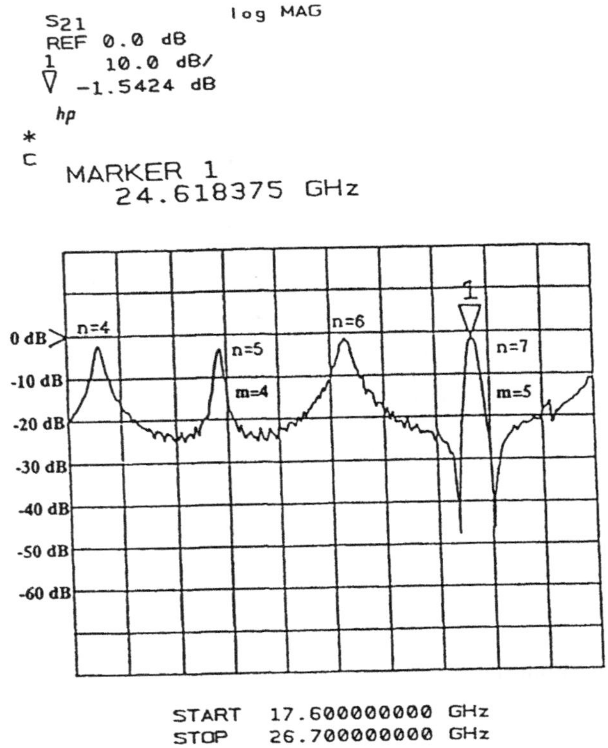

FIGURE 10.24 Measured frequency response for the single-cavity dual-mode filter with a tuning post at 135°.

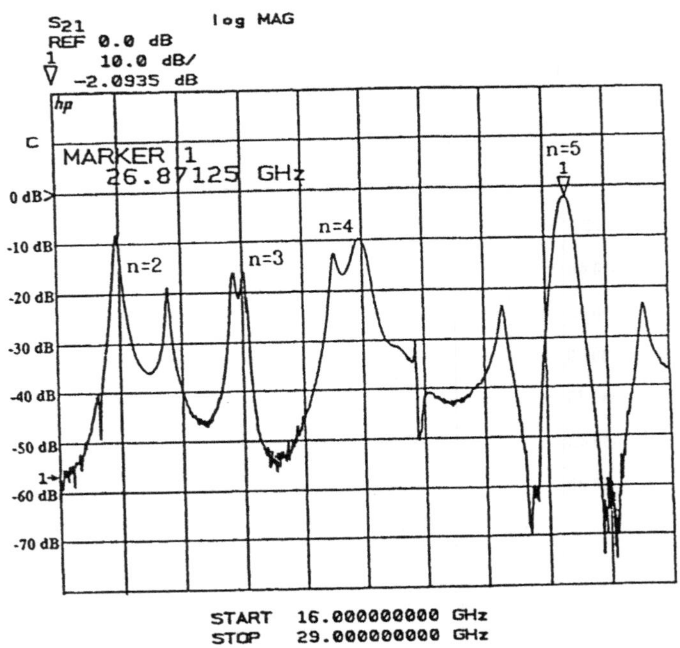

FIGURE 10.25 Measured frequency response for a two-cavity dual-mode filter.

figure, a two-cavity dual-mode filter was achieved for the $n = 5$ mode with the following results: (1) center frequency $f_0 = 26.82\,\text{GHz}$; (2) bandwidth $BW = 300\,\text{MHz}$; (3) insertion loss $IL = 2.09\,\text{dB}$; and (4) stopband attenuation $A = 70\,\text{dB}$.

REFERENCES

[1] T. C. Edwards, *Foundations for Microstrip Circuit Design*, Wiley, Chichester, England; 1981; 2d ed., 1992.

[2] K. Chang, F. Hsu, J. Berenz, and K. Nakano, "Find optimum substrate thickness for millimeter-wave GaAs MMICs," *Microwaves & RF*, Vol. 27, pp. 123–128, September 1984.

[3] W. Hoefer and A. Chattopadhyay, "Evaluation of the equivalent circuit parameters of microstrip discontinuities through perturbation of a resonant ring," *IEEE Trans. Microwave Theory Tech.*, Vol. MTT-23, pp. 1067–1071, December 1975.

[4] P. Toughton, "Measurement technique in microstrip," *Electron. Lett.*, Vol. 5, No. 2, pp. 25–26, January 23, 1969.

[5] J. Deutsch and J. J. Jung, "Microstrip ring resonator and dispersion measurement on microstrip lines from 2 to 12 GHz," *Nachrichtentech. Z.*, Vol. 20, pp. 620–624, 1970.

[6] I. Wolff and N. Knoppik, "Microstrip ring resonator and dispersion measurements on microstrip lines," *Electron. Lett.*, Vol. 7, No. 26, pp. 779–781, December 30, 1971.

[7] H. J. Finlay, R. H. Jansen, J. A. Jenkins, and I. G. Eddison, "Accurate characterization and modeling of transmission lines for GaAs MMICs," in *IEEE MTT-S Int. Microwave Symp. Dig.*, pp. 267–270, June 1986.

[8] P. A. Bernard and J. M. Gautray, "Measurement of relative dielectric constant using a microstrip ring resonator," *IEEE Trans. Microwave Theory Tech.*, Vol. MTT-39, pp. 592–595, March 1991.

[9] I. Wolff, "Microstrip bandpass filter using degenerate modes of a microstrip ring resonator," *Electron. Lett.*, Vol. 8, No. 12, pp. 302–303, June 15, 1972.

[10] M. Guglielmi and G. Gatti, "Experimental investigation of dual-mode microstrip ring resonators," *Proc. 20th Eur. Microwave Conf.*, pp. 901–906, September 1990.

[11] U. Karacaoglu, I. D. Robertson, and M. Guglielmi, "A dual-mode microstrip ring resonator filter with active devices for loss compensation," in *IEEE MTT-S Int. Microwave Symp. Dig.*, pp. 189–192, June 1993.

[12] A. Presser, "Varactor-tunable, high-Q microwave filter," *RCA Rev.*, Vol. 42, pp. 691–705, December 1981.

[13] M. Makimoto and M. Sagawa, "Varactor tuned bandpass filters using microstrip-line ring resonators," in *IEEE MTT-S Int. Microwave Symp. Dig.*, pp. 411–414, June 1986.

[14] K. Chang, T. S. Martin, F. Wang, and J. L. Klein, "On the study of microstrip ring and varactor-tuned ring circuits," *IEEE Trans. Microwave Theory Tech.*, Vol. MTT-35, pp. 1288–1295, December 1987.

[15] S. H. Al-Charchafchi and C. P. Dawson, "Varactor tuned microstrip ring resonators," *IEE Proc. H, Microwaves, Optics and Antennas*, Vol. 136, pp. 165–168, April 1989.

[16] S. Kumar, "Electronically tunable ring resonator microstrip and suspended-substrate filters," *Electron. Lett.*, Vol. 27, No. 27, pp. 521–523, March 14, 1991.

[17] T. S. Martin, F. Wang, and K. Chang, "Theoretical and experimental investigation of novel varactor-tuned switchable microstrip ring resonator circuits," *IEEE Trans. Microwave Theory Tech.*, Vol. MTT-36, pp. 1733–1739, December 1988.

[18] D. S. McGregor, C. S. Park, M. H. Weichold, H. F. Taylor, and K. Chang, "Optically excited microwave ring resonators in gallium arsenide," *Microwave Opt. Tech. Lett.*, Vol. 2, No. 5, pp. 159–162, May 1989.

[19] G. K. Gopalakrishnan, B. W. Fairchild, C. H. Yeh, C. S. Park, K. Chang, M. H. Weichold, and H. F. Taylor, "Microwave performance of nonlinear optoelectronic microstrip ring resonator," *Electron. Lett.*, Vol. 27, No. 2, pp. 121–123, January 17, 1991.

[20] G. K. Gopalakrishnan, B. W. Fairchild, C. H. Yeh, C. S. Park, K. Chang, M. H. Weichold, and H. F. Taylor, "Experimental investigation of microwave optoelectronic interactions in a microstrip ring resonator," *IEEE Trans. Microwave Theory Tech.*, Vol. MTT-39, pp. 2052–2060, December 1991.

[21] G. K. Gopalakrsihnan, "Microwave and optoelectronic performance of hybrid and monolithic microstrip ring resonator circuits," Ph.D. dissertation, Texas A&M University, College Station, May 1991.

[22] I. J. Bahl, S. S. Stuchly, and M. A. Stuchly, "A new microstrip radiator for medical applications," *IEEE Trans. Microwave Theory Tech.*, Vol. MTT-28, pp. 1464–1468, December 1980.

[23] J. A. Navarro and K. Chang, "Varactor-tunable uniplanar ring reseonators," *IEEE Trans. Microwave Theory Tech.*, Vol. 41, No. 5, pp. 760–766, May 1993.

[24] C. Ho, "Slotline, CPW ring circuits and waveguide ring cavities for coupler and filter applications," Ph.D. dissertation, Texas A&M University, College Station, May 1994.

[25] C. Ho, L. Fan, and K. Chang, "A new type of waveguide ring cavity for resonator and filter applications," *IEEE Trans. Microwave Theroy Tech.*, Vol. 42, No. 1, pp. 41–51, January 1994.

[26] L. Lewin et al., *Electromagnetic Waves and Curved Structures*, IEEE Press, New York, pp. 36–43, 1977, Chap. 4.

[27] R. G. Rogers, *Low Phase Noise Microwave Oscillator Design*, Artech House, Boston, pp. 74–82, 1991, Chap. 3.

[28] A. E. Atia and A. E. Williams, "New types of bandpass filters for satellite transponders," *COMSAT Tech. Rev.*, Vol. 1, No. 1, pp. 21–43, Fall 1971.

[29] A. E. Atia and A. E. Williams, "Narrow bandpass waveguide filters," *IEEE Trans. Microwave Theory Tech.*, Vol. MTT-20, pp. 258–265, April 1972.

[30] H. C. Chang and K. A. Zaki, "Evanescent-mode coupling of dual-mode rectangular waveguide filters," *IEEE Trans. Microwave Theory Tech.*, Vol. MTT-39, pp. 1307–1312, August 1991.

[31] K. A. Zaki, C. M Chen, and A. E. Atia, "A circuit model of probes in dual-mode cavity," *IEEE Trans. Microwave Theory Tech.*, Vol. MTT-36, pp. 1740–1745, December 1988.

CHAPTER ELEVEN

Ring Antennas and Frequency-Selective Surfaces

11.1 INTRODUCTION

The ring antenna has been used in many wireless systems. The ring resonator is constructed as a resonant antenna by increasing the width of the microstrip [1–4]. As shown in Figure 11.1, a coaxial feed with the center conductor extended to the ring can be used to feed the antenna. The ring antenna has been rigorously analyzed using Galerkin's method [5, 6]. It was concluded that the TM_{12} mode is the best mode for antenna applications, whereas TM_{11} mode is best for resonator applications. Another rigorous analysis of probe-feed ring antenna was introduced in [7]. In [7], a numerical model based on a full-wave spectral-domain method of moment is used to model the connection between the probe feed and ring antenna.

The slot ring antenna is a dual microstrip ring antenna. It has a wider impedance bandwidth than the microstrip antenna. Therefore, the bandwidth of the slot antenna is greater than that of the microstrip antenna [8–10]. By introducing some asymmetry to the slot antenna, a circular polarization (CP) radiation can be obtained. The slot ring antenna in the ground plane of a microstrip transmission line can be readily made into a corporate-fed array by implementing microstrip dividers.

Active antennas have received great attention because they offer savings in size, weight, and cost over conventional designs. These advantages make them desirable for possible application in microwave systems such as wireless communications, collision warning radars, vehicle identification transceiver, self-mixing Doppler radar for speed measurement, and microwave identification systems [11, 12].

Microwave Ring Circuits and Related Structures, Second Edition,
by Kai Chang and Lung-Hwa Hsieh
ISBN 0-471-44474-X

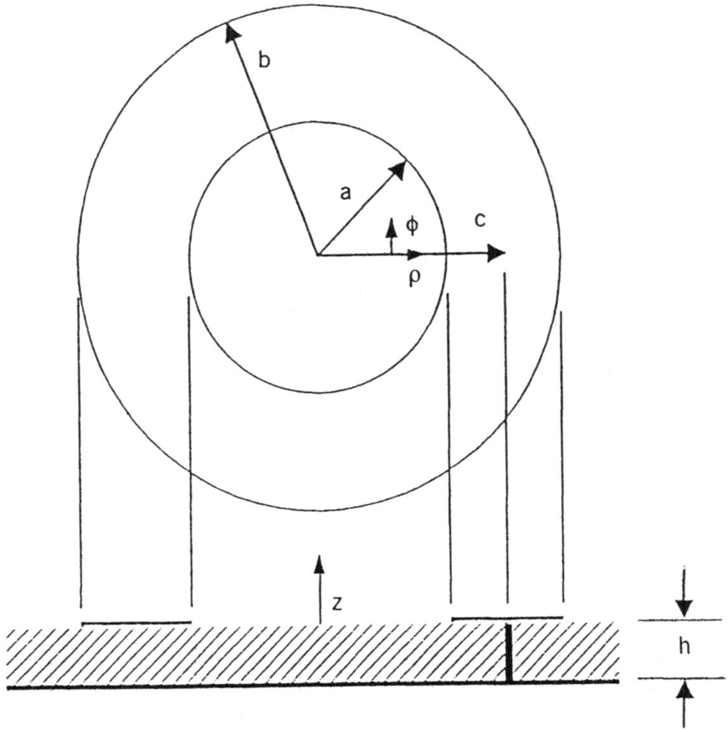

FIGURE 11.1 The annular ring antenna configuration.

Frequency-selective surfaces (FSSs) using circular or rectangular rings have been used as the spatial bandpass or bandstop filters. This chapter will briefly discuss these applications. Also, a reflectarray using ring resonators will be described in this chapter.

11.2 RING ANTENNA CIRCUIT MODEL

The annular ring antenna shown in Figure 11.1 can be modeled by radial transmission lines terminated by radiating apertures [13, 14]. The antenna is constructed on a substrate of thickness h and relative dielectric constant ε_r. The inside radius is a, the outside radius is b, and the feed point radius is c. This model will allow the calculation of the impedance seen from an input at point c. The first step in obtaining the model is to find the **E** and **H** fields supported by the annular ring.

11.2.1 Approximations and Fields

The antenna is constructed on a substrate of thickness h, which is very small compared to the wavelength (λ). The feed is assumed to support only a z-

directed current with no variation in the z direction ($\delta/\delta z = 0$). This current excitation will produce transverse magnetic (TM) to z-fields that satisfy the following equations in the (ρ, ϕ, z) coordinate system [15]:

$$E_z = \frac{k^2}{j\omega\varepsilon}\Psi \tag{11.1}$$

$$H_\rho = \frac{1}{\rho}\frac{\delta\Psi}{\delta\phi} \tag{11.2}$$

$$H_\phi = -\frac{\delta\Psi}{\delta\rho} \tag{11.3}$$

where

$$\psi = \frac{j\omega\varepsilon}{k^2}(A_n J_n(k\rho) + B_n Y_n(k\rho)) f_n(\phi)$$

$$k = \omega\sqrt{\mu_0\varepsilon_0\varepsilon_r}$$

$$\omega = \text{frequency in radians per second}$$

$$\mu_0 = \text{permeability of free space}$$

$$\varepsilon_0 = \text{permittivity of free space}$$

$$j = \sqrt{-1} \tag{11.4}$$

$f_n(\phi)$ is a linear combination of $\cos(n\phi)$ and $\sin(n\phi)$, A_n and B_n are arbitrary constants, J_n is the nth-order Bessel function, and Y_n is the nth-order Neumann function.

The equations for $E_z(\rho)$ and $H_\phi(\rho)$, without the ϕ dependence, are

$$E_z(\rho) = A_n J_n(k\rho) + B_n Y_n(k\rho) \tag{11.5}$$

$$H_\phi(\rho) = -\frac{jk}{\omega\mu_0}[A_n J_n'(k\rho) + B_n Y_n'(k\rho)] \tag{11.6}$$

where $J_n'(k\rho)$ is the derivative of the nth-order Bessel function and $Y_n'(k\rho)$ is the derivative of the nth-order Neumann function with respect to the entire argument $k\rho$.

These fields are used to define modal voltages and currents. The modal voltage is simply defined as $E_z(\rho)$. The modal current is $-\rho H_\phi(\rho)$ or $\rho H_\phi(\rho)$ for power propagating in the ρ or $-\rho$ direction, respectively. This results in the following expressions for the admittance at any point ρ:

$$Y(\rho) = \frac{\rho H_\phi(\rho)}{E_z(\rho)}, \qquad \rho < c \tag{11.7}$$

$$Y(\rho) = \frac{-\rho H_\phi(\rho)}{E_z(\rho)}, \qquad \rho > c \tag{11.8}$$

11.2.2 Wall Admittance Calculation

As shown in Figure 11.2 the annular ring antenna is modeled by radial transmission lines loaded with admittances at the edges. The *s* subscript is used to denote self-admittance while the *m* subscript is used to denote mutual admittance. The admittances at the walls ($Y_m(a, b)$, $Y_s(a)$, $Y_s(b)$) are found using two approaches. The reactive part of the self-admittances ($Y_s(a)$, $Y_s(b)$) is the wall susceptance. The wall susceptances $b_s(a)$ and $b_s(b)$ come from Equations (11.7) and (11.8), respectively. The magnetic-wall assumption is used to find the constants A_n and B_n in Equation (11.6). The $H_\phi(\rho)$ field is assumed to go to zero at the effective radius b_e and a_e. The effective radius is used to account for the fringing of the fields.

$$b_e = b\sqrt{1+\frac{2hx}{\pi b\varepsilon_r}}$$

$$a_e = a\sqrt{1-\frac{2hx'}{\pi a\varepsilon_r}}$$

$$x = \ln\left(\frac{b}{2h}\right)+1.41\varepsilon_r+1.77+\frac{h}{b}(0.268\varepsilon_r+1.65)$$

$$x' = \ln\left(\frac{a}{2h}\right)+1.41\varepsilon_r+1.77+\frac{h}{a}(0.268\varepsilon_r+1.65)$$

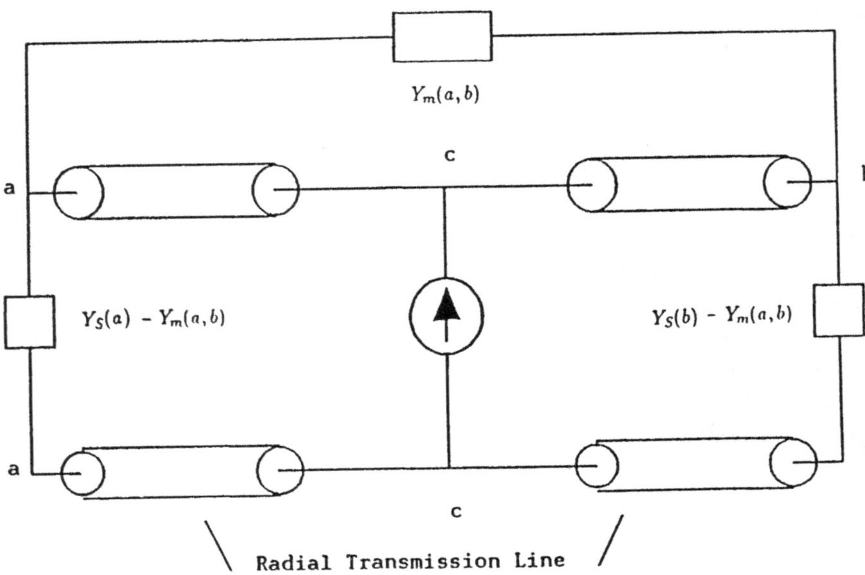

FIGURE 11.2 The annular ring antenna modeled as radial transmission lines and load admittances [13]. (Permission from IEEE.)

It is easily seen that Equations (11.7) and (11.8) will be purely reactive when the magnetic-wall assumption is used to calculate A_n and B_n. This results in the expressions

$$b_s(b)=\frac{kb}{\omega\mu_0}\frac{J_n'(kb)Y_n'(kb_e)-Y_n'(kb)J_n'(kb_e)}{J_n(kb)Y_n'(kb_e)-Y_n(kb)J_n'(kb_e)} \tag{11.9}$$

$$b_s(a)=-\frac{ka}{\omega\mu_0}\frac{J_n'(ka)Y_n'(ka_e)-Y_n'(ka)J_n'(ka_e)}{J_n(ka)Y_n'(ka_e)-Y_n(ka)J_n'(ka_e)} \tag{11.10}$$

The mutual admittance $Y_m(a, b)$ and wall conductances $g_s(a)$ and $g_s(b)$ are found by reducing the annular ring structure to two concentric, circular, coplanar magnetic line sources. The variational technique is then used to determine the equations [15].

The magnetic line current at $\rho = a$ was divided into differential segments and then used to generate the differential electric vector potential $\mathbf{dF}$. The electric field at an observation point is found from

$$\mathbf{dE} = -\nabla\times\mathbf{dF} \tag{11.11}$$

Then the magnetic field at $\rho = b$ and $z = 0$ can be found from Maxwell's equation:

$$\mathbf{dH} = -\frac{1}{j\omega\mu_0}\nabla\times\mathbf{dE} \tag{11.12}$$

The total H_ϕ component of the magnetic field at $\rho = b$ and $z = 0$ due to the current at $\rho = a$ is

$$H_\phi=\int_{\alpha=0}^{2\pi}dH_\phi \tag{11.13}$$

where α is the angle representation for the differential segments.

The mutual admittance will obey the reciprocity theorem, that is, the effect of a current at a on b will be the same as a current at b on a. The reaction concept is used to obtain

$$Y_m(a,b)=\frac{\int_0^{2\pi}H_\phi hbE_b\cos\phi\,d\phi}{\pi hE_aE_b}$$

where

E_a = the radial electric fringing aperture field at a

E_b = the radial electric fringing aperture field at b

The mutual admittance is then found to be

$$Y_m(a,b)=\frac{jabh}{2\pi^2\omega\mu_0}\int_0^{2\pi}\cos\phi\left[\int_0^{2\pi}\cos\alpha\left(\frac{e^{-jk_0r}}{r^3}\right)\{2\cos(\phi-\alpha)(1+jk_0r)\right.$$
$$\left.+\frac{(b\cos(\phi-\alpha)-a)(b-a\cos(\phi-\alpha))}{r^2}\times(k_0^2r^2-3jk_0r-3)\right\}da\Bigg]d\phi \tag{11.14}$$

where

$$r=\sqrt{a^2+b^2-2ab\cos\phi-\alpha}$$
$$k_0=\omega\sqrt{\varepsilon_0\mu_0}$$

This equation can be reduced to a single integral equation by replacing the coefficient of the cos ϕ term in the Fourier expansion of H_ϕ with the sum of all the coefficients and evaluating at $\phi = 0$:

$$Y_m(a,b)=\frac{jabh}{\pi\omega\mu_0}\int_0^{2\pi}\cos\alpha\frac{e^{-jk_0r}}{r^3}$$
$$\times\left[2(1+jk_0r)\cos\alpha+\frac{(b\cos\alpha-a)(b-a\cos\alpha)}{r^2}(k_0^2r^2-3jk_0r-3)\right]d\alpha \tag{11.15}$$

and

$$r=\sqrt{a^2+b^2-2ab\cos\alpha}$$

The self-conductance at a or b can be found by substituting $a = b$ in Equation (11.15) and extracting only the real part:

$$g_s(a)=\frac{a^2h}{2\pi\omega\mu_0}\int_0^{2\pi}\frac{\cos\alpha}{r_a^3}$$
$$\times\left[\left(1+\cos\frac{\alpha^2}{2}\right)(\sin k_0r_a-k_0r_a\cos k_0r_a)-k_0^2r_a^2\sin\frac{\alpha^2}{2}\sin k_0r_a\right]d\alpha \tag{11.16}$$

$$g_s(b)=\frac{b^2h}{2\pi\omega\mu_0}\int_0^{2\pi}\frac{\cos\alpha}{r_b^3}$$
$$\times\left[\left(1+\cos\frac{\alpha^2}{2}\right)(\sin k_0r_b-k_0r_b\cos k_0r_b)-k_0^2r_b^2\sin\frac{\alpha^2}{2}\sin k_0r_b\right]d\alpha \tag{11.17}$$

where

$$r_a=2a\sin\frac{\alpha}{2}$$

$$r_b=2b\sin\frac{\alpha}{2}$$

This completes the solutions for the admittances at the edges of the ring:

$$Y_s(a) = g_s(a) + jb_s(a)$$

$$Y_s(b) = g_s(b) + jb_s(b)$$

11.2.3 Input Impedance Formulation for the Dominant Mode

The next step is to transform the transmission lines to the equivalent π-network. This is accomplished by finding the admittance matrix of the two-port transmission line. The g-parameters of a π-network can then easily be found:

$$g_1(\rho) = \frac{-j}{\omega\mu_0\Delta(\rho_1, \rho_2)}\left[k\rho_1\Delta_1(\rho_1, \rho_2) + \frac{2}{\pi}\right]$$

$$g_2(\rho) = \frac{-2j}{\pi\omega\mu_0\Delta(\rho_2, \rho_1)}$$

$$g_3(\rho) = \frac{j}{\omega\mu_0\Delta(\rho_2, \rho_1)}\left[k\rho_2\Delta_1(\rho_2, \rho_1) + \frac{2}{\pi}\right]$$

where

$$\Delta(\rho_1, \rho_2) = J_n(k\rho_1)Y_n(k\rho_2) - Y_n(k\rho_1)J_n(k\rho_2)$$

$$\Delta_1(\rho_1, \rho_2) = J_n'(k\rho_1)Y_n(k\rho_2) - Y_n'(k\rho_1)J_n(k\rho_2)$$

For $\rho = a$, ρ_1 is replaced by c and ρ_2 is replaced with a. When $\rho = b$, ρ_1 is replaced with b and ρ_2 by c. Figure 11.3 shows the equivalent circuit and the simplified circuit.

From simple circuit theory, the input impedance is seen to be:

$$Z_{\text{in}} = \frac{h}{\pi\sigma_n}\frac{(Z_A + R_A)(Z_B + R_B)Z_C + R_C Z_C(Z_A + R_A + Z_B + R_B)}{\pi\sigma_n(Z_A + R_A)(Z_B + R_B) + (Z_C + R_C)(Z_A + R_A + Z_B + R_B)} \quad (11.18)$$

where

h = thickness of the substrate

$\sigma_n = 2$ *for* $n = 0$; 1 *for* $n > 0$

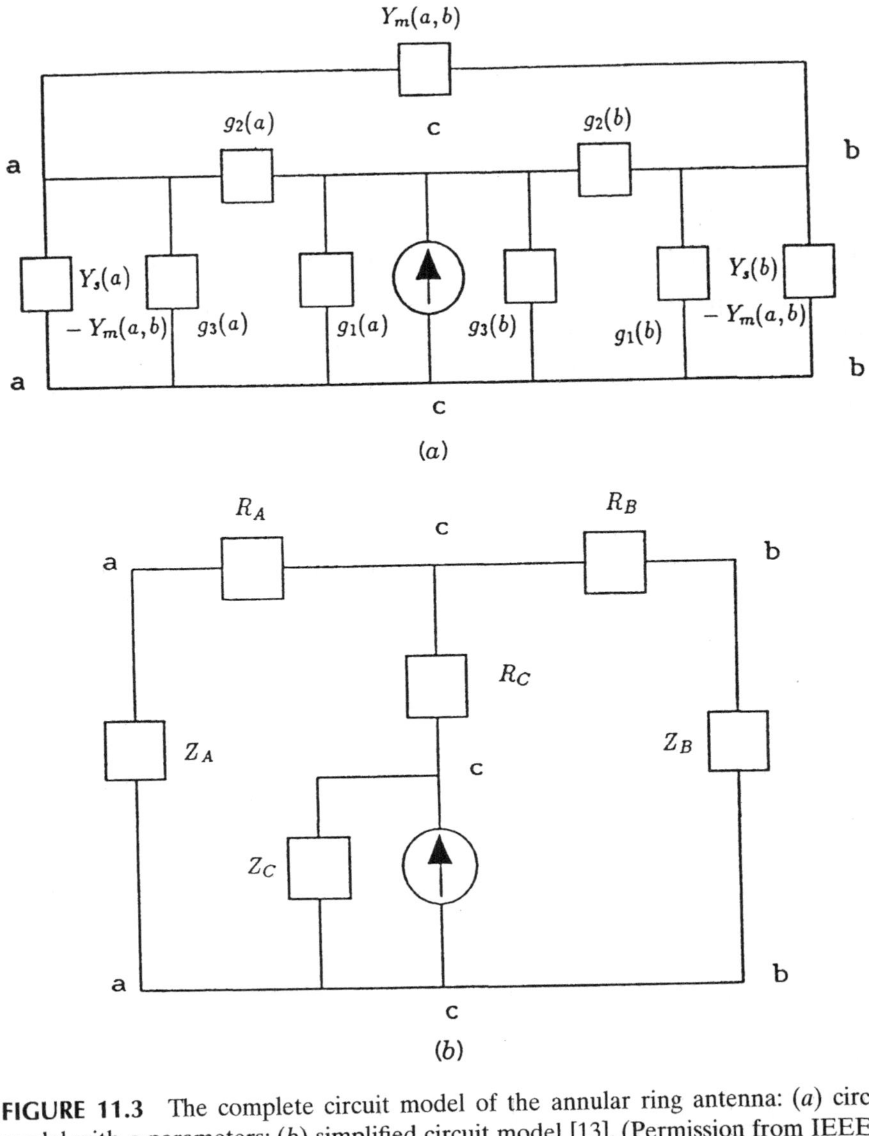

FIGURE 11.3 The complete circuit model of the annular ring antenna: (*a*) circuit model with *g*-parameters; (*b*) simplified circuit model [13]. (Permission from IEEE.)

$$Z_A = \frac{1}{Y_s(a) - Y_m(a,b) + g_3(a)}$$

$$Z_B = \frac{1}{Y_s(a) - Y_m(a,b) + g_1(a)}$$

$$Z_C = \frac{1}{g_s(b) + g_1(a)}$$

$$R_A = \frac{1}{2}\left(\frac{1}{\dfrac{g_2(b)g_2(a)}{g_2(b)+g_2(a)} + Y_m(a,b)} - \frac{1}{g_2(b) + \dfrac{Y_m(a,b)g_2(a)}{g_2(a)+Y_m(a,b)}} + \frac{1}{g_2(a) + \dfrac{Y_m(a,b)g_2(b)}{g_2(b)+Y_m(a,b)}} \right)$$

$$R_B = \frac{1}{2}\left(\frac{1}{\dfrac{g_2(b)g_2(a)}{g_2(b)+g_2(a)} + Y_m(a,b)} + \frac{1}{g_2(b) + \dfrac{Y_m(a,b)g_2(a)}{g_2(a)+Y_m(a,b)}} - \frac{1}{g_2(a) + \dfrac{Y_m(a,b)g_2(b)}{g_2(b)+Y_m(a,b)}} \right)$$

$$R_c = \frac{1}{2}\left(\frac{1}{g_{2(b)} + \dfrac{Y_m(a,b)g_2(a)}{g_2(b)+Y_m(a,b)}} + \frac{1}{g_2(a) + \dfrac{Y_m(a,b)g_2(b)}{g_2(b)+Y_m(a,b)}} - \frac{1}{\dfrac{g_2(b)g_2(a)}{g_2(b)+g_2(a)} + Y_m(a,b)} \right)$$

The $h/(\pi\sigma_n)$ term arises from the discontinuity of the H_ϕ field at c.

11.2.4 Other Reactive Terms

The equation for Z_{in}, Equation (11.18), given earlier assumes that the dominant mode is the only source of input impedance. The width of the feed probe and nonresonant modes contribute primarily to a reactive term. The wave equation is solved using the magnetic walls, as stated earlier, to find the nonresonant mode reactance:

$$X_M = \sum_{\substack{m=0 \\ m\neq n}}^{a} \frac{\omega\mu_0 h}{2\sigma_m}\left[\frac{J_m(kc)Y_m'(ka_e) - Y_m(kc)J_m'(ka_e)}{J_m'(kb_e)Y_m'(ka_e) - Y_m'(kb_e)J_m'(ka_e)}\right] \times [J_m(kc)Y_m'(kb_e) - Y_m(kc)J_m'(kb_e)]\left[\frac{\sin(md/2c)}{(md/2c)}\right]^2 \qquad (11.19)$$

$$\sigma_m = 2 \text{ for } m = 0; 1 \text{ for } m > 0$$
$$d = \text{the feed width}$$
$$n = \text{the resonant mode number}$$

The reactance due to the probe is approximated from the dominate term of the reactance of a probe in a homogeneous parallel-plate waveguide [16]:

$$X_p = \frac{\omega\mu_0 h}{2\pi}\ln\frac{4v_c}{1.781\omega\sqrt{\varepsilon_r}d} \tag{11.20}$$

where v_c is the speed of light.

11.2.5 Overall Input Impedance

The complete input impedance is found by summing the reactive elements given earlier. The final form of Z_{input} is

$$Z_{\text{input}} = \text{Re}\{Z_{\text{in}}\} + j[\text{Im}\{Z_{\text{in}}\} + X_M + X_P] \tag{11.21}$$

where Re and Im represent the real and imaginary parts of Z_{in}, respectively. The reactive terms are summed because X_M and X_p contribute very little to the radiated fields.

11.2.6 Computer Simulation

A computer program was written in Fortran to find the input impedance. The program followed the steps shown in Figure 11.4. The results shown in Figure 11.5 were checked well with the published results of Bhattacharyya and Garg [13].

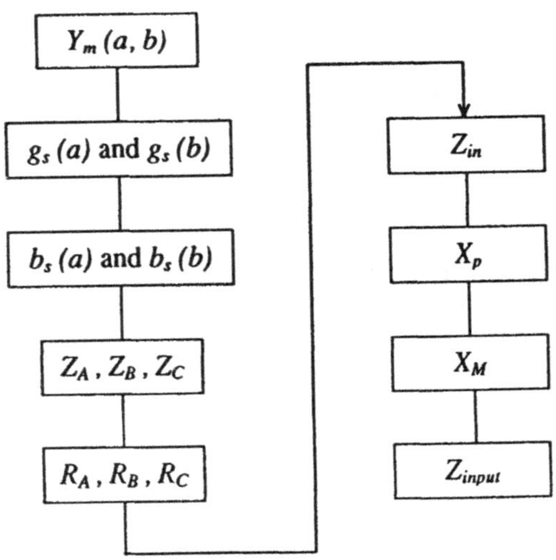

FIGURE 11.4 Flow chart of the input impedance calculation.

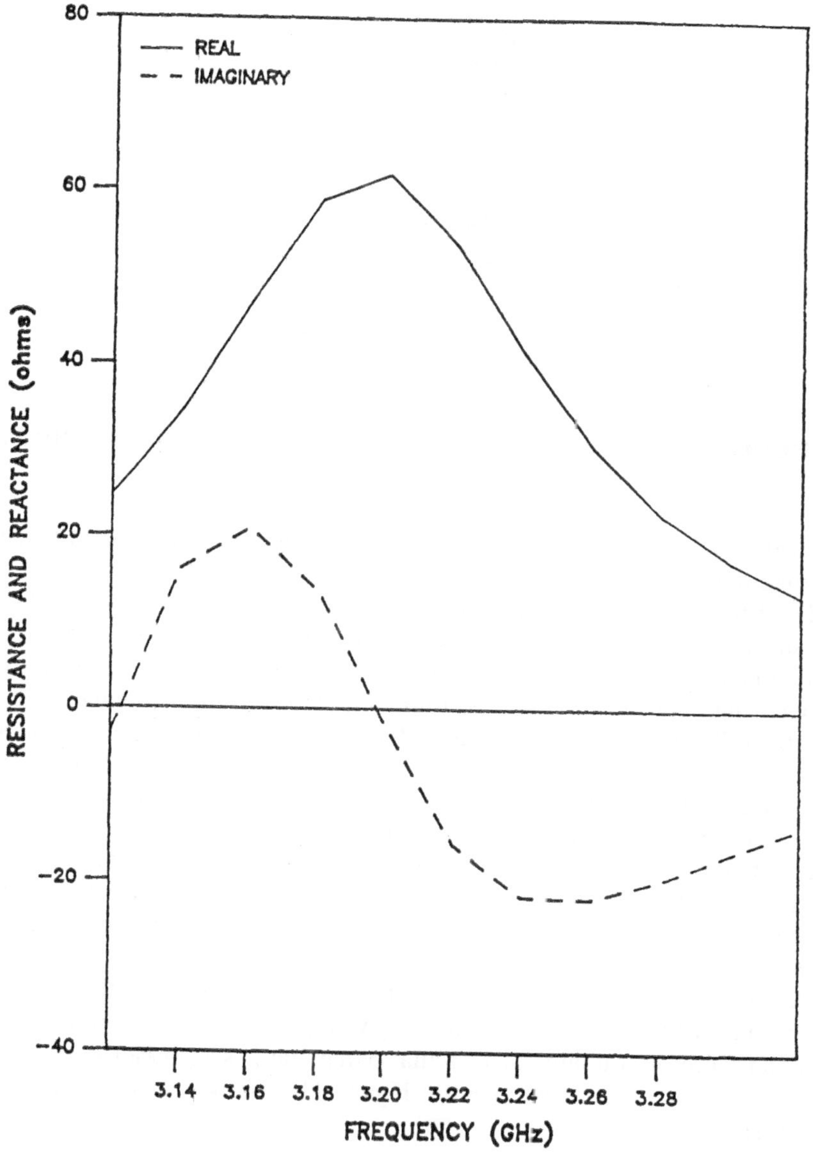

FIGURE 11.5 Input impedance of the TM_{12} mode. $a = 3.0$ cm; $b = 6.0$ cm; $\epsilon_r = 2.2$.

11.3 CIRCULAR POLARIZATION AND DUAL-FREQUENCY RING ANTENNAS

A method for circular polarized ring antennas has been proposed in which an ear is used at the outer periphery [17]. The ear is used as a perturbation to

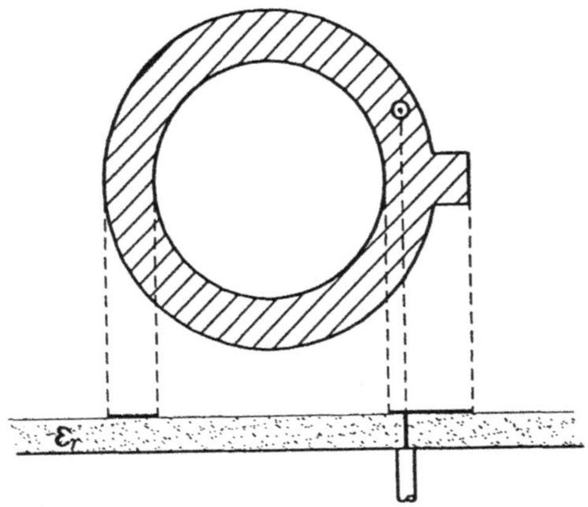

FIGURE 11.6 Circular polarized ring antenna [17]. (Permission from IEEE.)

separate two orthogonal degenerate modes. Figure 11.6 shows the circuit arrangement.

Dual-frequency operation can be achieved using stacked structures [18]. As shown in Figure 11.7, the inner conductor of the coaxial probe passes through a clearance hole in the lower ring and is electrically connected to the upper ring. The lower ring is only coupled by the fringing field and the overall structure can be viewed as two coupled ring cavities. Since the fringing fields are different for the two cavities, their effective inner and outer radii are different even though their physical dimensions are the same. Two resonant frequencies are thus obtained. The separations of the two resonant frequencies ranging from 6.30 to 9.36 percent for the first three modes have been achieved. The frequency separation can be altered by means of an adjustable air gap between the lower ring and the upper substrate.

A shorted annular ring antenna that was made by shorting the inner edge of the ring with a cylindrical conducting wall [19] was recently reported. This antenna therefore radiates as a circular patch, but has a smaller stored energy that allows for a larger bandwidth. Figure 11.8 shows the geometry of the arrangement.

11.4 SLOTLINE RING ANTENNAS

The slotline ring antenna is the dual of the microstrip ring antenna. The comparison is given in Figure 11.9 [20]. Analyses of slot ring antenna can be found in [20, 21]. To use the structure as an antenna, the first-order mode is excited as shown in Figure 11.10, and the impedance seen by the voltage source will be real at resonance. All the power delivered to the ring will

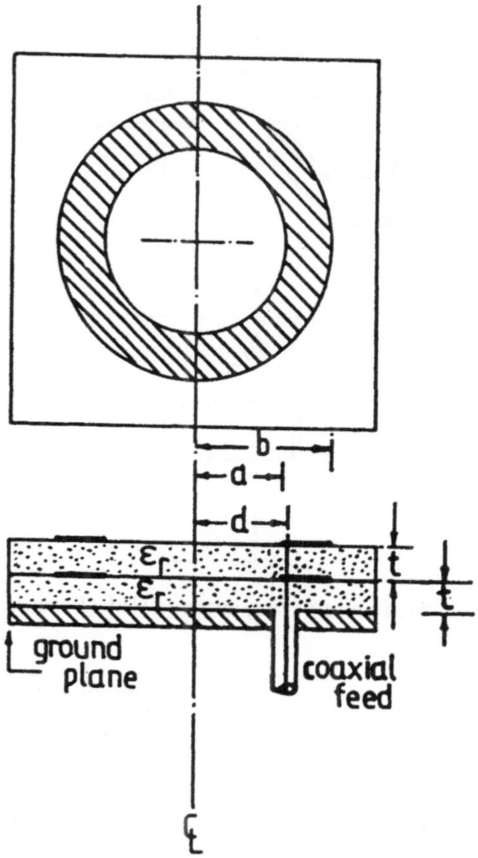

FIGURE 11.7 Dual-frequency stacked annular ring microstrip antenna [18]. (Permission from IEEE.)

be radiated [20]. The resonant frequency, which is the operating frequency, can be calculated using the transmission-line model discussed earlier in the previous chapters. Following the analysis by Stephan et al. [20], the far-field radiation patterns and the input impedance at the feed point can be calculated.

Using the standard spherical coordinates r, θ, and ϕ to refer to the point at which the field are measured, the far-field equations are [20]

$$E_\theta(r,\theta,\phi) = -k_0 \frac{e^{-jk_0 r}}{r} \frac{j^n e^{jn\phi}}{2} [\tilde{E}_0(k_0 \sin\theta)] \tag{11.22}$$

$$E_0(r,\theta,\phi) = +k_0 \frac{e^{-jk_0 r}}{r} \frac{j^{n+1} e^{jn\phi}}{2} \cos\theta [\tilde{E}_e(k_0 \sin\theta)] \tag{11.23}$$

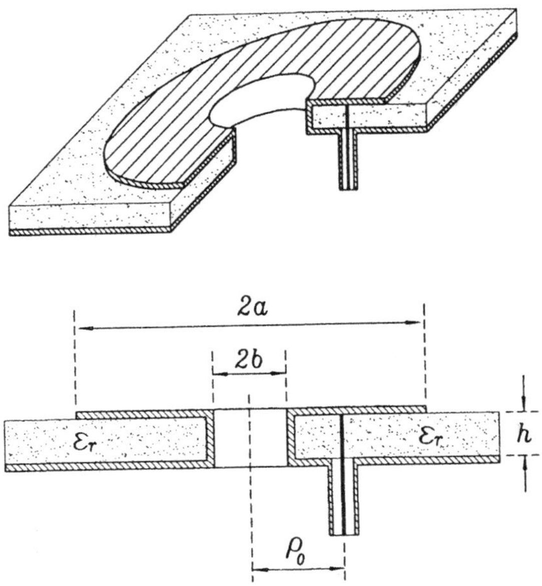

FIGURE 11.8 Shorted annular ring antenna [19]. (Permission from Wiley.)

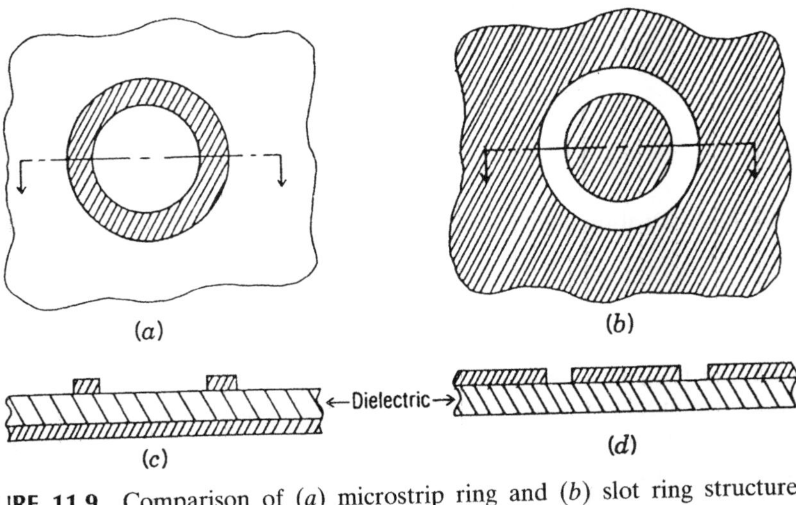

FIGURE 11.9 Comparison of (*a*) microstrip ring and (*b*) slot ring structures. (*c*) Ground plane. (*d*) No ground plane [20]. (Permission from IEEE.)

where $k_0 = \omega\sqrt{\mu_0\varepsilon_0}$ and the linear combinations of the Hankel-transformed estimates are used

$$\tilde{E}_0(k_0 \sin\theta) = \tilde{E}_{(+)}(k_0 \sin\theta) - \tilde{E}_{(-)}(k_0 \sin\theta) \tag{11.24}$$

$$\tilde{E}_e(k_0 \sin\theta) = \tilde{E}_{(+)}(k_0 \sin\theta) + \tilde{E}_{(-)}(k_0 \sin\theta) \tag{11.25}$$

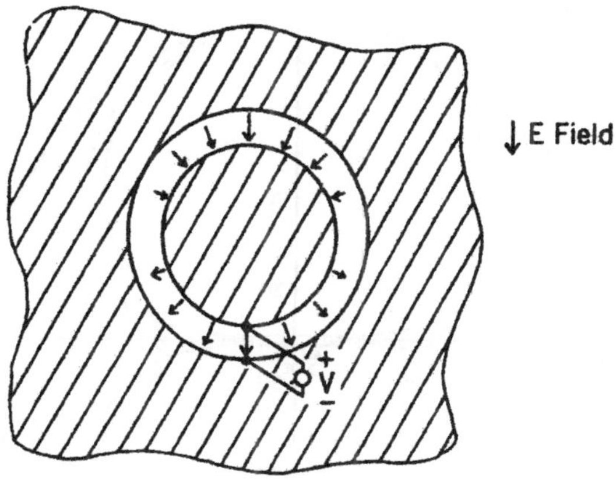

FIGURE 11.10 Slot ring feed method showing electric field [20]. (Permission from IEEE.)

where the $(n \pm 1)$th-order Hankel transforms are defined by

$$\tilde{E}_{(\pm)}(a) = \int_{r_i}^{r_a} J_{n\pm 1}(\alpha r)dr \tag{11.26}$$

where $J_n(\alpha r)$ is the nth-order Bessel function of the first kind, α is the Hankel-transform variable, and r_i and r_a are the inner and outer ring radii, respectively. These integrals can be evaluated analytically using tables. At the center of the ring, $r = 0$, n is the order of resonance being analyzed. In the case of interest, $n = 1$ and $\omega = \omega_0$ = the resonant frequency.

For the finite thickness of the dielectric substrate, the preceding equations for field patterns need to be modified for better accuracy [20]. The input impedance at the feed point can be calculated by [20]:

$$Z_{\text{in}} = \frac{\left[\ln(r_a/r_i)^2\right]}{P} \tag{11.27}$$

where P is the power given by

$$P = \iint_{\text{sphere}} \frac{\frac{1}{2}\sqrt{|E_\theta|^2 + |E_\phi|^2}}{Z_{\text{fs}}} ds \tag{11.28}$$

where Z_{fs} is the intrinsic impedance of free space. An example of calculated and measured E and H-plane patterns is given in Figure 11.11.

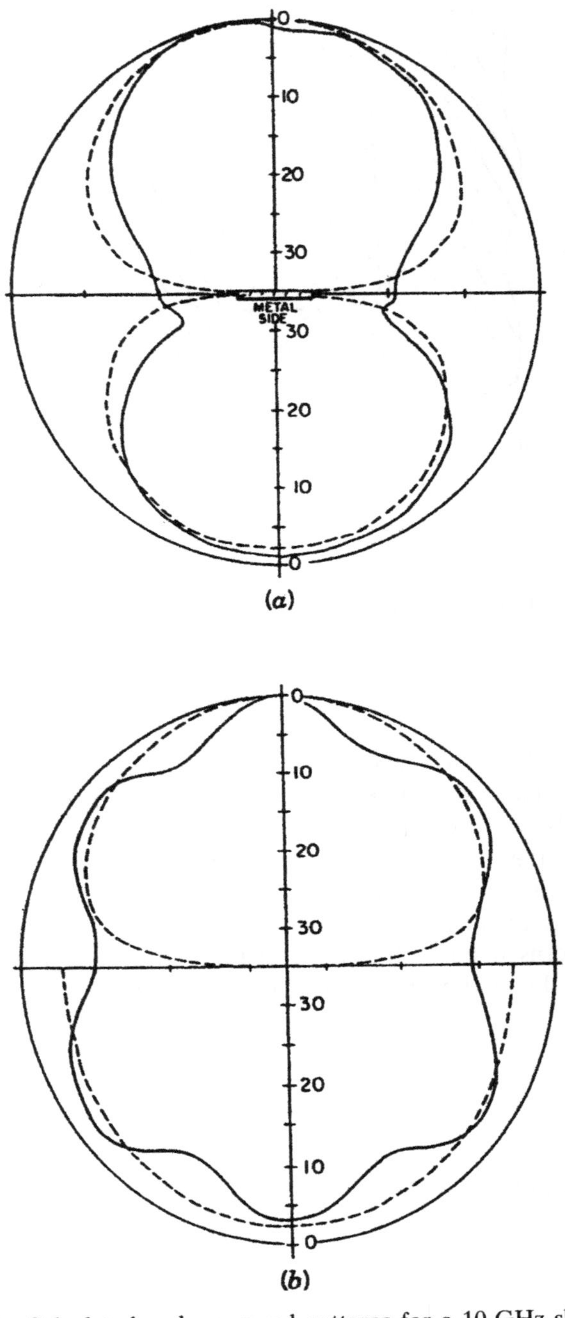

FIGURE 11.11 Calculated and measured patterns for a 10-GHz slot ring antenna. Inner ring radius = 0.39 cm, outer ring radius = 0.54 cm, dielectric $\varepsilon_r = 2.23$, thickness = 0.3175 cm. All patterns are decibels down from maximum. (*a*) *H*-plane; (*b*) *E*-plane. Key:– –calculated;—measured [20]. (Permission from IEEE.)

Figure 11.12 shows a multifrequency annular slot antenna [8, 9]. A 50-ohm microstrip feed is electromagnetically coupled to the slot ring at point A and is extended to the point C. The circuit was etched on a Keene Cor-poration substrate with relative dielectric constant of 2.45 and height of 0.762 mm. The widths of the microstrip (w_m) and slot ring (w_s) were 2.16 mm and 2.9 mm, respectively. The mean circumference of the slot ring is 93.3 mm.

Ignoring the microstrip feed and treating the slot-ring antenna as a transmission line, one expects the operating frequency to be the frequency at which the circumference of the slot-ring antenna becomes one guided wavelength of the slot (λ_{gs}). Slot-guided wavelength for the frequency range of interest can be obtained from [22]

$$\lambda_{gs} = \lambda_o \left\{ 1.045 - 0.365 \ln \varepsilon_r + \frac{6.3(w_s / h)\varepsilon_r^{0.945}}{(238.64 + 100 w_s / h)} - \left[0.148 - \frac{8.81(\varepsilon_r + 0.95)}{100 \varepsilon_r} \right] \ln(h / \lambda_o) \right\} \tag{11.29}$$

where λ_o is the free-space wavelength and h is the thickness of the substrate. At 2.97 GHz, λ_{gs} is equal to the mean circumference of the antenna (93.3 mm).

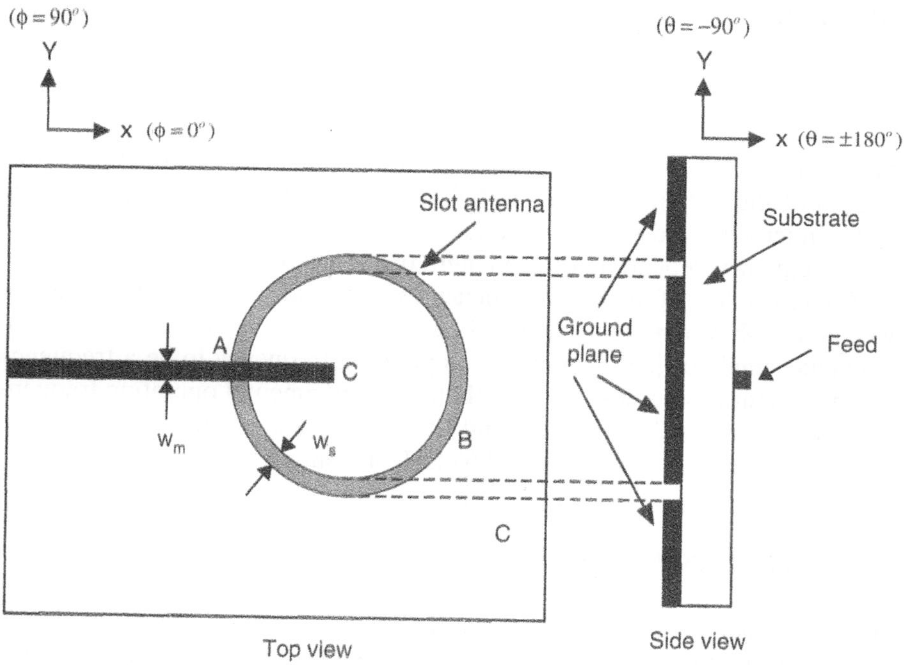

FIGURE 11.12 The configuration of the multifrequency annular antenna.

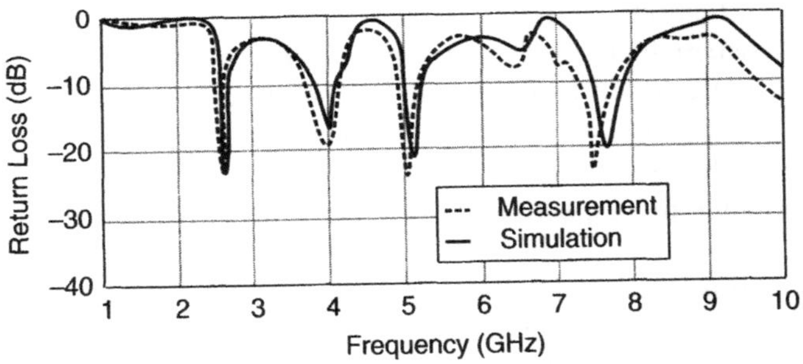

FIGURE 11.13 Measured and simulated return loss of the multifrequency antenna with AC = 46.85 mm [8]. (Permission from IEEE.)

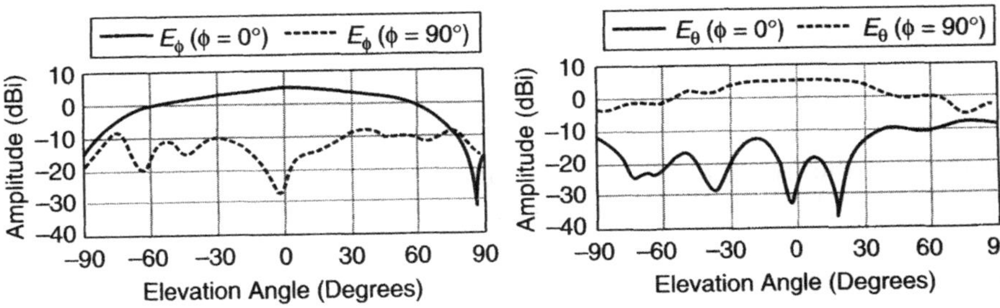

FIGURE 11.14 Radiation patterns of the multifrequency antenna with microstrip stub length AC = 46.85 mm at 2.65 GHz [8]. (Permission from IEEE.)

From this information, as a first-order approximation, first-operating frequency of the slot-ring antenna is 2.97 GHz. The actual operating frequency of the microstrip-fed slot-ring antenna can be above or below this approximate frequency depending on the length of the microstrip stub.

The return loss of the multifrequency antenna and simulation results agree well and as shown in Figure 11.13. The simulation was carried out by electromagnetic simulator [23]. Defining the operating frequency to be a frequency at which return loss is less than 10 dB, these experimental operating frequencies are centered at 2.58, 3.9, 5.03, and 7.52 GHz. The measured patterns of the antenna at resonant frequency of 2.65 GHz are shown in Figure 11.14.

11.5 ACTIVE ANTENNAS USING RING CIRCUITS

An active antenna was developed by the direct integration of a Gunn device with a ring antenna as shown in Figure 11.15 [24]. The radiated output power level and frequency response of the active antenna are shown in Figure 11.16.

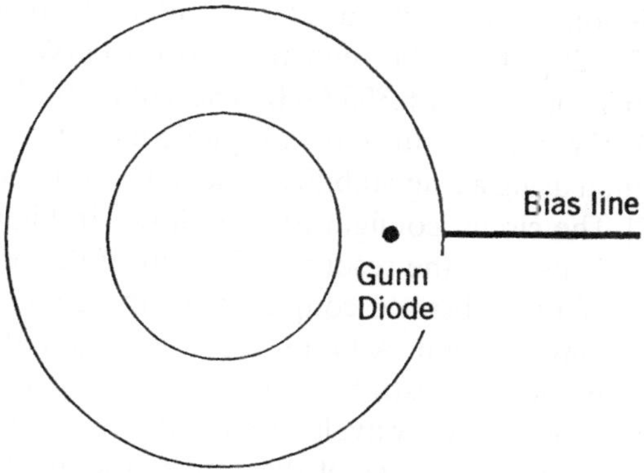

FIGURE 11.15 The active annular ring antenna integrated with Gunn diode [24]. (Permission from Wiley.)

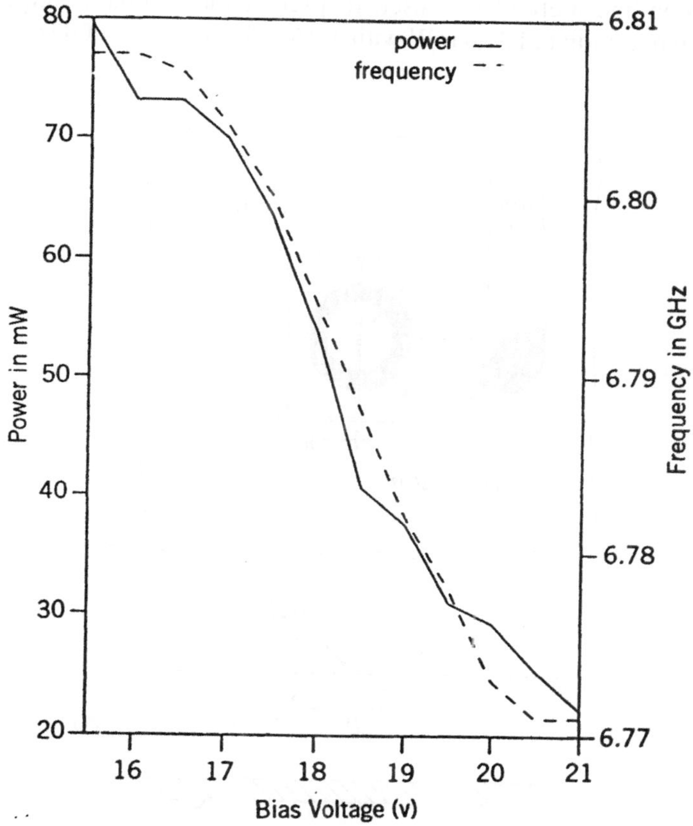

FIGURE 11.16 Power output and frequency vs. bias voltage [24]. (Permission from Wiley.)

The Friis transmission equation was used to calculate the power radiated from the active ring [25, 26]. It can be seen that over 70-mW output power was achieved with a bias of 16 V at 6.805 GHz. The Gunn diode used produced a maximum of 100 mW in an optimized waveguide circuit.

An active antenna using a ring-stabilized oscillator coupled to a slot antenna was reported [27]. The circuit configuration is shown in Figure 11.17. A circular microstrip ring is used as the resonant element of the oscillator. A slot on the ground plane of the substrate coupled with the circular microstrip ring served as the radiating element. A Gunn diode is mounted between the ring and the ground plane of the substrate at either side of the ring. A metal mirror block is introduced one-quarter wavelength behind the ring to avoid any back scattering. The operating frequency of the active antenna was designed to be close to the first resonant frequency of the circular microstrip ring. A radiated power of +16 dBm at 5.5 GHz occurred at the bias level of 12.6 V. The radiation patterns are shown in Figures 11.18 and 11.19.

An active slotline ring antenna integrated with an FET oscillator was also developed [28]. Figure 11.20 shows the physical configuration. A simple transmission-line method was used to predict the resonant frequency. The active antenna radiated 21.6 mW with 18% efficiency at 7.7 GHz.

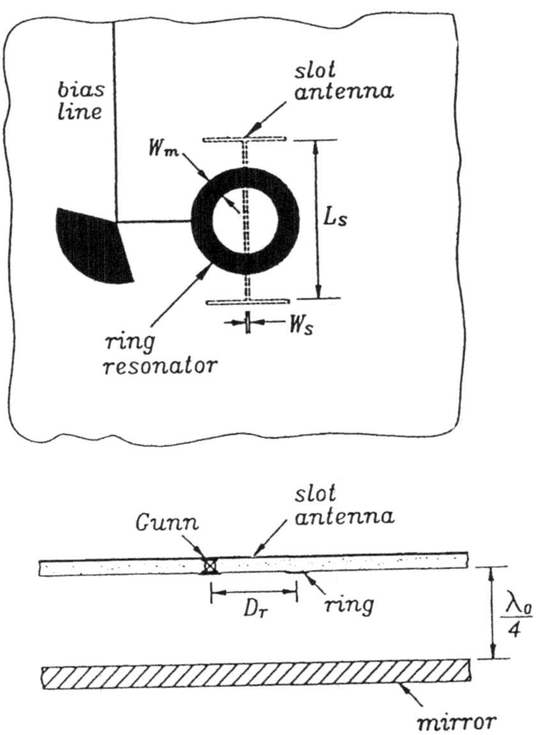

FIGURE 11.17 Circuit configuration [27]. (Permission from IEEE.)

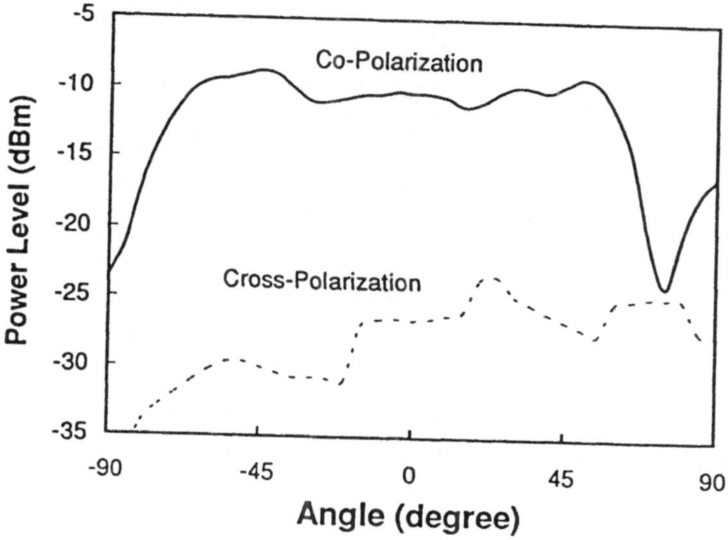

FIGURE 11.18 *E*-plane pattern [27]. (Permission from IEEE.)

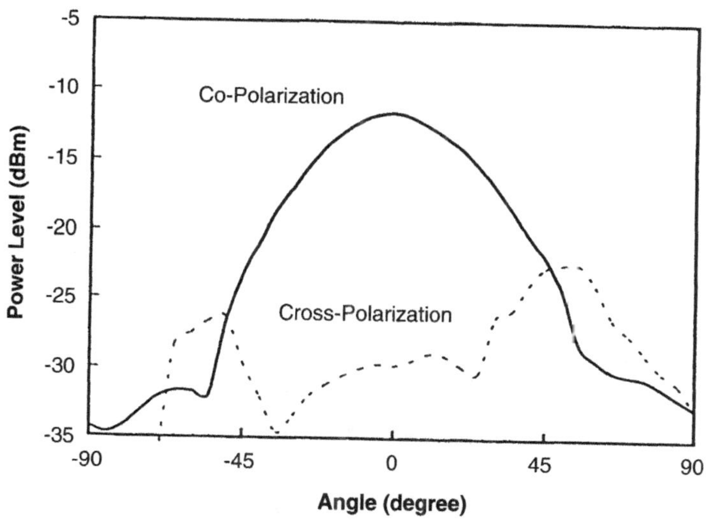

FIGURE 11.19 *H*-plane pattern [27]. (Permission from IEEE.)

The active antenna shown in Figure 11.21 consists of a Gunn diode and a slotline-notch antenna stabilized with a slotline-ring resonator [29]. The Gunn diode is placed across the ring resonator at a low-impedance point to meet the conditions for oscillation [30]. The slotline ring's resonant wavelength can be determined from

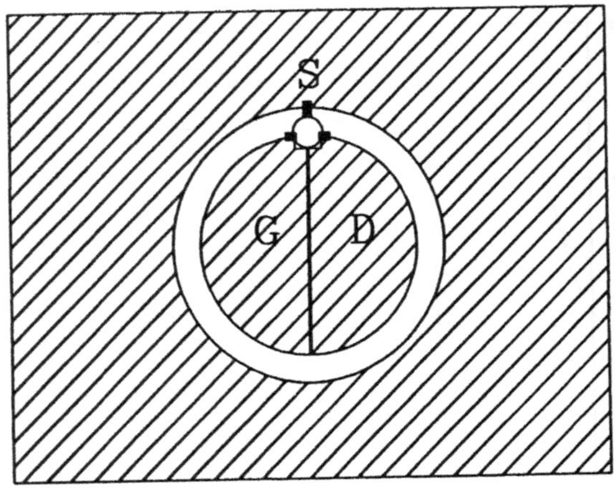

FIGURE 11.20 Circuit configuration of an FET active slotline ring antenna [28]. (Permission from *Electronics Letters.*)

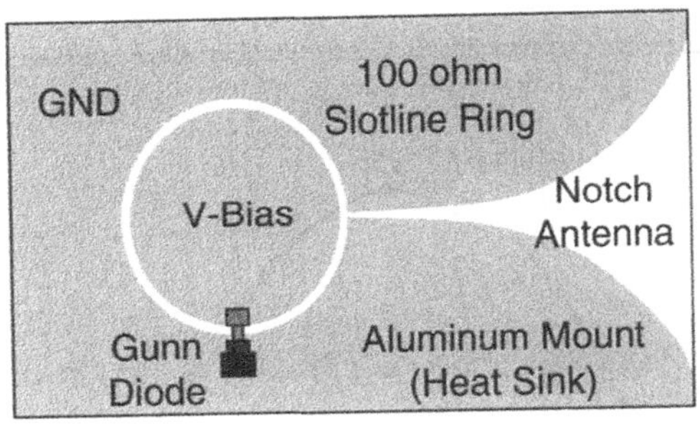

(a)

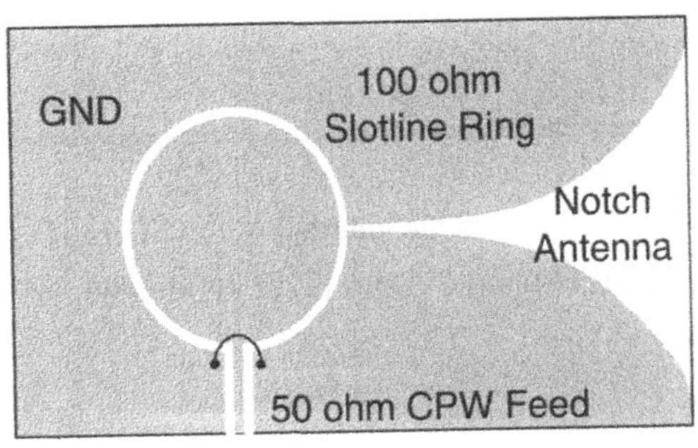

(b)

FIGURE 11.21 Configuration of active antenna: (a) Gunn-diode active-notch antenna using a slotline-ring resonator (the wire for dc bias to the center of the ring is not shown) and (b) CPW-fed passive antenna (some dimensions in the figures are exaggerated to enhance detail) [29].

$$2\pi r = n\lambda_g \qquad for\ n = 1, 2, 3, \ldots \tag{11.30}$$

where r is the mean radius of the slotline ring, λ_g is the guide wavelength in the slotline ring and n is the mode number.

Figure 11.21*a* shows the circuit configuration of the active antenna, whereas Figure 11.21*b* shows the coplanar waveguide (CPW)-fed passive antenna developed for radiation pattern comparison. The antennas were etched on Duroid 5870 board with a relative dielectric constant of 2.33, substrate thickness of 62 mils (1.575 mm), and 1-oz copper metallization. The antennas are truly uniplanar, requiring no backplane for excellent performance. The slotline ring has a mean radius of 3.81 mm and a slot width of 0.18 mm. The CPW feedline in the passive antenna has a center strip width of 1.2 mm and symmetrical side gaps of 0.08 mm. A thin-wire air bridge is used to operate the CPW line in the even mode. The slotline-notch antenna uses an exponential taper to match the impedance of the ring to free-space. The antenna length is 6 cm, and the gap at the feed point is 0.18 mm. The aluminum mount used in the active antenna also serves as the heat sink for the Gunn diode. The dc bias is provided directly to the center of the slotline ring. The active antenna radiates a clean spectrum at 9.26 GHz with a bias voltage of 10.0 V and draws 410 mA. The active antenna produces an effective power output of 27.1 mW and an effective isotropic radiated power (EIRP) of 720.0 mW. The spectrum has a phase noise of −95.33 dBc/Hz at 100 KHz from the carrier, and the second harmonic radiation produced by the active antenna is 26.16 dB below the fundamental frequency. Figure 11.22 shows the radiation patterns of the active antenna.

The E-plane and H-plane patterns are smooth with cross-polarization levels of 13.18 and 6.69 dB below copolarization. The radiation E-plane and H-plane are 33° and 47°, respectively. The radiation pattern of the CPW-fed passive antenna is essentially similar with the exception that the cross-polarization levels are 18.74 and 16.51 or the E- and H-planes, respectively.

11.6 FREQUENCY-SELECTIVE SURFACES

Frequency-selective surfaces (FSSs) have found many applications in quasi-optical filters, diplexers, and multiplexers. Many different element geometries have been used for FSSs [31]. They include dipole, square patch, circular patch, cross dipole, Jerusalem cross, circular ring, and square loop. Figure 11.23 shows these elements. A number of representative techniques for analyzing FSSs have been reviewed in a paper by Mittra, Chan, and Cwik [31].

FFSs using circular rings or square loops have been studied extensively [32–46]. For square loops, closed-form equations are available to design the elements [39]. For example, the gridded square-loop element shown in.

Figure 11.24 can be represented by an equivalent circuit given in Figure 11.25. For a vertically incident electric field, an inductance L_2 represents the

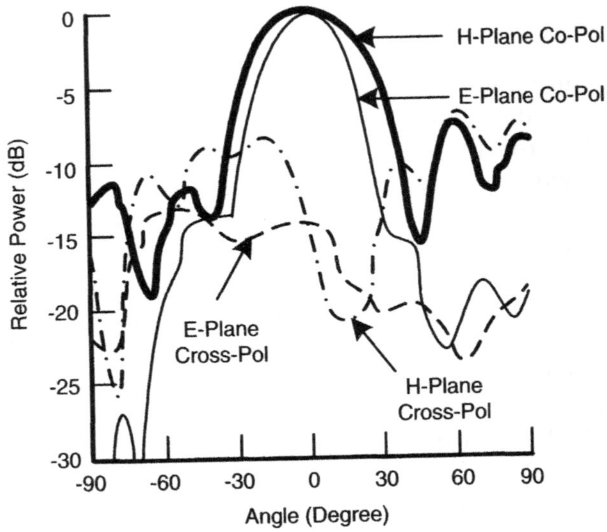

FIGURE 11.22 E-plane radiation pattern: HPBW = 33°, cross-polarization = 13.18 dB below copolarization and H-plane radiation pattern: HPBW = 47°, cross-polarization = 6.69 dB below copolarization [29].

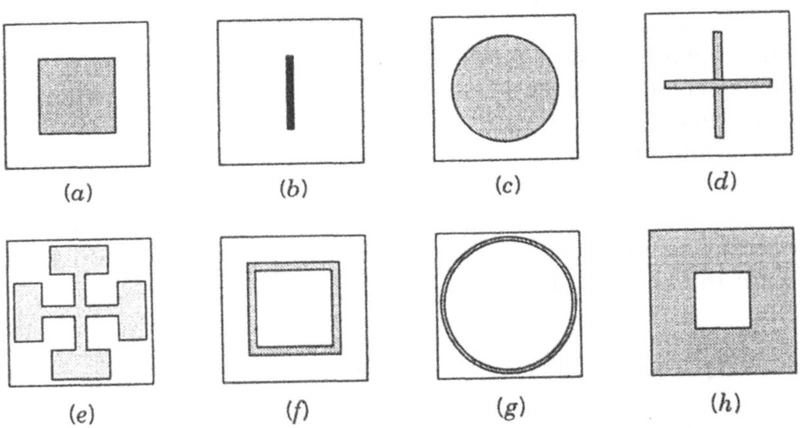

FIGURE 11.23 Some typical FSS unit cell geometries [31]. (*a*) Square patch. (*b*) Dipole. (*c*) Circular patch. (*d*) Cross dipole. (*e*) Jerusalem cross. (*f*) Square loop. (*g*) Circular loop. (*h*) Square aperture. (Permission from IEEE.)

grid and a series-resonant inductance L_1 and capacitance C represent the squares.

The equations given below are used to design for the transmission and rejection bands [39]. Solving for the circuit admittance, the transmission coefficient is given by

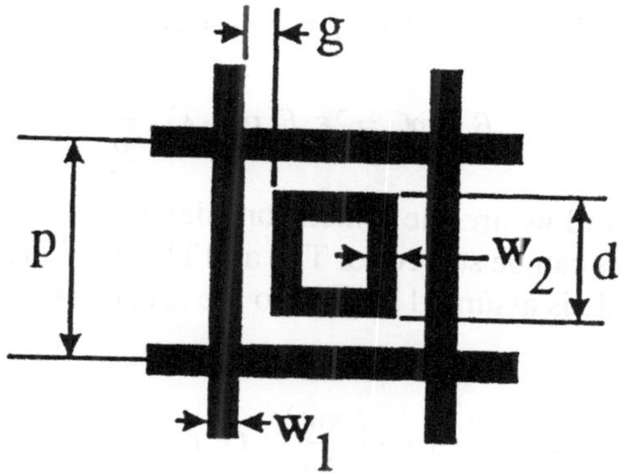

FIGURE 11.24 Unit cell for gridded square frequency-selective surface.

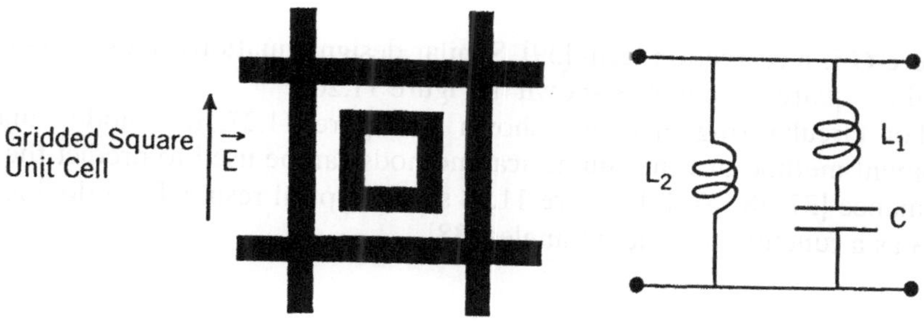

FIGURE 11.25 Equivalent circuit for the gridded FSS element.

$$|\tau|^2 = \frac{4}{4 + |Y|^2}$$

where

$$Y = \frac{X_1 + X_2 - (1/B)}{X_2(X_1 - (1/B))}$$

The values of L_1, L_2, and C are given as

$$X_1 = \omega L_1 = 2(X_2 \| X_3)$$

where

$$X_3 = F(p, 2w_2, \lambda) \cdot \frac{d}{p}$$

$$X_2 = \omega L_2 = F(p, w_1, \lambda)$$

$$B = \omega C = 2\varepsilon_r F(p, g, \lambda) \cdot \frac{d}{p}$$

where p, g, d, w_1, and w_2 are the dimensions defined in Figure 11.24.

The admittances can be solved for TE- and TM-incidence at oblique angles. If the incident angle is assumed normal to the array plane, the resulting equations then become

$$F(p, w, \lambda) = \frac{p}{\lambda}\left[\ln \csc\left(\frac{\pi w}{2p}\right) + G(p, w, \lambda, \theta \,\&\, \phi = 0^\circ)\right]$$

$$F(p, g, \lambda) = \frac{p}{\lambda}\left[\ln \csc\left(\frac{\pi g}{2p}\right) + G(p, g, \lambda, \theta \,\&\, \phi = 0^\circ)\right]$$

where G is a correction term [39]. Similar design equations can be applied to double-square elements as shown in Figure 11.26.

For circular ring elements shown in Figure 11.27, the model analysis moment method or other numerical methods can be used to predict the performance [37, 38, 40, 41]. Figure 11.28 shows typical results for a double-ring FSS as a function of incident angles [38].

11.7 REFLECTARRAYS USING RING RESONATORS

Microstrip reflectarrays have been shown to be good candidates for replacing conventional parabolic reflector antennas. The technologies involve a flat printed array integrated with space-feeding horn and have the advantages of low cost, low profile, easy fabrication, and flat surface [47].

Figure 11.29 shows the configuration of the reflectarray with a Ka-band feed horn [48]. The reflectarray is constructed by ring resonators with two slits at the top and the bottom. Comparing with the reflectarray using square patch antennas, this configuration using ring antennas has wider bandwidth and can reduce blockage when using in a dual-layer and dual-frequency application.

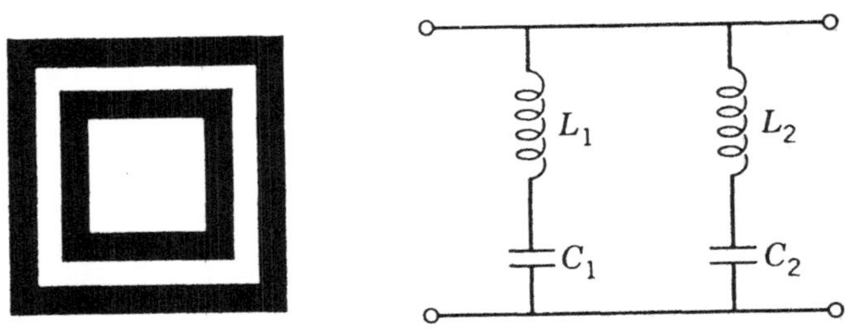

FIGURE 11.26 Double-square element and its equivalent circuit.

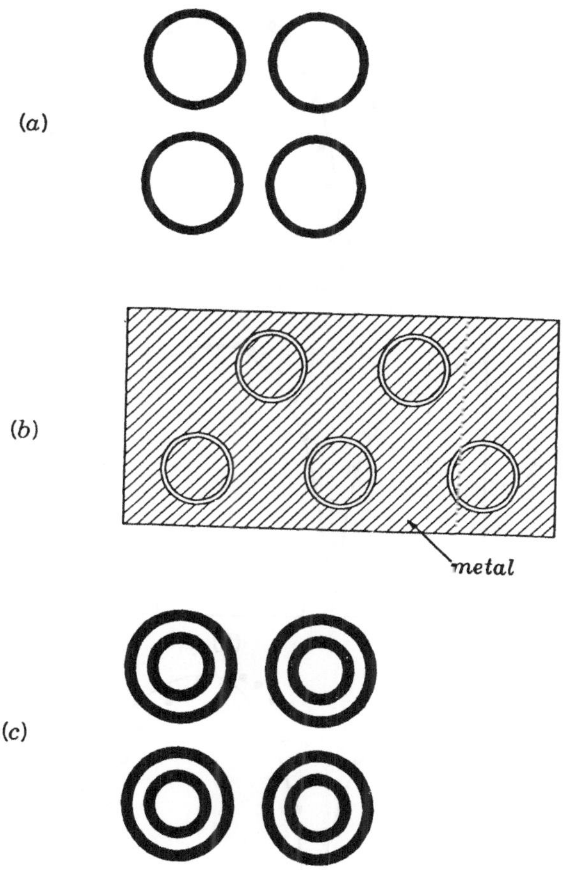

FIGURE 11.27 FSSs use circular ring elements. (*a*) Single ring. (*b*) Slot ring. (*c*) Double ring.

Due to the nature of reflection, an incident right-hand circularly polarized wave is reflected as a left-hand circularly polarized wave. However, by using these two slits on the ring, the reversal of polarization can be canceled out. Also, a simple trigonometric relation is used to calculate the required additional path lengths to create a parabolic phase front across an array's surface. Each ring antenna is rotated counterclockwise by ϕ radians at each position when 2ϕ radians are needed for compensating these additional path lengths. For the whole reflectarray, the repetitious behaviors of array configurations are found to be more than four times of a 2π radian.

A 0.5-m reflectarray has been designed for broadside radiation at a focal distance of 350 mm corresponding to a *f*/D ratio of 0.7. The ring antennas are fabricated on an RT/Duroid 5870 substrate with $\varepsilon_r = 2.33$ and 0.508 mm thickness. The space between rings is 0.5 free-space wavelength, which is selected to avoid grating lobes. The measured results at 31.75 GHz are shown in Figure

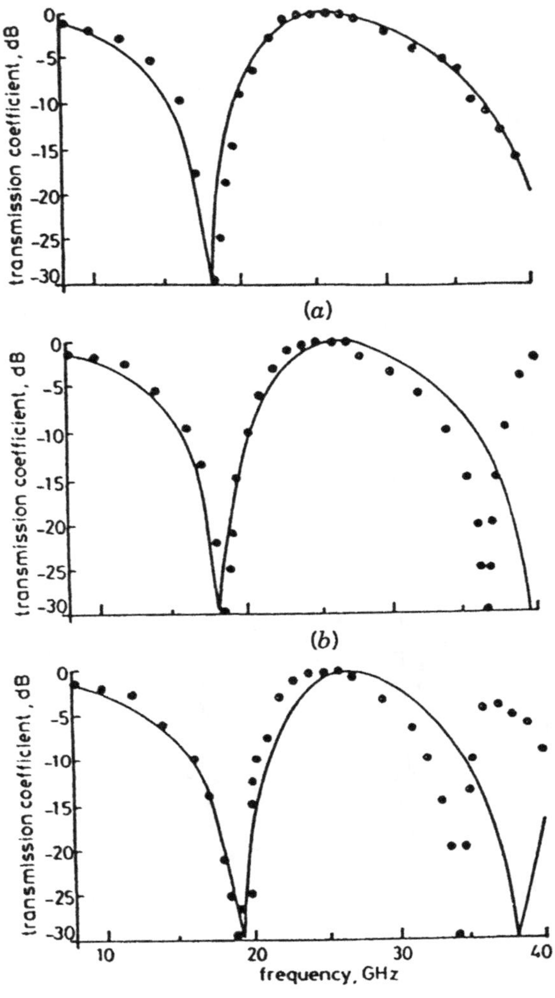

FIGURE 11.28 *H*-plane transmission coefficients as a function of incident angles [38]. (*a*) $\theta = 0°$. (*b*) $\theta = 30°$. (*c*) $\theta = 45°$. Key: — calculated values; ••• experimental results. (Permission from *Electronics Letters.*)

11.30. The main beam has a beam width of 1.3°. The cross-polarization level is 40.71 dB down at broadside, and the sidelobe suppression is greater than 19.47 dB occurring at 2°. This cross-polarization level corresponds to an axial ration of less than 0.5 dB. The relative high peak sidelobe is due to feed horn blockage effects. The measured efficiency of the reflectarray is shown in Figure 11.31. The efficiency is better than 40% from 31.5 to 32.5 GHz. The highest efficiency is 50% at 31.75 GHz.

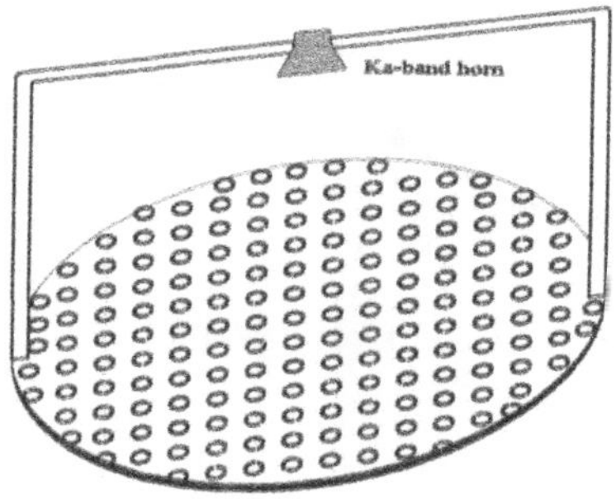

(a)

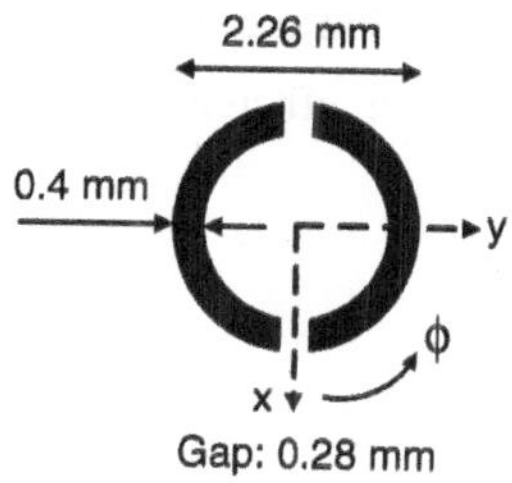

(b)

FIGURE 11.29 Configuration of Ka-band reflectarray (a) setup and (b) ring antenna [48]. (Permission from *Electronics Letters*.)

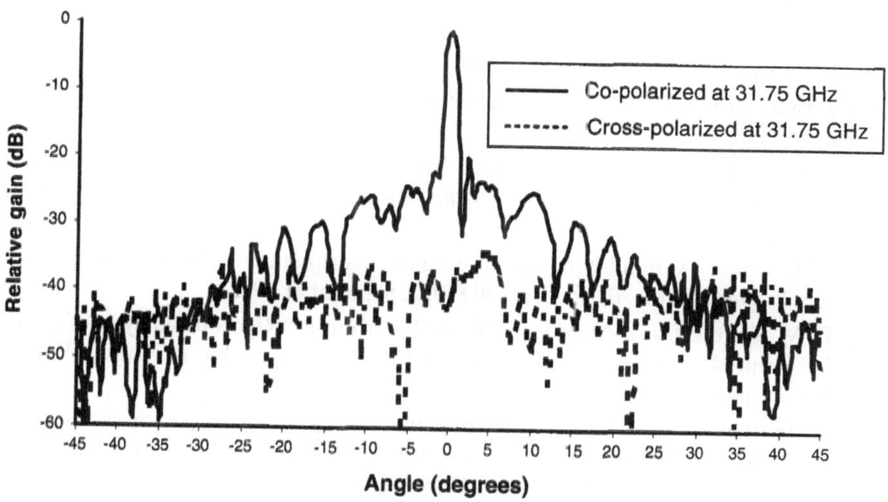

FIGURE 11.30 Measured radiation patterns [48]. (Permission from *Electronics Letters.*)

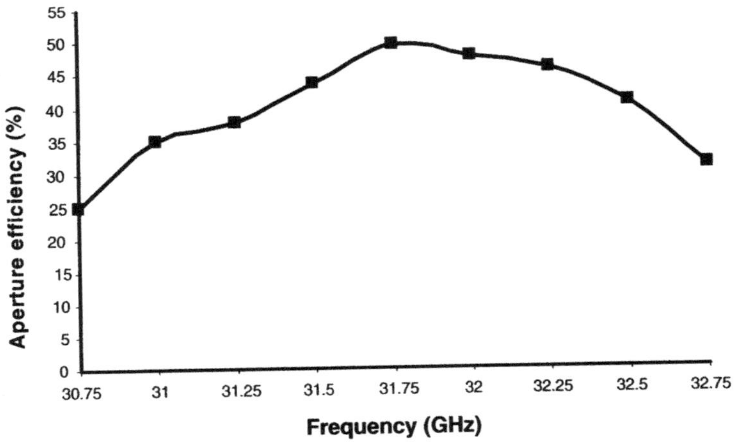

FIGURE 11.31 Measured aperture efficiency [48]. (Permission from *Electronics Letters.*)

REFERENCES

[1] R. R. Ramirez, F. De Flaviis, and N. G. Alexopoules, "Single-feed circularly polarized microstrip ring antenna and arrays," *IEEE Trans. Antennas Propagat.*, Vol. 48, No. 7, pp. 1040–1047, July 2000.

[2] W. -S. Chen, C. -K. Wu, and K. -L Wong, "Square-ring microstrip antenna with a cross strip for compact circular polarization operation," *IEEE Trans. Antennas Propagat.*, Vol. 47, No. 10, pp. 1566–1568, October 1999.

[3] P. M. Bafrooei and L. Shafai, "Characteristics of single- and double-layer microstrip saqaure-ring antennas," *IEEE Trans. Antennas Propagat.*, Vol. 47, No. 10, pp. 1633–1639, October 1999.

[4] J. Yang, J. Litva, C. Wu, and T. Tam, "Practical microstrip dual-ring antenna," *Electron. Lett.*, Vol. 32, No. 6, pp. 511–512, March 1996.

[5] S. M. Ali, W. C. Chew, and J. A. Kong, "Vector Hankel transform analysis of annual-ring microstrip antenna," *IEEE Trans. Antennas Propagat.*, Vol. 30, No. 4, pp. 637–644, July 1982.

[6] W. C. Chew, "A broad-band annual-ring microstrip antenna," *IEEE Trans. Antennas Propagat.*, Vol. 30, No. 5, pp. 918–922, September 1982.

[7] D. M. Kokotoff, J. T. Aberle, and R. B. Waterhouse, "Rigorous analysis of probe-fed printed annular ring antenna," *IEEE Trans. Antennas Propagat.*, Vol. 47, No. 2, pp. 384–388, February 1999.

[8] H. Tehrani and K. Chang, "Multifrequency operation of microstrip-fed slot-ring antennas on thin low-dielectric permittivity substrates," *IEEE Trans. Antennas Propagat.*, Vol. 50, No. 9, pp. 1299–1308, September 2002.

[9] H. Tehrani and K. Chang, "A multi-frequency microstrip-fed annular slot antenna," *Proc. IEEE AP-S Int. Symp. Dig.*, pp. 632–636, July 2000.

[10] K. -L. Wong, C. -C. Huang, and W. -S. Chen, "Printed ring slot antenna for circular polarization," *IEEE Trans. Antennas Propagat.*, Vol. 50, No. 1, pp. 75–77, January 2002.

[11] J. A. Navarro and K. Chang, *Integrated Active Antennas and Spatial Power Combining*, Wiley, New York, 1996.

[12] J. A. Hagerty and Z. Popovic, "A 10 GHz active annular ring antenna," *IEEE AP-S Int. Symp. Dig.*, Vol. 2, pp. 284–287, 2002.

[13] A. K. Bhattacharyya and R. Garg, "Input impedance of annual ring microstrip antenna using circuit theory approach," *IEEE Trans. Antennas Propagat.*, Vol. AP-33, No. 4, pp. 369–374, April 1985.

[14] R. E. Miller, "Electronically tunable active antenna using microstrip annular ring," M. S. thesis, Texas A&M University, College Station, August 1989.

[15] A. K. Bhattacharyya and R. Garg, "Self and mutual admittance between two concentric, coplanar, circular radiating current sources," *Proc. IEE*, Vol. 130, Pt. H, No. 6, pp. 217–219, June 1984.

[16] E. Lier, "Rectangular microstrip patch antenna," Ph.D. dissertation, University Trondheim, Trondheim, Norway, 1982.

[17] A. K. Bhattacharyya and L. Shafai, "A wider band microstrip antenna for circular polarization," *IEEE Trans. Antennas Propagat.*, Vol. AP-36, pp. 157–163, February 1988.

[18] J. S. Dahele, K. F. Lee, and D. P. Wong, "Dual-frequency stacked annular-ring microstrip antenna," *IEEE Trans. Antennas Propagat.*, Vol. AP-35, pp. 1281–1285, November 1987.

[19] G. D. Massa and G. Mazzarella, "Shorted annular patch antenna," *Microwave Opt. Technol. Lett.*, Vol. 8, No. 4, pp. 222–226, March 1995.

[20] K. D. Stephan, N. Camilleri, and T. Itoh, "A quasi-optical polarization-duplexed balanced mixer for millimeter-wave applications," *IEEE Trans. Microwave Theory Tech.*, Vol. MTT-31, No. 2, pp. 164–170, February 1983.

[21] C. E. Tong and R. Blundell, "An annular slot antenna on a dielectric half-space," *IEEE Trans. Antennas Propagat.*, Vol. AP-42, No. 7, pp. 967–974, July 1994.

[22] R. Janaswamy and D. H. Schaubert, "Characteristic impedance of a wide slotline on low-permittivity substrate," *IEEE Trans. Microwave Theory Tech.*, Vol. 34, pp. 900–902, August 1986.

[23] IE3D Version 8.0, Zeland Software Inc., Fremont, CA, January 2001.

[24] R. E. Miller and K. Chang, "Integrated active antenna using annular ring microstrip antenna and Gunn diode," *Microwave Opt. Technol. Lett.*, Vol. 4, No. 2, pp. 72–75, January 1991.

[25] K. A. Hummer and K. Chang, "Spatial power combining using active micostrip patch antennas," *Microwave Opt. Technol. Lett.*, Vol. 1, No. 1, pp. 8–9, March 1988.

[26] K. A. Hummer and K. Chang, "Microstrip active antennas and arrays," *1998 IEEE Int. Microwave Symp. Dig.*, pp. 963–966, 1988.

[27] Z. Ding, L. Fan, and K. Chang, "A new type of active antenna for coupled Gunn oscillator driven spatial power combining arrays," *IEEE Microwave Guided Wave Lett.*, Vol. 5, No. 8, pp. 264–266, August 1995.

[28] C. H. Ho, L. Fan, and K. Chang, "New FET active slotline ring antenna," *Electron. Lett.*, Vol. 29. No. 6, pp. 521–522, August 1993.

[29] C. M. Montiel, L. Fan, and K. Chang, "Active-notch antennas stabilized with a slotline-ring resonator for wireless applications," *IEEE Trans. Antennas Propagat.*, Vol. 46, pp. 945–946, June 1998.

[30] K. Chang, *Microwave Solid-State Circuits and Applications*, Wiley, New York, 1994.

[31] R. Mittra, C. H. Chan, and T. Cwik, "Techniques for analyzing frequency selective surfaces—A review," *Proc. IEEE*, Vol. 76, No. 12, pp. 1593–1615, December 1988.

[32] E. A. Parker and S. M. A. Hamdy, "Rings as elements for frequency selective surfaces," *Electron. Lett.*, Vol. 17, No. 17, pp. 612–614, August 1981.

[33] R. Cahill and E. A. Parker, "Concentric ring and Jerusalem cross arrays as frequency selective surfaces for a 45° incidence diplexer," *Electron. Lett.*, Vol. 18, No. 8, pp. 313–314, April 1982.

[34] S. M. A. Hamdy and E. A. Parker, "Current distribution on the elements of a square loop frequency selective surface," *Electron. Lett.*, Vol. 18, No. 14, pp. 642–626, 1982.

[35] R. Cahill and E. A. Parker, "Crosspolar levels of ring arrays in reflection at 45° incidence influence of lattice spacing," *Electron. Lett.*, Vol. 18, No. 24, pp. 1060–1061, 1982.

[36] R. J. Langley and E. A. Parker, "Double-square frequency-selective surfaces and their equivalent circuit," *Electron. Lett.*, Vol. 19, No. 17, pp. 675–677, August 1983.

[37] E. A. Parker and J. C. Vardaxoglou, "Plane-wave illumination of concentric-ring frequency-selective surfaces," *IEE Proc.*, Vol. 132, Pt. H, No. 3, pp. 176–180, June 1985.

[38] E. A. Parker, S. M. A. Hamdy, and R. J. Langley, "Arrays of concentric rings as frequency selective surfaces," *Electron. Lett.*, Vol. 17, No. 23, pp. 880–881, November 12, 1981.

[39] C. K. Lee and R. J. Langley, "Equivalent-circuit models for frequency-selective surfaces at oblique angles of incidence," *IEE Proc.*, Vol. 132, Pt. H, pp. 395–399, October 1985.

[40] A. Roberts and R. C. McPhedran, "Bandpass grids with annular apertures," *IEEE Trans. Antennas Propagat.*, Vol. AP-36, No. 5, pp. 607–611, May 1988.

[41] A. Kondo, "Design and characteristics of ring-slot type FSS," *Electron. Lett.*, Vol. 27, pp. 240–241, January 31, 1991.

[42] R. Cahill and E. A. Parker, "Performance of millimeter-wave frequency selective surfaces in large incident angle quasi-optical systems," *Electron. Lett.*, Vol. 28, No. 8, pp. 788–789, April 9, 1992.

[43] R. Cahill, E. A. Parker, and C. Antonopoulos, "Design of multilayer frequency selective surface for diplexing two closely spaced channels," *Microwave Opt. Technol. Lett.*, Vol. 8, No. 6, pp. 293–296, April 20, 1995.

[44] T. K. Wu, "Single-screen triband FSS with double-square-loop elements," *Microwave Opt. Technol. Lett.*, Vol. 5, No. 2, pp. 56–59, February 1992.

[45] T. K. Wu, S. W. Lee, and M. L. Zimmerman, "Evaluation of frequency-selective reflector antenna systems," *Microwave Opt. Technol. Lett.*, Vol. 6, No. 3, pp. 175–179, March 5, 1993.

[46] J. D. McSpadden, T. Yoo, and K. Chang, "Theoretical and experimental investigation of a rectenna element for microwave power transmission," *IEEE Trans. Microwave Theory Tech.*, Vol. MTT-40, No. 12, pp. 2359–2366, December 1992.

[47] J. Huang and R. Pogorzelski, "A Ka-band microstrip reflectarray with elements having variable rotation angles," *IEEE Trans. Antennas Propagat.*, Vol. 46, No. 5, pp. 650–656, May 1998.

[48] C. Han and K. Chang, "Ka-band reflectarray using ring elements," *Electron. Lett.*, Vol. 39, No. 6, pp. 491–493, March 2003.

CHAPTER TWELVE

Ring Mixers, Oscillators, and Other Applications

12.1 INTRODUCTION

The previous chapters discuss the use of rings for resonators, measurements, filters, couplers, magic-Ts, cavities, antennas, and frequency-selective surface applications. This chapter introduces some ring mixers, oscillators, and other applications.

12.2 RAT-RACE BALANCED MIXERS

The hybrid couplers described in Chapters 8 and 9 can be used to build balanced mixers. One example is the microstrip rat-race balanced mixer using the rat-race coupler discussed in Section 8.2. Figure 12.1 shows the physical configuration of the microstrip rat-race hybrid-ring coupler. When a unit amplitude wave is incident at port 4 of the hybrid coupler, this wave is divided into two equal components at the ring junction. The two component waves arrive in phase at ports 2 and 3, and 180° out of phase at port 1. Therefore, ports 1 and 4 are isolated. Similarly, a wave incident at port 1 is divided equally and coupled to ports 2 and 3. The two component waves that both arrive at ports 2 and 3 are combined in phase. The combined wave at port 2 has a phase difference of 180° with the combined wave at port 3. The wave incident at port 1 will not be coupled to port 4 since the two component waves are 180° out of phase. These properties are used to build the balanced mixer.

Microwave Ring Circuits and Related Structures, Second Edition,
by Kai Chang and Lung-Hwa Hsieh
ISBN 0-471-44474-X

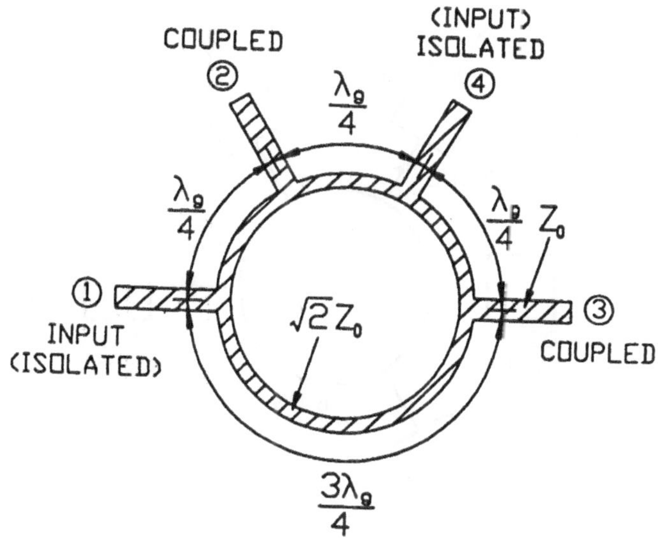

FIGURE 12.1 Physical layout of the microstrip rat-race hybrid-ring coupler.

The single-balanced mixer consists of two diodes arranged so that the local oscillator (LO) pump is 180° out of phase and the radio frequency (RF) signal is in phase at the diodes, or vice versa. The balanced operation results in LO noise suppression and provides a larger dynamic range and better intermodulation suppression compared with the single-ended mixer [1]. Figure 12.2 shows a rat-race hybrid-ring mixer. It consists of a hybridring coupler, two dc blocks, two mixer diodes, two RF chokes, and a low-pass filter. The RF input is split equally into two mixer diodes. The LO input is also split equally but is 180° out of phase at the mixer diodes. Both the LO and RF are mixed in these diodes, which generate signals that are then combined through the ring and taken out through a low-pass filter. The LO and RF ports are isolated. The RF chokes provide a tuning mechanism and prevent the RF signal from leaking into ground.

Because the microstrip hybrid ring coupler is bandwidth limited, only a 10 to 20% bandwidth has been achieved using rat-race mixers. Rat-race mixers have been demonstrated up to 94 GHz. Figure 12.3 shows the circuit of a 94-GHz rat-race mixer. A conversion loss of less than 8 dB was achieved over a 3-GHz RF bandwidth using LO pump power of +8 dBm, and less than 6.5 dB with LO pump power of +10 dBm [2]. The results are given in Figure 12.4. Wide-band mixers can be constructed using the broadband coplanar waveguide hybrid-ring couplers and magic-Ts described in Chapters 8 and 9.

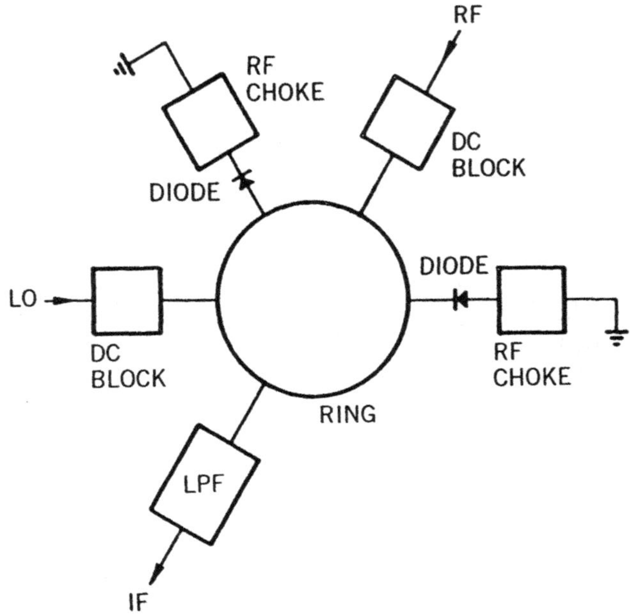

FIGURE 12.2 Rat-race mixer configuration.

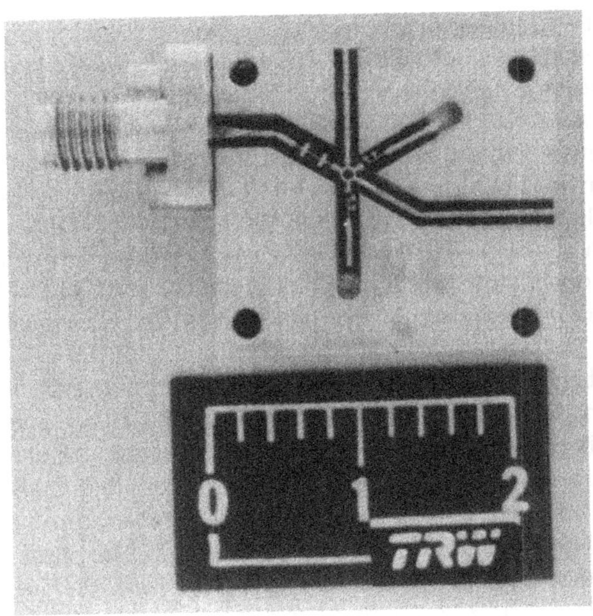

FIGURE 12.3 94-GHz rat-race mixer [2]. (Permission from IEEE.)

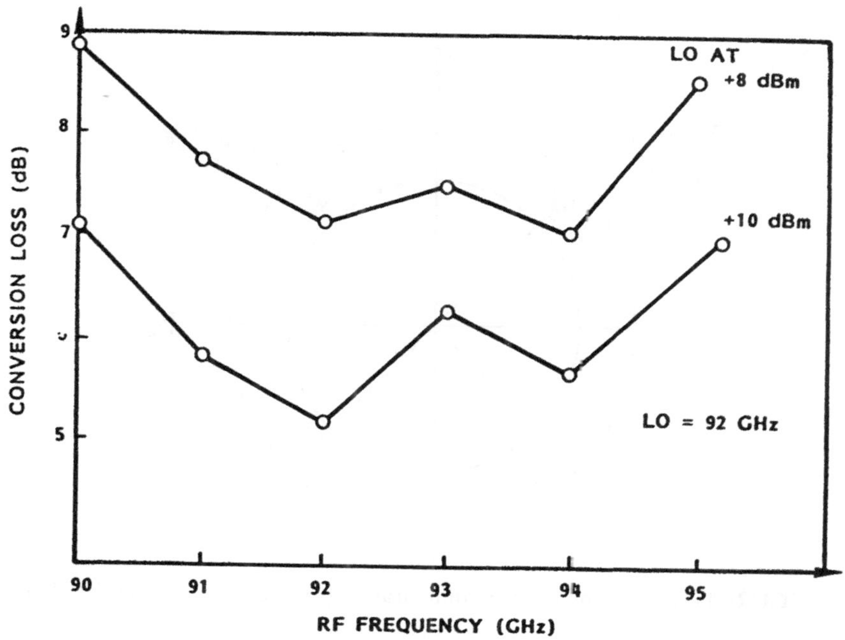

FIGURE 12.4 Performance of a 94-GHz rat-race mixer [2]. (Permission from IEEE.)

12.3 SLOTLINE RING QUASI-OPTICAL MIXERS

The slotline ring antenna discussed in Chapter 11 was used to build a quasi-optical mixer [3]. Figure 12.5 shows the circuit arrangement. The RF signal arrives as a horizontally polarized plane wave incident perpendicular to the antenna. The LO signal is vertically polarized, and can arrive from either side of the structure. V_{LO} and V_{RF} are the electric field vectors on the antenna plane. By resolving each vector into two perpendicular components, it is easy to see that the mixer diode D_1 receives

$$\frac{V_{LO} - V_{RF}}{\sqrt{2}}$$

while D_2 receives

$$\frac{V_{LO} + V_{RF}}{\sqrt{2}}$$

In effect, each diode has its own independent mixer circuit, with the intermediate frequency (IF) outputs added in parallel. The IF signal appears as a voltage between the central metal disk and the surrounding ground plane, and

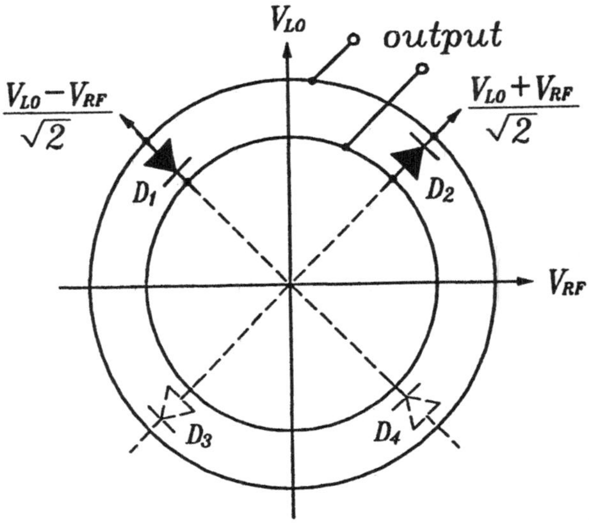

FIGURE 12.5 Antenna-mixer configuration [3]. (Permission from IEEE.)

is removed through an RF choke. A double-balanced mixer with improved isolation can be made by adding two additional diodes D_3 and D_4, as indicated.

The antenna-mixer has good LO-to-RF isolation, because of the symmetry provided by the balanced configuration. A conversion loss of 6.5 dB was measured for this quasi-optical mixer operating at X-band [3]. Similar circuits were recently analyzed using a nonlinear analysis [4].

12.4 RING OSCILLATORS

Since a ring circuit is a resonator, it can be used to stabilize an oscillator. Figure 12.6 shows a high-temperature superconductor ring-stabilized FET oscillator built on $LaAlO_3$ substrate [5]. The circuit exhibited an output power of 11.5 dBm and a maximum efficiency of 11.7%. At 77 K, the best phase noise of the superconductor oscillator was –68 dBc/Hz at an offset frequency of 10 kHz. This phase noise level is 12 dB and 26 dB less than the copper oscillator at 77 K and 300 K, respectively. A similar circuit was demonstrated using a high-electron mobility transistor (HEMT) device giving a phase noise of –75 dBc/Hz at 10 kHz from the carrier [6].

A voltage-tuned microstrip ring-resonator oscillator was reported to have a tuning bandwidth of 30% [7]. The circuit employed two microwave monolithic integrated circuit (MMIC) amplifiers as the active devices, and a tunable microstrip ring resonator in the feedback path was designed to operate over the frequency range of 1.5–2.0 GHz and fabricated with all the components mounted inside the ring as shown in Figure 12.7. A varactor diode was

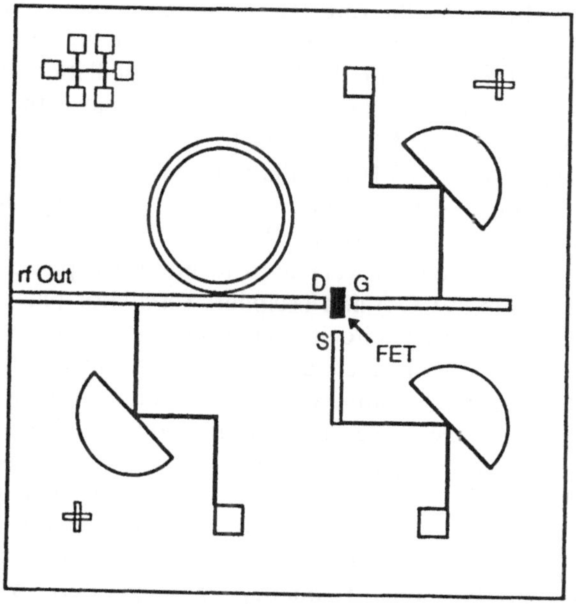

FIGURE 12.6 The physical layout of the reflection-mode oscillator on a 1-cm^2 $LaAlO_3$ substrate [5]. (Permission from IEEE.)

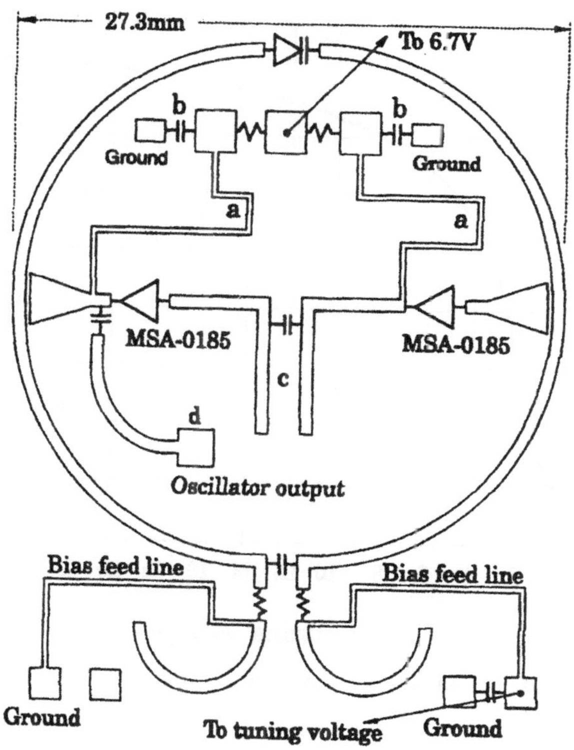

FIGURE 12.7 Layout of the microstrip ring resonator oscillator [7]. (Permission from *Electronics Letters.*)

mounted across the gap in the ring. By adjusting the bias voltage to the varactor, the resonant frequency of the ring was varied and the oscillation frequency was thus tuned. Figure 12.8 shows the oscillation frequency as a function of tuning varactor voltage, and Figure 12.9 shows the output power. The frequency was adjusted from 1.533 to 1.92 GHz with the capacitance changed from 0.44 to 3.69 pF. The oscillation frequency can be tuned down to 1.44 GHz, corresponding to a tuning range of 28.8% by slightly forward biasing the diode with 1-mA current [7].

Dual-mode ring resonators were used to build low phase noise voltage-controlled oscillators (VCOs) and oscipliers (oscillator plus multiplier) [8]. Figure 12.10 shows the VCO circuit configuration. Circuit 1 covers the lower frequency band ranges, while circuit 2 covers the higher frequency band ranges. Both oscillators are composed of a common dual-mode resonator and two identical negative resistance circuits. Using a dual-mode resonator reduces the variable frequency range to about one-half of the conventional one. As a result, the phase noise of the oscillators are significantly improved. Figure 12.11 shows the circuit configuration of osciplier [8]. The dual-mode resonator can be used to obtain two outputs of the fundamental frequency f_o and its second harmonic frequency $2f_o$, separately, with high isolation between them. An osciplier with an output signal of 1.6 GHz was demonstrated with a fundamental suppression level of 18 dB [8].

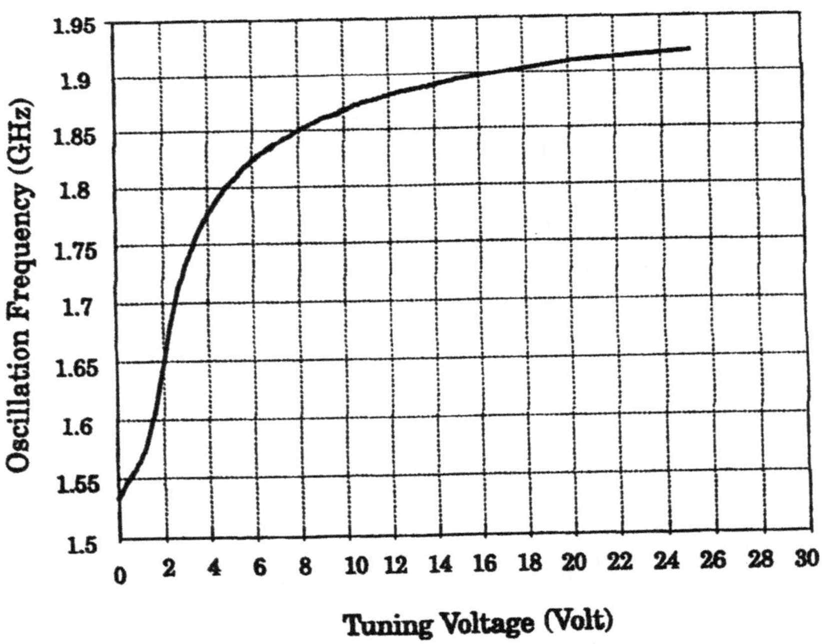

FIGURE 12.8 Oscillation frequency vs. tuning voltage [7]. (Permission from *Electronics Letters.*)

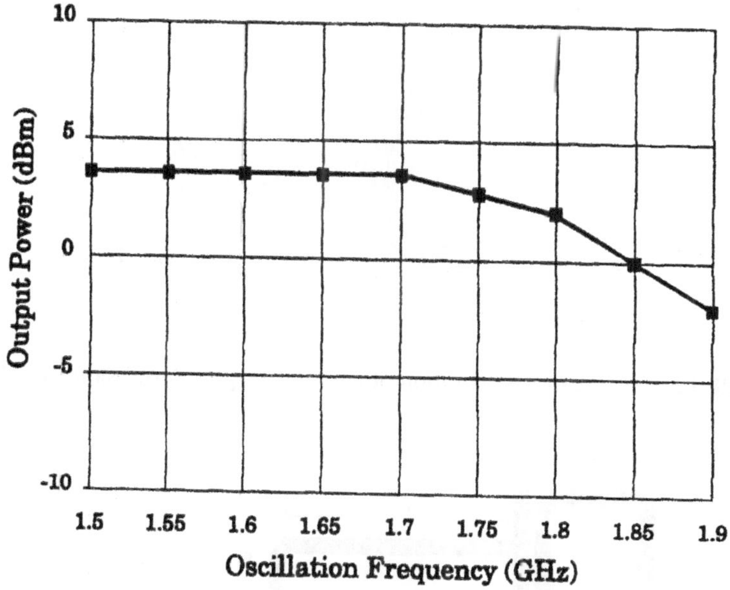

FIGURE 12.9 Output power vs. oscillation frequency [7]. (Permission from *Electronics Letters.*)

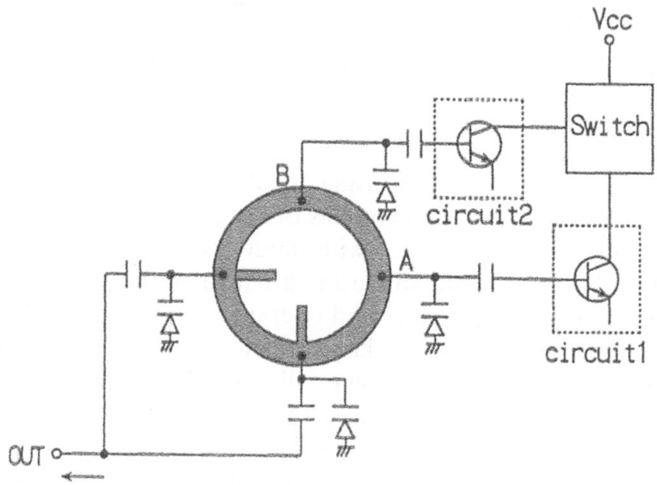

FIGURE 12.10 Circuit configuration of a low phase noise VCO [8]. (Permission from IEEE.)

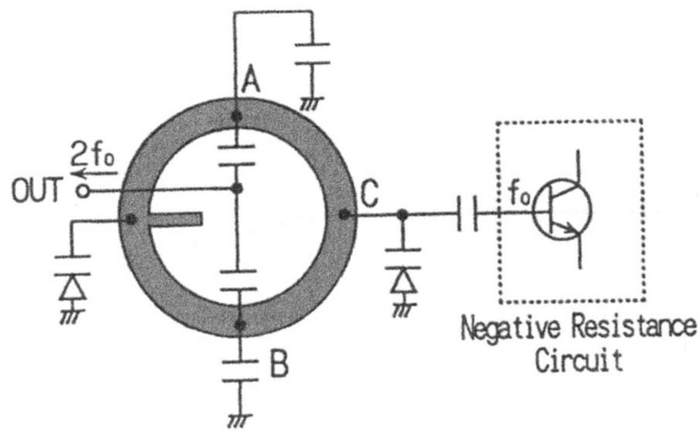

FIGURE 12.11 Circuit configuration of an osciplier [8]. (Permission from IEEE.)

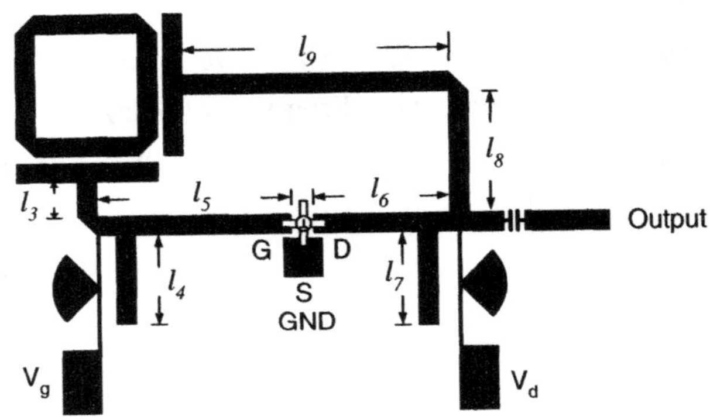

FIGURE 12.12 Feedback ring resonator oscillator [9]. (Permission from IEEE.)

Another type ring oscillator using feedback configuration is shown in Figure 12.12. This configuration consists of a feedback ring circuit and a two-port negative-resistance oscillator with input and output matching networks [9]. The close-loop ring resonator using a pair of orthogonal feed lines suppresses odd resonant frequencies and operates at even resonant frequencies. This operation has a similar characteristic of high operating resonant frequencies as that of the push-push oscillators [10, 11]. Also, the high Q ring resonator is used to reduce the noise of the two-port negative-resistance oscillator.

To investigate the high-frequency operation of the ring circuit, an orthogonal feed ring resonator is shown in Figure 12.13. As seen in Figure 12.13, the closed-loop ring resonator with total length of $l = n\lambda_g$ is fed by two orthogonal feed lines, where n is the mode number and λ_g is the guided-wavelength.

The ring resonator fed by the input and output feed lines represents a shunt circuit, which consists of the upper and lower sections of $l_1 = 3n\lambda_g/4$ and $l_2 =$

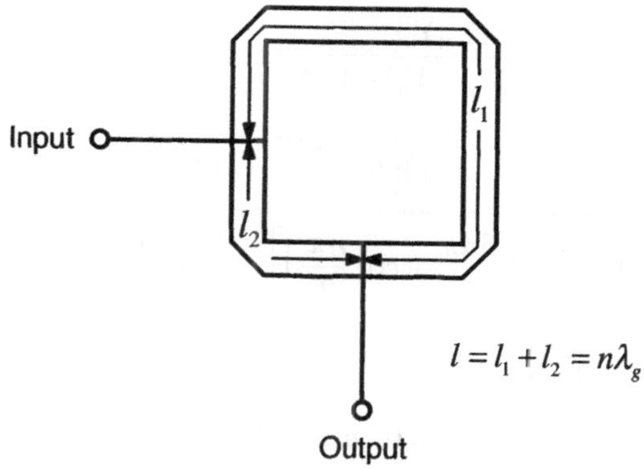

FIGURE 12.13 Configuration of the ring resonator fed by two orthogonal feed lines [9]. (Permission from IEEE.)

$n\lambda_g/4$, respectively. The *ABCD* matrixes of the upper and lower sections of the lossless ring circuit are given by

$$\begin{bmatrix} A & B \\ C & D \end{bmatrix}_{upper} = \begin{bmatrix} \cos \beta l_1 & jZ_o \sin \beta l_1 \\ jY_o \sin \beta l_1 & \cos \beta l_1 \end{bmatrix} \tag{12.1a}$$

and

$$\begin{bmatrix} A & B \\ C & D \end{bmatrix}_{lower} = \begin{bmatrix} \cos \beta l_2 & jZ_o \sin \beta l_2 \\ jY_o \sin \beta l_2 & \cos \beta l_2 \end{bmatrix} \tag{12.1b}$$

where β is the propagation constant and $Z_o = 1/Y_o$ is the characteristic impedance of the ring resonator. The *Y* parameters of the upper and lower sections are obtained from (12.1a) and (12.1b) and given by

$$\begin{bmatrix} Y_{11} & Y_{12} \\ Y_{21} & Y_{22} \end{bmatrix}_j = \begin{bmatrix} D_j / B_j & (B_j C_j - A_j D_j) / B_j \\ -1 / B_j & A_j / B_j \end{bmatrix} \tag{12.2}$$

where j = *upper* or *lower* is for upper or lower sections. In addition, the total Y parameter of the whole circuit is expressed as

$$\begin{aligned} \begin{bmatrix} Y_{11} & Y_{12} \\ Y_{21} & Y_{22} \end{bmatrix} &= \begin{bmatrix} Y_{11} & Y_{12} \\ Y_{21} & Y_{22} \end{bmatrix}_{upper} + \begin{bmatrix} Y_{11} & Y_{12} \\ Y_{21} & Y_{22} \end{bmatrix}_{lower} \\ &= \begin{bmatrix} -jY_o(\cos \beta l_1 + \cot \beta l_2) & jY_o(\csc \beta l_1 + \csc \beta l_2) \\ jY_o(\csc \beta l_1 + \csc \beta l_2) & -jY_o(\cos \beta l_1 + \cot \beta l_2) \end{bmatrix} \end{aligned} \tag{12.3}$$

Furthermore, S_{21} of the ring circuit can be found from (12.3) and is expressed as

$$S_{21} = \frac{-2Y_{21}Y_o}{(Y_{11}+Y_o)(Y_{22}+Y_o)-Y_{12}Y_{21}}$$

$$= \frac{-j2\left(\csc\frac{3n\pi}{2}+\csc\frac{n\pi}{2}\right)}{\left[1-j\left(\cot\frac{3n\pi}{2}+\cot\frac{n\pi}{2}\right)\right]^2+\left[\csc\frac{3n\pi}{2}+\csc\frac{n\pi}{2}\right]^2} \tag{12.4}$$

For odd-mode excitation

$$|S_{21}| = 0, \; n = 1, 3, 5. \ldots \tag{12.5a}$$

and for even-mode excitation

$$|S_{21}| = 1, \; n = 2, 4, 6. \ldots \tag{12.5b}$$

The calculated results in (12.5) show that the ring resonator fed by two orthogonal fed lines can suppress the odd mode resonant frequencies and operate at even mode resonant frequencies only. This operation has a similar characteristic of high operating resonant frequencies as that of the push-push oscillator [10, 11]. Figure 12.14 shows the layout of the ring circuit using two orthogonal feed lines with coupling gap size of s. This ring circuit was designed at the fundamental mode of 6 GHz and fabricated on a 20-mil-thick RT/Duroid 5870 substrate with a relative dielectric constant of $\varepsilon_r = 2.33$. The dimensions of the ring circuit are $l_1 = 27.38$ mm, $l_2 = 9.13$ mm, $l_f = 8$ mm, $w = 1.49$ mm, and $s = 0.2$ mm.

The measured and simulated results of this circuit are shown in Figure 12.15. The simulation is performed using an IE3D EM simulator [12]. Observing the measured and simulated results, they agree well with each other. The results also agree with the predictions given by (12.5). The measured unloaded Q of the ring resonator is 125.2.

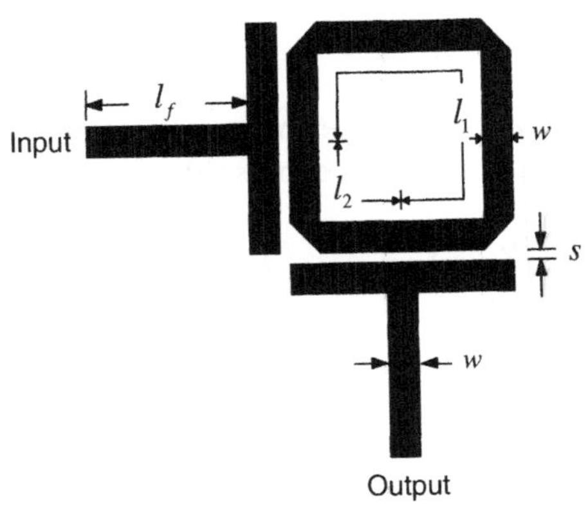

FIGURE 12.14 Configuration of the ring resonator using enhanced orthogonal feed lines [9]. (Permission from IEEE.)

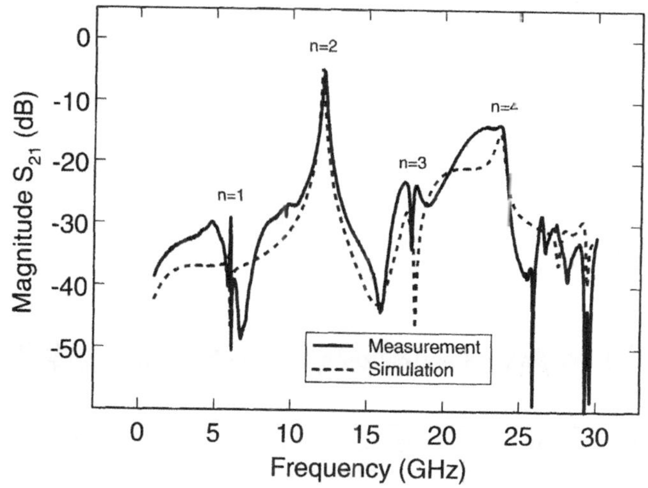

FIGURE 12.15 Simulated and measured results for the ring resonator using enhanced orthogonal feed lines [9]. (Permission from IEEE.)

By using the characteristic of the high resonant frequency operation shown in Figure 12.15, the feedback oscillator shown in Figure 12.12 can oscillate at high oscillation frequency. The active device used in the oscillator is a NE 32484A HEMT. The dimensions of the oscillator are $l_3 = 3$ mm, $l_4 = 6.95$ mm, $l_5 = 15.19$ mm, $l_6 = 10.69$ mm, $l_7 = 7.3$ mm, $l_8 = 9.47$ mm, and $l_9 = 21.19$ mm. The two-port negative-resistance oscillator uses the one-open-end S terminal as a series-feedback element to obtain a potential instability. Also, with the input and output matching network, the two-port oscillator with an applied bias of of $V_{gs} = -0.65$ V and $V_{ds} = 1$ V has a negative resistance around 12 GHz. Inspecting the equation of the DC-to-RF efficiency in Equation (12.6), if the decreasing rate of $I_{ds}V_{ds}$ is faster than that of the RF output power, P_{out}, then oscillators can possibly research to a high DC-to-RF efficiency.

$$\text{Efficiency} = \eta(\%) = \frac{P_{out}}{I_{ds}V_{ds}} \times 100\% \qquad (12.6)$$

Observing Equation (12.6), the maximum efficiency can be obtained by selecting a low V_{gs} and V_{ds}. The highest DC-to-RF efficiency of the oscillator of 41.4% is obtained with output power of 6.17 dBm at the oscillation frequency of 12.104 GHz.

Figure 12.16 shows the measured spectrum of the oscillator with applied voltages of $V_{gs} = -0.65$ V and $V_{ds} = 1$ V. Also, as shown in Figure 12.16, the oscillator is operated at the second harmonic of the ring resonator. The oscillator has the efficiency of 48.7% with output power of 3.41 mW at 12.09 GHz. The phase noise of the oscillator is −96.17 dBc/Hz at offset frequency of 100 KHz. The second and third harmonics of the oscillator are 22.8 dB and 15.1 dB down from the fundamental oscillation frequency.

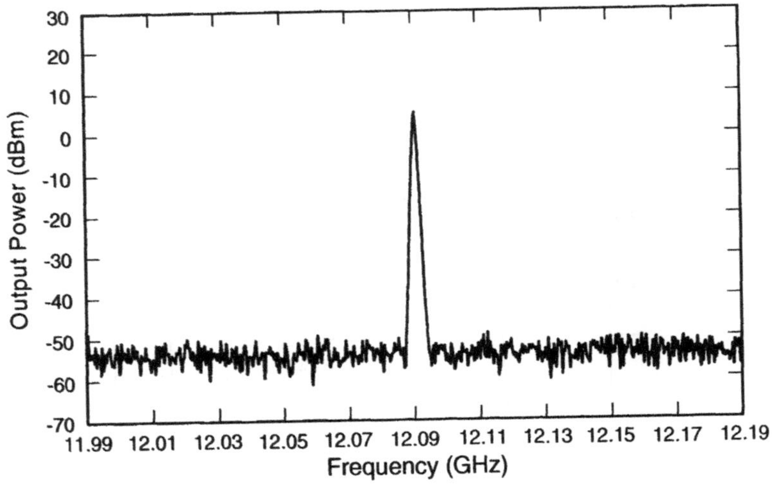

FIGURE 12.16 Output power for the feedback ring resonator oscillator operated at the second harmonic of the ring resonator [9]. (Permission from IEEE.)

These harmonics have less effect on the fundamental oscillation frequency. Comparing with other oscillators [13], this oscillator provides a high DC-to-RF efficiency.

Figure 12.17 shows the configuration of the ring resonator oscillator integrated with a piezoelectric transducer (PET) with an attached dielectric perturber. When applying a DC voltage to the PET, the PET move the perturber up or down vertically to change the effective dielectric constant of the ring resonator [9, 14], and thus vary the resonant frequency of the ring resonator.

Figure 12.18 shows the measured results of the oscillator using the PET tuning. The perturber attached on the PET has a dielectric constant of $\varepsilon_r = 10.8$ and a thickness of h = 50 mil. The tuning range of the oscillator is from 11.49 GHz (+90 V) with a power output of 3.17 dBm to 12 GHz (0 V) with a power output of 5.33 dBm.

Figure 12.19 shows the tunable oscillation frequencies and output power levels versus PET tuning voltages. As seen in Figure 12.19, the PET tuning range is about 4.25% around the oscillation frequency of 12 GHz, and the output power is varied from 2.67 to 5.33 dBm. This good tuning rage is due to a large area perturbation on the whole ring that significantly tunes the resonant frequency of the ring. In addition, by using a higher dielectric perturber, a wider tuning range and a lower DC applied voltage could be achieved [15].

12.5 MICROWAVE OPTOELECTRONICS APPLICATIONS

An optical control in microwave ring devices has been developed for its potential applications in signal switching, mixing, and frequency modulation. Fur-

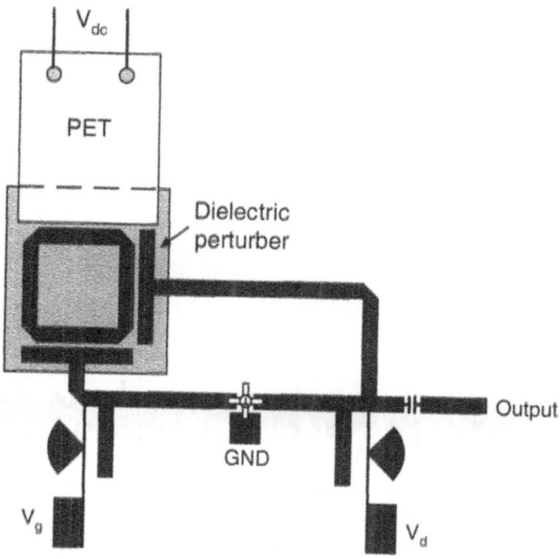

(a)

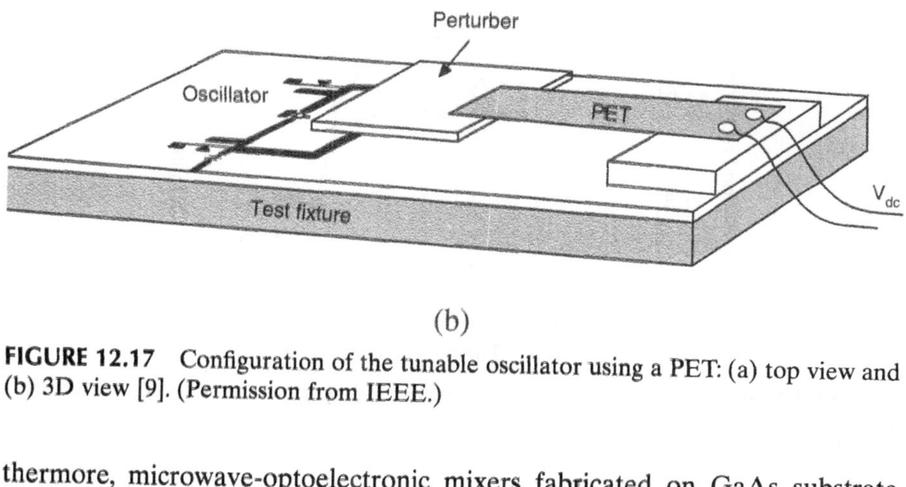

(b)

FIGURE 12.17 Configuration of the tunable oscillator using a PET: (a) top view and (b) 3D view [9]. (Permission from IEEE.)

thermore, microwave-optoelectronic mixers fabricated on GaAs substrate have been reported [16–19]. The layout of the circuit is illustrated in Figure 12.20. Since the Q-factor of the ring resonator is better than that of the linear resonator, the ring was chosen for experiments. The circuit is fabricated on semi-insulating GaAs.

Resonances were measured to occur at 3.467 GHz, 7.18 GHz, and 10.4 GHz. Corresponding loaded Q-factors are 45, 58, and 74. Two 4-μm slits are introduced at diametrically opposite locations of the ring for optical excitation. These slits are designed to be collinear with the feed lines so that mode

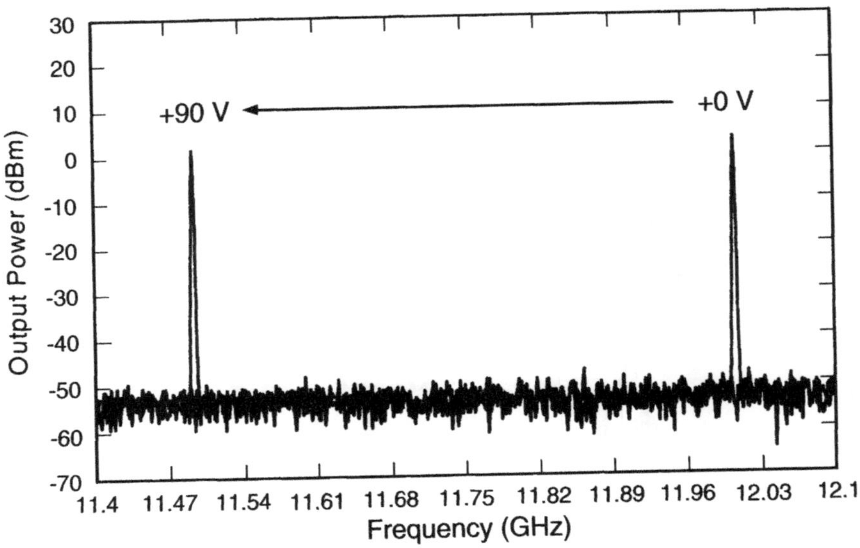

FIGURE 12.18 Measured tuning range of 510 MHz for the tunable oscillator using a PET [9]. (Permission from IEEE.)

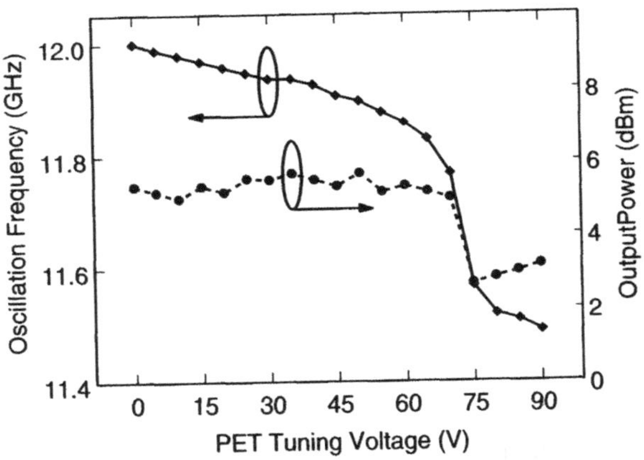

FIGURE 12.19 Tuning oscillation frequencies and output power levels versus PET tuning voltages [9]. (Permission from IEEE.)

configuration of this resonator is identical to that of the completely closed ring. The dimensions of the coupling gaps between the feed lines and the resonator were chosen to be 30 μm and 100 μm, respectively. In this configuration, the microwave LO excitation is applied via the more loosely coupled 100-μm gap and the output signal is extracted across the 30-μm gap. It is thus ensured that whereas the LO signal is loosely coupled into the resonator, extraction of the output signal is more efficient due to the tighter coupling associated with the 30-μm gap.

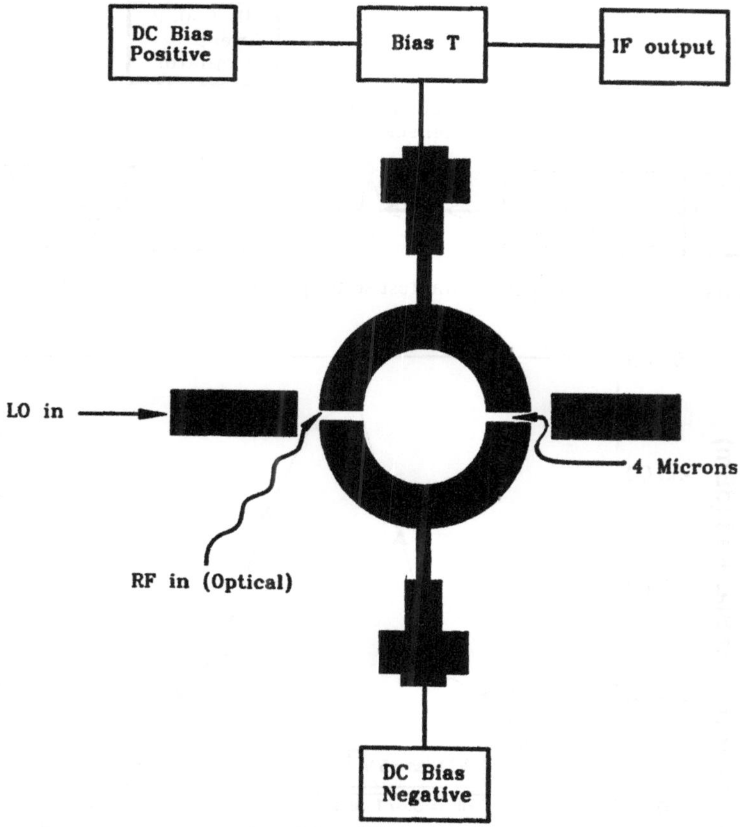

FIGURE 12.20 Layout of ring resonator circuit [19]. (Permission from IEEE.)

The test setup is illustrated in Figure 12.21. When a modulated optical signal from a laser diode is applied to one of the slits of the ring resonator, an RF voltage is induced. By virtue of the ring's moderately high Q-factor, the manifestation of this phenomenon is enhanced when the circumference of the ring becomes an integral multiple of the wavelength corresponding to the RF signals. The RF signal is the modulating signal to the optical carrier. When a larger amplitude LO microwave signal is applied to the feed line of the circuit, this signal is mixed with the RF optical signal if both the LO and RF frequencies are at the ring's resonance; the down-converted IF difference signal is obtained from the bias pad of the circuit. When the IF signal at basehand is extracted from the bias pad, the circuit is said to be operated in the "resistive mixing" mode, as the circuit operation in this case involves the modulation of the conductance of the detector diodes. For operation in this mode, the RF and LO ports are mutually isolated and the low-pass filter automatically suppresses the image frequency.

The Ortel SL 1010 laser diode, with an operating wavelength of 0.84 μm

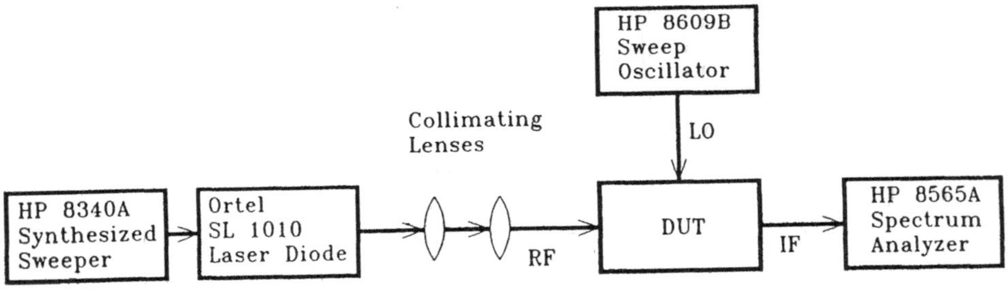

FIGURE 12.21 Experimental test setup [19]. (Permission from IEEE.)

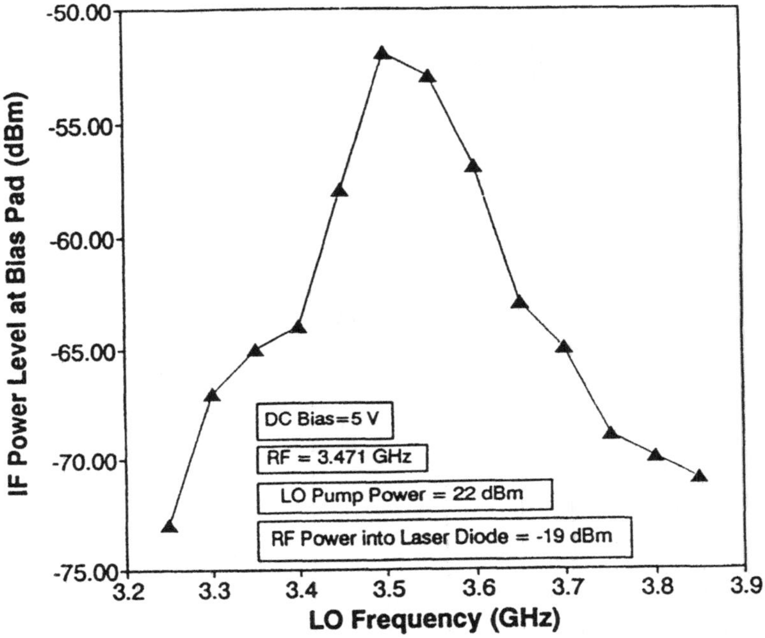

FIGURE 12.22 IF power output vs. LO frequency [19]. (Permission from IEEE.)

and a threshold current of 6.6 mA, is biased at 9 mA and operated with an input-modulated power of –14 dBm at 3.467 GHz. If either one of the RF or LO frequencies is tuned away from resonance, the IF signal strength at the bias pad gradually decreases. This is illustrated in Figure 12.22. As can be seen, the peak of the IF signal output occurs when the LO is close to the ring's resonance; when tuned out of resonance, the strength goes down. Similar effects were observed in varying the RF.

In the "parametric mode," sum and difference frequencies in the microwave band are extracted from the feed line of the circuit. For operation in this mode, the ring should resonate at the RF, LO, and IF frequencies. Both degenerate and nondegenerate parametric amplification of the optical carrier signal can take place [19].

12.6 METAMATERIALS USING SPLIT-RING RESONATORS

The metamaterials with simultaneously negative permittivity and permeability ($\varepsilon < 0$ and $\mu < 0$) were proposed by Veselago in the late 1960s [20]. He termed the metamaterial with simultaneously negative permittivity and permeability as "left-handed material" (LHM) because the vectors E, H, and k form a left-handed triplet. Also, the wave vector k and Poynting vector are anti-parallel, which shows a reversal of Snell's law [21]. However, these simultaneously negative permittivity and permeability were only derived from mathematics without any experimental proofs because the negative permittivity and permeability do not exist in the nature world.

Recently, many papers have been published for the matematerials [21–25]. By using a periodic split-ring resonator array, a negative permeability can be obtained [22]. Also, some propose the negative refraction index by using periodically L-C loaded transmission line [24, 25]. However, despite those incredible reports in LHM, there are some attempts to debunk all of these experiments [26–29].

Figure 12.23 shows the one unit of split-ring resonator arrays. The unit

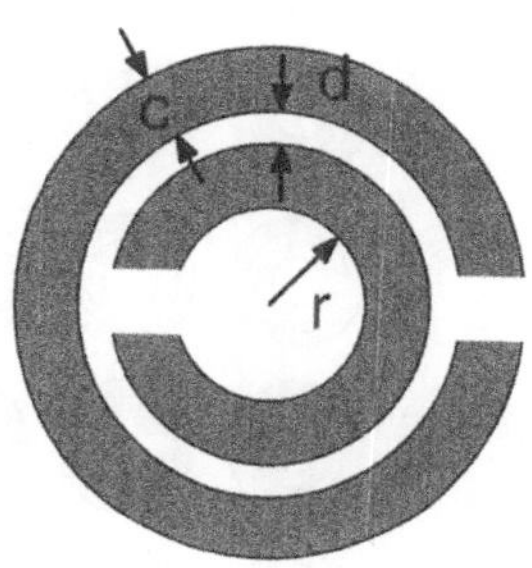

(a)

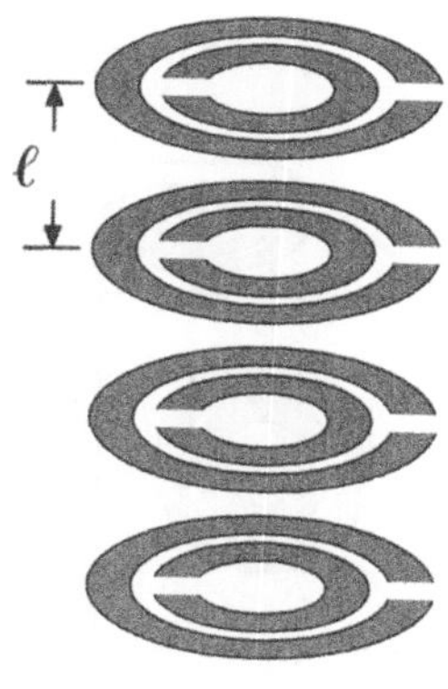

(b)

FIGURE 12.23 (a) Plan view of a split ring showing definitions of distances and (b) sequence of split rings shown in their stacking sequence [22]. (Permission from IEEE.)

resonator consists of two concentric rings, and each has a split that is used to prevent current from flowing around any ring. The inside ring is used to induce capacitances to make current flow to the ring. The capacitance between tow rings is given by [22]

$$C_1 = \frac{\varepsilon_o}{\pi}\ln\frac{2c}{d} = \frac{1}{\pi\mu_o c_o^2}\ln\frac{2c}{d} \tag{12.7}$$

where c is the width of the ring, d is the gap size between tow rings, and c_o is the speed of light in free space. Also, the effective permeability is given by [22]

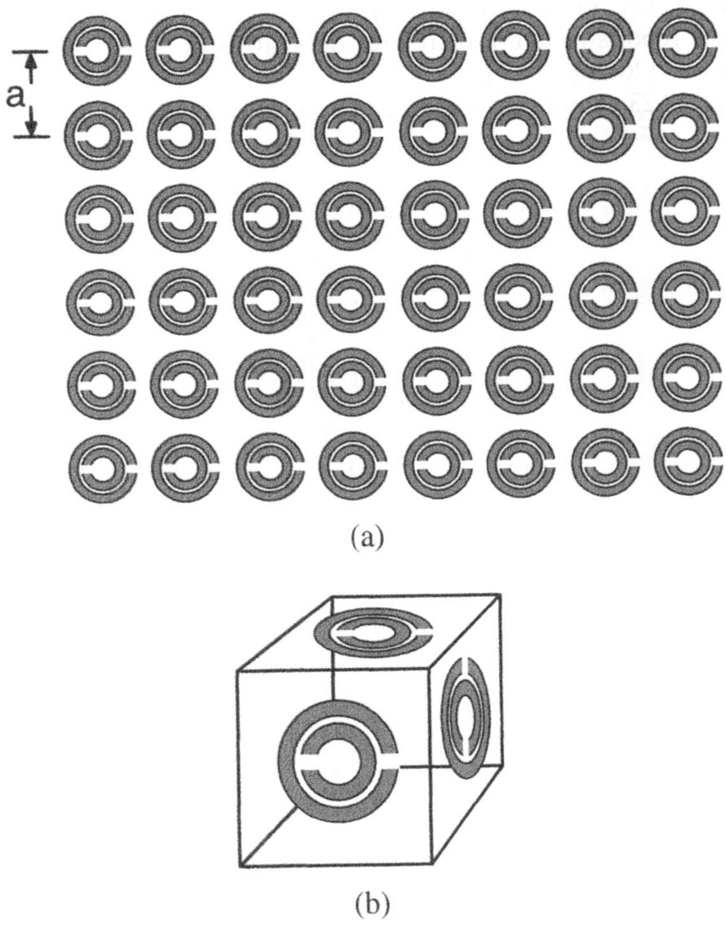

FIGURE 12.24 (a) Plain view and (b) 3D view of a split rings structure in an array (lattice spacing a) [22]. (Permission from IEEE.)

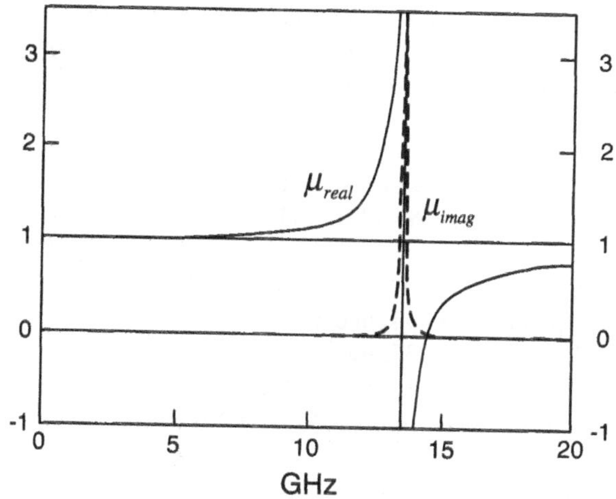

FIGURE 12.25 Plot of μ_{eff} for the cubic split ring structure [22]. (Permission from IEEE)

$$\mu_{eff} = 1 - \frac{\frac{\pi r^2}{a^2}}{1 + \frac{2l\sigma_1}{\omega r \mu_o} i - \frac{3\ell c_o^2}{\pi \omega^2 \ln \frac{2c}{d} r^3}} \tag{12.8}$$

where σ_1 is the resistance of unit length of the sheets measured around the circumference, r is the radii of the inside ring, and a is the distance between two split-ring resonators (SRR), as shown in Figure 12.24. The plotting of μ_{eff} is shown in Figure 12.25 by using parameters of $a = 1.0 \times 10^{-2}$ m, $c = 1.0 \times 10^{-3}$ m, $d = 1.0 \times 10^{-4}$ m, $l = 2.0 \times 10^{-3}$ m, and $r = 2.0 \times 10^{-3}$ m. It can be found the effective negative permeability is around 13.6 GHz within a narrow band.

REFERENCES

[1] K. Chang, *Microwave Solid-State Circuits and Applications*, Wiley, New York, 1994, Chap. 6.

[2] K. Chang, D. M. English, R. S. Tahim, A. J. Grote, T. Phan, C. Sun, G. M. Hayashibara, P. Yen, and W. Piotrowski, "W-band (75–110 GHz) microstrip components," *IEEE Trans. Microwave Theory Tech.*, Vol. MTT-33, No. 12, pp. 1375–1382, December 1985.

[3] K. D. Stephan, N. Camilleri, and T. Itoh, "A quasi-optical polarization-duplexed balanced mixer for millimeter-wave applications," *IEEE Trans. Microwave Theory Tech.*, Vol. MTT-31, No. 2, pp. 164–170, February 1983.

[4] S. K. Masarweh, T. N. Sherer, K. S. Yngvesson, R. L. Gingras, C. Drubin, A. G. Cardiasmenos, and J. Wolverton, "Modeling of a monolithic slot ring quasi-optical mixer," *IEEE Trans. Microwave Theory Tech.*, Vol. MTT-42, No. 9, pp. 1602–1609, September 1994.

[5] N. J. Rohrer, G. J. Valco, and K. B. Bhasin, "Hybrid high temperature superconductor/GaAs 10 GHz microwave oscillator: Temperature and bias effects," *IEEE Trans. Microwave Theory Tech.*, Vol. MTT-41, No. 11, pp. 1865–1871, November 1993.

[6] D. Chauvel, Y. Crosnier, J. C. Carru, and D. Chambonnet, "A 12-GHz high-temperature superconducting semiconductor oscillator," *Microwave Opt. Technol. Lett.*, Vol. 9, No. 5, pp. 235–237, August 5, 1995.

[7] P. Gardner, D. K. Paul, and K. P. Tan, "Microwave voltage tuned microstrip ring resonator oscillator," *Electron. Lett.*, Vol. 30, No. 21, pp. 1770–1771, October 134, 1994.

[8] H. Yabuki, M. Matsuo, M. Sagawa, and M. Makimoto, "Miniaturized stripline dual-mode ring resonators and their application to oscillating devices," *1995 IEEE Int. Microwave Symp. Dig.*, Orlando, Fla., pp. 1313–1316, 1995.

[9] L.-H. Hsieh and K. Chang, "High efficiency piezoelectric transducer tuned feedback microstrip ring resonator oscillators operating at high resonant frequencies," *IEEE Trans. Microwave Theory Tech.*, Vol. 51, No. 4, pp. 1141–1145, April 2003.

[10] Y. D. Lee, M. H. Lee, K. H. Lee, W. P. Hong, and U. S. Hong, "Voltage-controlled hair-pin resonator oscillator with new tuning mechanism," *Electron. Lett.*, Vol. 36, No. 17, pp. 1470–1471, August 2000.

[11] J. H. Hwang, J. S. Cho, J. H. Kim, J. C. Lee, B. K. Kim, N. Y. Kim, and U. S. Hong, "Effective structure of a push-push oscillator using a hair-pin resonator for K-band application," *IEEE MTT-S Int. Microwave Symp. Dig.*, pp. 452–455, 1999.

[12] IE3D Version 8.0, Zeland Software Inc., Fremont, CA, January 2001.

[13] O. Ishihara, T. Mori, H. Sawano, and M. Nakatani, "A highly stabilized GaAs FET oscillation using a dielectric resonator feedback circuit in 9–14 GHz," *IEEE Trans. Microwave Theory Tech.*, Vol. 28, No. 8, pp. 817–824, August 1980.

[14] L.-H. Hsieh and K. Chang, "Tunable microstrip bandpass filters with two transmission zeros," *IEEE Trans. Microwave Theory Tech.*, Vol. 51, No. 2, pp. 520–525, Feburary 2003.

[15] Y. M. Poplavko, Y. V. Prokopenko, V. I. Molchanov, and A. Dogan, "Frequency-tunable microwave dielectric resonator," *IEEE Trans. Microwave Theory Tech.*, Vol. 49, No. 6, pp. 1020–1026, June 2001.

[16] J.-C. Lee, H. F. Taylor, and K. Chang, "Degenerate parametric amplification in an optoelectronic GaAs CPW-to-slotline ring resonator," *IEEE Microwave Guided Wave Lett.*, Vol. 7, No. 9, pp. 267–269, September 1997.

[17] G. K. Gopalakrishnan, "Microwave and optoelectronic performance of hybrid and monolithic microstrip ring resonator circuits," Ph.D. dissertation, Texas A&M University, College Station, May 1991.

[18] G. K. Gopalakrishnan, B. W. Fairchild, C. L. Yeh, C. S. Park, K. Chang, M. H. Weichold, and H. F. Taylor, "Microwave performance of nonlinear optoelectronic microstrip ring resonator," *Electron. Lett.*, Vol. 27, No. 2, pp. 121–123, January 17, 1991.

[19] G. K. Gopalakrishnan, B. W. Fairchild, C. L. Yeh, C. Park, K. Chang, M. H. Weichold, and H. F. Taylor, "Experimental investigation of microwave-optoelectronic interactions in a microstrip ring resonator," *IEEE Trans. Microwave Theory Tech.*, Vol. MTT-39, No. 12, pp. 2052–2060, December 1991.

[20] V. G. Veselago, "The electrodynamics of substances with simultaneously negative values of ε and μ," *Sov. Phys. –Usp.*, Vol. 10, No. 4, pp. 509–514, January–February 1968.

[21] J. B. Pendry, "Negative refraction makes a perfect lens," *Phys. Rev. Lett.*, Vol. 85, No. 18, pp. 3966–3969, October 2000.

[22] J. B. Pendry, A. J. Holden, D. J. Robins, and W. J. Stewart, "Magnetism from conductors and enhanced nonlinear phenomena," *IEEE Trans. Microwave Theory Tech.*, Vol. 47, No. 11, pp. 2075–2084, November 1999.

[23] R. W. Ziolkowski, "Double negative metamaterial design, experiments, and applications," *IEEE MTT-S Int. Microwave Symp. Dig.*, pp. 396–399, 2002.

[24] G. V. Eleftheriades, A. K. Iyer, and P. C. Kremer, "Planar negative refractive index media using periodically L-C loaded transmission lines," *IEEE Trans. Microwave Theory Tech.*, Vol. 50, No. 12, pp. 2702–2712, December 2002.

[25] G. V. Eleftheriades, O. Siddiqui, and A. K. Iyer, "Transmission line models for negative refractive index media and associated implementations without excess resonators," *IEEE Wireless Compon. Lett.*, Vol. 13, No. 2, pp. 51–53, February 2003.

[26] G. W. 't Hooft, "Comment on 'negative refraction makes a perfect lens'," *Phys. Rev. Lett.*, Vol. 87, p. 249701, December 2001.

[27] J. M. Williams, "Some problems with negative refraction," *Phys. Rev. Lett.*, Vol. 87, p. 249703, December 2001.

[28] P. M. Valanju, R. M. Walser, and A. P. Valanju, "Wave refraction in negative-index media: always positive and very inhomogeneous," *Phys. Rev. Lett.*, Vol. 88, pp. 187041–187044, May 2002.

[29] N. Garcia and M. Nieto-Vesperinas, "Left-handed materials do not make a perfect lens," *Phys. Rev. Lett.*, Vol. 88, pp. 207403–207406, May 2002.

Index

Microwave Ring Circuits and Related Structures, Second Edition,
by Kai Chang and Lung-Hwa Hsieh
ISBN 0-471-44474-X

Printed and bound by CPI Group (UK) Ltd, Croydon, CR0 4YY

16/09/2026

14963442-0003